Reinbert Krol
Germany's Conscience

The series is edited by Egon Flaig, Daniel Fulda, Petra Gehring, Friedrich Jaeger, Jörn Rüsen, Jürgen Straub and Manfred Sicking.

Reinbert Krol, born 1979, teaches history at the University of Groningen. His research interests are German political and cultural history, with a specific focus on philosophy of history.

Reinbert Krol

Germany's Conscience

Friedrich Meinecke: Champion of German Historicism

The translation and printing of this book was supported by the Friedrich-Meinecke-Gesellschaft, a subdivision of the Ernst-Reuter-Gesellschaft Freie Universität Berlin and by Groningen Institute for the Study of Culture, University of Groningen, Faculty of Arts.

Bibliographic information published by the Deutsche Nationalbibliothek
The Deutsche Nationalbibliothek lists this publication in the Deutsche National-bibliografie; detailed bibliographic data are available in the Internet at http://dnb.d-nb.de

© 2021 transcript Verlag, Bielefeld

Cover layout: Kordula Röckenhaus, Bielefeld
Proofread by Uta Zeller and Reinbert Krol
Translated by Reinbert Krol
Printed by Majuskel Medienproduktion GmbH, Wetzlar
Print-ISBN 978-3-8376-5135-5
PDF-ISBN 978-3-8394-5135-9
https://doi.org/10.14361/9783839451359

Contents

Preface

Conscience is perhaps an obvious concept that comes to mind for a study on the German historian Friedrich Meinecke (1862-1954). For, there can be few historians whose concern with the reflection on the past and informing the present was so prominent and prudent. Many works on Meinecke acknowledge this aspect; however, none of them explain what role conscience exactly plays in Meinecke's philosophy of history. Existing studies even suggest the need for an explanation. This study offers such an explanation, and thereby it also offers a coherent interpretation of his work and thought. This interpretation goes further than a mere explanation of the role conscience plays in Meinecke's thought, for it reveals his, until now, unnoticed reconciliatory philosophy of history in which conscience is encapsulated.

Reconciliation is thus another key concept in Meinecke's thought. This book analyzes how Meinecke's apparently ambivalent attitude on political issues as well as the contrasts that are central to his work become comprehensible against the background of this notion of reconciliation. A thorough analysis of Meinecke's intellectual development and his role as a champion of German historicism will not only provide insight into the much-discussed question of the coherence of Meinecke's oeuvre and its importance for the historicist tradition, but can also be a source of inspiration for many today. For, in addition to historical and philosophical insight, it gives insight into (historical) questions of truth, ethics, state power and propaganda, political and ethical relativism, and questions on how we can have knowledge of the past, the difficult relationship between power and ethics in power political decision making, and how to render account of catastrophes and reconcile oneself with one's past. These issues are all present in Meinecke's philosophy of history.

Almost every historian will have to render account of ethical, cultural, political, and intellectual questions, but probably no generation of historians as did Friedrich Meinecke's lived through a more deeply unsettling transformation of life that all these questions together where most pressing. This generation was indeed witness to major political, socio-economic, cultural and intellectual crises and had to, one way or the other, try to cope with them.

Beginning in *fin-de-siècle* Germany, different processes coincided: industrialization, socialism, mass democracy, class conflicts, socio-economic pluralism, urbanization, state and nation building.[1] Germany was thus faced with the task of simultaneously addressing crises of identity, legitimacy, participation and integration.[2] From 1890 the various crises intensified. It is the year that Chancellor Otto von Bismarck was dismissed by Emperor Wilhelm II, which put an end to *Realpolitik* and was the advent of an impulsive and aggressive *Weltpolitik*.[3] Wilhelm II's imperialist policy was focussed on the future, on increasing the power and territory of Germany, but was also used to defuse internal tensions. Moreover, it was a policy characterized by unrest, impatience, irritability and nervousness.[4]

After the forced resignation of Bismarck, in addition to the political, social and economic changes, there was also a major cultural change. Various associations and societies were established, which Bismarck had previously banned or kept small. According to some, there was cultural prosperity and political decline. There is much to be said for this, especially with regard to the works of poets and thinkers such as Friedrich Nietzsche, Richard Wagner, Stefan George, Max Weber and Thomas Mann. The rise of all kinds of artistic circles and youth movements are also evidence of a cultural boom. At the same time, all kinds of politico-cultural and especially utopian-oriented movements emerged such as Marxism, Social Darwinism, eugenics and racial theory. The combination of optimism, utopianism, despondency and nervousness, and the multitude of crises and changes in the political, economic, cultural and social spheres; the dynamics, expansion, speed, optimism and future-orientation as expressed in technical innovations, means of transport and mass cultural developments – such as the car, tram, train, ships, but also the bicycle, telegraph, telephone, radio, gramophone, film – also led to a darker side of that modernity: a sense of alienation and fragmentation, which manifested itself in the phenomena of hypersensitivity, fatigue, decadence, irrationalism, de-

---

1   Brian Ladd, 'Berlin' in: John Merriman and Jay Winter eds., *Europe 1789 to 1914 Encyclopedia of the Age of Industry and Empire* (Detroit enz 2006) 215-220; Thomas Nipperdey, 'Probleme der Modernisierung in Deutschland' in: idem, *Kann Geschichte Objektiv sein? Historische Essays* (Munich 2013) 84-104, there 98.

2   Nipperdey, "Probleme der Modernisierung in Deutschland," 98.

3   Volker Ullrich, *Die Nervöse Grossmacht 1871-1918: Aufstieg und Untergang des deutschen Kaiserreichs* (Frankfurt am Main 1997), passim. Wilhelm II caused for even more crises, like the Morocco crises, the infamous Krüger-Telegram, the Daily Telegraph affair. On his mental condition: Christopher M. Clark, *Kaiser Wilhelm II. A Life in Power* (London 2009) 29-34.

4   Joachim Radkau, 'Die wilhelminische Ära als nervöses Zeitalter, oder: Die Nerven als Netz zwischen Tempo- und Körpergeschichte', *Geschichte und Gesellschaft* 20 No. 2 (April/June 1994) 211-241, there 238; Ullrich, *Die Nervöse Grossmacht*, 581, 720.

cay and degeneration.[5] And this was all going on before the advent of two world wars, socio-economic crises, and many other political and intellectual crises.

Despite or probably thanks to this turbulent as well as euphoric time, an explosion of intellectual creativity took place. This was of course not limited to Germany, for its thinkers were present all over Europe; one can think of Dilthey, Simmel, Proust, Freud, Bergson, Collingwood, Cassirer, Mannheim, Whitehead, Ortega y Gasset, Huizinga, Croce, Heidegger, Wittgenstein, Einstein, and of course Meinecke, to name a few. These thinkers, and many more, have laid the foundation for our time, and essentially, we still encounter problems similar to those they addressed: problems of the mind, language, being, time, ethics, aesthetics, relativism, individuality and the historicity of all the aforementioned. Even more so, those who also had to endure the atrocities of the First and the Second World War had to take this into account, or one might expect they did. Meinecke was one of those thinkers who did. He reflected on his time and the history of his country.

The landscape of the historical sciences in the first half of twentieth-century Germany was indeed shaped to a significant extent by Friedrich Meinecke. Not only did his most important works appear during this period, he also exerted considerable influence on current political and historical-theoretical debates as editor-in-chief of the prestigious *Historische Zeitschrift*. The many political and intellectual crises that Germany faced during this period forced Meinecke to constantly reflect on the past. Moreover, where others became entangled in despair, he relied on the resilience of the German spirit.

This study is a thoroughly revised and updated translation of my Dutch dissertation, but it is essentially a second and more profound book on Meinecke. The Dutch version was written under supervision of Frank Ankersmit and Herman Paul in the context of the project 'Beyond Historicism' at the University of Groningen. This program focussed on historicism, its history and viability and has led to three studies on this subject.[6]

---

5   Nipperdey, 'Probleme der Modernisierung in Deutschland', 95, 99-100; Joachim Radkau, *Das Zeitalter der Nervosität: Deutschland zwischen Bismarck und Hitler* (Munich and Vienna 1998) passim; Max Nordau, *Entartung*, 2 vols (Berlin 1892-1893); Andreas Killen, *Berlin Electropolis. Shock, Nerves, and German Modernity* (Berkely, Los Angeles and London 2006) 2, 4-5, 33; Wessel Krul, 'Tweemaal fin-de-siècle: van de negentiende naar de twintigste eeuw', *Groniek: Historisch Tijdschrift* 143 (1998) 155-167, there 155; David Baneke, *Synthetisch denken. Natuurwetenschappers over hun rol in een moderne maatschappij, 1900-1940* (Hilversum 2008) 24; R.A. Krol, 'Hyper Duitsland. Tijdservaring en nervositeit in het fin de siècle', *Groniek: Historisch Tijdschrift* 219 (voorjaar 2019) 143-156.

6   Herman Paul, *Het moeras van de geschiedenis. Nederlandse debatten over historisme* (Amsterdam 2012); Frank Ankersmit, *Meaning, Truth, and Reference in Historical Representation* (Ithaca 2012);

Acknowledgements

This book would not have been written if it wasn't for Rik Peters. He encouraged me to translate and revise my dissertation for an international audience. His continuing support and enthusiasm kept me going, for which I am now deeply grateful.

I am obliged to Bernd Sösemann and the *Friedrich-Meinecke-Gesellschaft* for their confidence and support, especially in the final stages of this project. Likewise, I owe a debt of gratitude to Daniel Fulda for recommending me to *transcript* in the first place; thanks to him this study is now published at this renowned publishing house. The collaboration with *transcript* was very friendly and productive; in particular I am very thankful for the trust and patience of Annika Linnemann. For the careful correction of my English and a friendly exchange of thought on this topic I am obliged to Uta Zeller.

In the course of this project several people provided me with scholarly advice for which I am very grateful. First and foremost, I am indebted to my mentor and friend Frank Ankersmit. Over the years we exchanged ideas on almost every subject concerning philosophy of history as well as music and literature; I am very grateful for our friendship. For encouragement, inspiration and also a shared interest in music I am indebted to Wessel Krul. For ideas to improve my research I offer my thanks to the late J.J.A. Mooij in Groningen and the late H.W. von der Dunk in Utrecht. For support and interest in my project, and an exchange of ideas on Meinecke I thank Gisela Bock, Thijs Bogers and Herman Paul. To my colleagues at the Department of History I offer my thanks for a stimulating environment, an exchange of ideas, and support over the years, in particular many thanks to: Rik Peters, Jeremia Pelgrom, Leonieke Vermeer, Babette Hellemans, Clemens Six, Dorien Daling, Maarten Zwiers, Hans van Koningsbrugge, and Dirk Jan Wolffram.

I want to thank my close friends for showing an interest in my work, supporting what I do and most of all for their friendship: Annelies Noordhof-Hoorn, Yvar de Boer, Jeroen Willems, and in particular Eva Rovers: our conversations are uplifting and give depth to life.

Last but certainly not least I thank my family and most of all my parents for their unconditional support. I would not have been able to bring this work to a successful conclusion without them. For this and many other reasons I dedicate this book to them.

Reinbert Krol, *Het geweten van Duitsland. Friedrich Meinecke als pleitbezorger van het Duitse historisme* (Groningen 2013).

Introduction

Friedrich Meinecke's oeuvre is permeated with a desire for reconciliation. A desire which was deeply connected to the history Meinecke witnessed, for his life spanned the rise and primarily the fall of the German state. As an eight-year-old boy he saw the Prussian soldiers entering Berlin after their victory over the French in 1871. And as a witness to the First World War, the Weimar Republic, the Third Reich, the Second World War, and the subsequent partition of Germany Meinecke lived through all major political events of modern Germany.[1] These events left an indelible mark on his philosophy of history, which he wrote to get a grip on present political and intellectual debates. Confronted with these political and intellectual upheavals he reflected on their past. In order to counterbalance the many crises his country went through, he constructed a view of history with reconciliation at its heart.

At first glance, Meinecke's work is focussed on contrasts rather than reconciliation. As editor-in-chief of the *Historische Zeitschrift*, for example, Meinecke was involved in the so-called *Methodenstreit* (1891-1899), which, revolving around the ideas of Karl Lamprecht, was a dispute about the 'right' form of historiography.[2] Furthermore, Meinecke's well-known trilogy *Weltbürgertum und Nationalstaat* (1908), *Die Idee der Staatsräson* (1924) and *Die Entstehung des Historismus* (1936) contain many apparently contradicting notions and ideas. In all three studies and in his theoretical articles he is concerned with these (apparent) opposites of power and ethics, individuality and universality, Enlightenment and historicism.

---

1   Friedrich Meinecke, 'Erlebtes 1862-1901' in: idem, *Autobiographische Schriften* (Stuttgart 1969) 1-134, there 22-23; E. H. Kossmann, 'Friedrich Meinecke (1862-1954)' in: idem, *Vergankelijkheid en continuïteit. Opstellen over geschiedenis* (Amsterdam 1995) 209-224, there 211.

2   Ernst Schulin, 'Friedrich Meinecke', in: Hans-Ulrich Wehler ed., *Deutsche Historiker* 9 Bd (Göttingen 1971) Volume 1, 39-57, there 39. Since 1893 Meinecke was co-editor with Heinrich von Sybel and from 1896 untill 1935 he was editor-in-chief. Reinbert Krol, 'De "scheppende spiegel" van Friedrich Meinecke. Over geschiedschrijving en de taak van de historicus', *Leidschrift* 25 (2010) 73-94, there 75-84; Frits Boterman, *Duitse dichters en denkers. Het belang van cultuur in de moderne Duitse geschiedenis* (Amsterdam en Antwerpen 2008) 209.

Finally, with regard to political events, Meinecke's attitude was also ambivalent. In August 1914 he was enthusiastic about the outbreak of the war, but after a while he longed for peace. He 'converted' to the Weimar Republic; he became a *Vernunftre-publikaner*, but remained a *Herzensmonarchist*.[3] That is, he supported the Republic on rational grounds, but in his heart he remained an advocate of the monarchy. His attitude towards the Second World War was likewise ambivalent. At first, he was mildly enthusiastic, but later on he again longed for peace and even hoped Germany would lose the war.[4] In sum, it would seem that contrasts and opposites are dominant in Meinecke's thought.

The thesis of this book is, in contrast to the above, that reconciliation is the key to understanding Meinecke's oeuvre. His supposedly ambivalent attitude in political issues and the apparent contradictions that are central to his work are only understandable against the background of his desire for reconciliation. In particular, Meinecke's panentheistic philosophy enables him to reconcile opposites. This philosophy will make us understand Meinecke's world view and philosophy of history. Panentheism draws a distinction between reality and God – 'God' not in the sense of a personal God, but an abstract principle. In this respect it differs from (Spinoza's) pantheism in the sense that reality or the world is perceived as *not* identical with God. In panentheism reality is seen as an *emanation* of God. The world is 'in' God, while God also transcends the world. That means that God leaves room for individualities to develop at will. It follows that every emanation has a 'spark' of God in itself; man, as an emanation of God, understands the good, true and beautiful, which is in its perfect form in God. Furthermore, opposites like 'good and evil', 'light and dark', 'spirit and matter' share the same origin, all being emanations of God. This panentheistic philosophy enables Meinecke to reconcile polarities such as 'power and ethics', 'politics and culture', 'individuality and the absolute' 'objectivity and subjectivity', and the 'historical and ahistorical'. In other words, Meinecke's panentheistic philosophy of history gave him the tools to withstand the many political and intellectual crises of his age.

Until now, this reconciliatory view has been neglected in the complex effective history (*Wirkungsgeschichte*) of Meinecke's oeuvre. I will discuss this reception of Meinecke's thought in some detail in the next section, for now, I will briefly give a short thematic overview of the different questions that were raised over the years with regard to Meinecke's oeuvre. During the fifties, Meinecke was admired and highly praised. Most commentators were focussed on defending and explaining his political views and his philosophy of history. These writers were chiefly 'followers' and admirers who wanted to preserve and promote Meinecke's legacy. At the end

---

3   Friedrich Meinecke, 'Verfassung und Verwaltung der deutschen Republik' in: idem, *Politische Schriften und Reden* (Darmstadt 1958) 280-298, there 281.

4   Kossmann, 'Friedrich Meinecke', 211; Ernst Schulin, 'Friedrich Meinecke', 44.

of the fifties they also initiated the publication of Meinecke's collected works. At the same time, especially during the sixties, admiration turned into severe critique. Meinecke's historicism was labelled as a politically conservative ideology, and the question arose to what extent (Meinecke's) historicism was related to National Socialism. Next, in the early seventies several publications appeared which sought to explain the principles of Meinecke's philosophy of history from a more appreciative point of view. Again, accompanying this positive tone, there was also strong criticism, which focussed on his studies on historicism, and especially on Meinecke's view of history and his political theory. During the eighties and nineties, the emphasis shifted to questions of ethics. In this context, Meinecke's alleged ethical relativism was questioned in relation to the so-called 'crisis of historicism' at the beginning of the twentieth century. Meinecke was accused of keeping aloof of the problems of relativism, but it was also argued that he did indeed engage in the debate on this crisis, and that he even came up with a proposal for a solution. From the year 2000 onwards a gradual revaluation took place, which suggests there was now room for a historicization of Meinecke's oeuvre and life.[5] As is clearly expressed by the editors of a commemorative volume (2006) on Meinecke's life and work: 'As has been the case for quite some time, praise and reproach won't be prominent, neither the urgent appeal to either identify oneself with or reject Meinecke's historiographical and political position'.[6] The same counts for the studies published from 2006 onwards. Like all the aforementioned studies – with a few exceptions – all these new views on Meinecke mostly discuss only certain elements of his conception of history, not his whole philosophy of history, which Meinecke developed throughout his oeuvre. These new views are sometimes reminiscent of positions taken between the fifties up to and including the nineties, albeit in a historicized manner.

In the following detailed historiographic analysis – which, apart from German literature, also comprises Dutch, American, and Swiss views on Meinecke[7] – two

---

5   Gisela Bock and Daniel Schönpflug ed., *Friedrich Meinecke in seiner Zeit. Studien zu Leben und Werk* (Stuttgart 2006); Friedrich Meinecke, *Akademischer Lehrer und emigrierte Schüler. Briefe und Aufzeichnungen 1910-1977*, ed. Gerhard A. Ritter (Munich 2006). Jens Fabian Pyper, 'Meinecke, Croce, and the Individual: The Moral Foundations of the Study of History, 1918-1946' (Florence 2008) unpublished dissertation. Pyper's study describes the history of the exchange of ideas between Croce and Meinecke. Pyper focusses on the 'individual' within historicism, in relation to Hegel.

6   Gisela Bock and Daniel Schönpflug, 'Vorwort' in: Gisela Bock and Daniel Schönpflug ed., *Friedrich Meinecke in seiner Zeit. Studien zu Leben und Werk* (Stuttgart 2006) 7-8. 'Im Vordergrund stehen nicht, wie lange Zeit der Fall war, Lob oder Tadel und nicht die Aufforderung, sich mit Meineckes historiographischen und politischen Positionen entweder zu identifizieren oder sie abzulehnen'. All translations are mine, unless otherwise noted.

7   Due to the language barrier I will leave the Italian reception of Meinecke's thought for the most part out of consideration. In chapter 4 I will discuss Benedetto Croce (1866-1952)

issues are essential. First, at the expense of a historicization, both controversy and apologia have dominated the reception of Meinecke's oeuvre. Second, the emphasis was time and again on the (dis)continuity of Meinecke's intellectual development (*Werdegang*). Both these issues are connected insofar as praise and criticism are often aimed at different aspects of Meinecke's work, instead of being focussed on the fundamental principles of his thought. For that reason, this study will disclose these fundamental principles of Meinecke's oeuvre. As indicated, I will identify this fundamental principle of Meinecke's oeuvre with a panentheistic desire for reconciliation. At the end of this introduction, I will further elaborate on the aims, method and structure of this book.

Meinecke studies

The following comprehensive as well as concise critical discussion of Meinecke studies is not only aimed at providing the dominant views in relation to the problem stated above, but also serves to introduce an Anglophone audience to the predominantly German and Dutch discussions on Meinecke's philosophy of history. Moreover, it is an introduction to the many aspects of Meinecke's thought, which will be discussed and analysed in detail in the following chapters.

Monism and dualism

The first major study on Meinecke's philosophy of history was written by the Swiss historian Walther Hofer (1920-2013), and was entitled *Geschichtsschreibung*

---

with regard to Meinecke's *Die Entstehung des Historismus*. Croce exchanged many letters with Meinecke, and he also discussed Meinecke's historicism at length (which is available in English and German). Cf. Pyper, 'Meinecke, Croce, and the Individual'. On Croce's thought: Rik Peters, *History as Thought and Action. The Philosophies of Croce, Gentile, de Ruggiero, and Collingwood* (Exeter and La Vergne 2013) 17-25, 40-69, 301-304. For an overview of the Italian reception in general: 'Michael Erbe, 'Zur Meinecke-Rezeption im Ausland' in: idem ed., *Friedrich Meinecke heute. Bericht über ein Gedenk-Colloquium zu seinem 25. Todestag am 5. und 6. April 1979* (Berlin 1981) 147-165; Hans Herzfeld, 'Friedrich-Meinecke-Renaissance im Ausland' in: *Festschrift für Hermann Heimpel. Zum 70. Geburtstag am 19. September 1971. Erster Band*, ed. Max-Planck-Instituts für Geschichte (Göttingen 1971) 42-62; Fulvio Tessitore, 'Meinecke in Italien' in: Gisela Bock and Daniel Schönpflug ed., *Friedrich Meinecke in seiner Zeit. Studien zu Leben und Werk* (Stuttgart 2006) 227-256. Further I will not discuss, also due to the language barrier, the following recent studies: Mihael Antolović, *Historiography and Politics. The Intellectual Biography of Friedrich Meinecke (1862-1954)* (Belgrade 2017), this study is written in Serbian. Marcelo Durão Rodrigues Da Cunha, *A História Em Tempos De Crise: Friedrich Meinecke (1862-1954) E Os Problemas Do Historicismo Alemão* (Vitória 2017).

und Weltanschauung.[8] Hofer draws a distinction between two stages in Meinecke's thought that has been dominant ever since. According to Hofer, the First World War was a rupture in Meinecke's life and thought. Hofer characterizes Meinecke's pre-war attitude as 'monistic', and the time after the war as 'dualistic'.[9] Hofer argues that in Meinecke's monistic thought 'idea' and 'reality' are in harmony, or better yet: the desire for, or the possibility of such a harmony is fundamental to his thought. Hofer claims that Meinecke's pre-war studies clearly echo German Idealism – he even speaks of a renaissance of German Idealism.[10]

Though Meinecke's thought was certainly rooted in German Idealism, and deeply influenced by Romanticism, Hofer's idea of monism – which is essentially pantheistic – neither applies to Meinecke's pre-war thought, nor does it fit his post-war thought. Contrary to Hofer's interpretation, I will argue that Meinecke's philosophy of history leaves more room between 'creator' and 'creation' than Hofer's pantheistic interpretation allows for.

According to Hofer, after the First World War Meinecke's optimistic and monistic idealism was replaced by a pessimism which manifested itself in a dualistic world view. 'A strictly enforced dualism', says Hofer, 'is from now on the program that defined the historian's [Meinecke, RK] historical thought, world view, and political convictions'.[11] The difference between monism and dualism, according to Hofer, is 'the withdrawal of the 'idea' from 'reality'.[12] This change in Meinecke's

---

8   This study is based on his PhD-thesis: Walther Hofer, *Friedrich Meinecke als geschichtlicher Denker. Untersuchungen über die Bedeutung der Weltanschauung für die historische Begriffsbildung* (Munich 1949). Walther Hofer, *Geschichtsschreibung und Weltanschauung. Betrachtungen zum Werk Friedrich Meineckes* (Munich 1950). Other important studies that agree with Hofer or at least are written in a similar vein: Walter Goetz, 'Die Entstehung des Historismus' in: idem, *Historiker in meiner Zeit* (Cologne-Graz 1957) 351-360; Heinrich Ritter von Srbik, *Geist und Geschichte vom deutschen Humanismus bis zur Gegenwart* volume 2 (Munich 1961) 279-293; Ludwig Dehio, *Friedrich Meinecke. Der Historiker in der Krise. Festrede, gehalten am Tage des 90. Geburtstags* (Berlin 1953); Hans Rothfels, *Friedrich Meinecke – Ein Rückblick auf sein wissenschaftliches Lebenswerk. Trauerrede, gehalten in Berlin am 27. Februar 1954* (Berlin 1954); Philip J. Wolfson, 'Friedrich Meinecke (1862-1954)', *Journal of the History of Ideas* 17 (1956) 511-525. Also: Ernst Schulin, 'Das Problem der Individualität. Eine kritische Betrachtung des Historismus-Werkes von Friedrich Meinecke', in: idem, *Traditionskritik und Rekonstruktionsversuch. Studien zur Entwicklung von Geschichtswissenschaft und historischem Denken* (Göttingen 1979) 97-116, there 98. An exception during this period (because of its severe criticism) is: Gerald Strauss, 'Meinecke, Historismus, and the Cult of the Irrational', *German Quarterly* 26 (1953) 107-114.

9   Hofer, *Geschichtsschreibung*, 25-26.

10   Ibidem, 27.

11   Ibidem, 29. 'Ein streng durchgeführter Dualismus ist das Programm, das von jetzt an das geschichtliche und weltanschauliche Denken wie das politische Wollen unseres Historikers bestimmt'.

12   Ibidem, 32. 'der Rückzug der "Idee" aus der "Wirklichkeit"'.

world view and political theory is, in Hofer's opinion, clearly visible in Meinecke's replacement of notions like 'harmony', 'synthesis' and 'unity' with concepts such as 'antinomy', 'polarity', 'discrepancy', 'tragedy' and 'demonic' (*Dämonie*).[13]

Hofer's interpretation was first questioned by the American historian Richard Sterling, who claimed that Meinecke's oeuvre is characterized by continuity instead of discontinuity.[14] This continuity is, according to Sterling, expressed in Meinecke's use of dualistic concepts in all of his studies.[15] Sterling refers to conceptual opposites like 'power and ethics', 'state and culture', 'spirit and nature', 'individuality and universality'. One quick glance at a few titles of Meinecke's studies and articles clearly show these contrasting concepts: *Weltbürgertum und Nationalstaat*, 'Kausalitäten und Werte', 'Geschichte und Gegenwart', 'Persönlichkeit und geschichtliche Welt', and 'Ranke und Burckhardt'.[16] Sterling thus challenges Hofer's idea of a rupture in Meinecke's thought by emphasizing the continuity of dualisms in his work. Put differently, Sterling claims that the First World War, from a political viewpoint, was a definite break, however, intellectually Meinecke did not change at all.[17]

Next to Sterling the American historian Frederick Kreiling (1923-2004) was convinced of a certain consistency in Meinecke's thought. In his unpublished dissertation he argues that the search for harmony or an overarching idea or concept is key to Meinecke's thought.[18] Kreiling, therefore, does not agree with Hofer's idea of a rupture in Meinecke's thought, for the First World War did not undo

---

13   Walther Hofer, 'Friedrich Meinecke als politischer Denker' in: idem, *Geschichte zwischen Philosophie und Politik. Studien zur Problematik des modernen Geschichtsdenkens* (Basel 1956) 69-97, there 83-84; Hofer, *Geschichtsschreibung*, 34.

14   Richard W. Sterling, *Ethics in a World of Power: The Political Ideas of Friedrich Meinecke* (Princeton 1958) 29; Also: Wolfson, 'Friedrich Meinecke (1862-1954)', 520-521 and note 25. Stefan Meineke is also critical of Hofer's view: Stefan Meineke, *Friedrich Meinecke. Persönlichkeit und politisches Denken bis zum Ende des Ersten Weltkrieges* (Berlin and New York 1995) 20 and 298. Also: Ernst Schulin, 'Meineckes Leben und Werk. Versuch einer Gesamtcharakteristik' (1971) in: idem, *Traditionskritik und Rekonstruktionsversuch. Studien zur Entwicklung von Geschichtswissenschaft und historischem Denken* (Göttingen 1979) 117-132, there 120; Robert A. Pois, *Friedrich Meinecke and German Politics in the Twentieth Century* (Berkeley and Los Angeles 1972) 86-87.

15   Sterling, *Ethics in a World of Power*, 29. Masur argues that Meinecke's view is not concerned with antinomies or polarities, but contrasts. Georg Masur, "Ethics in a World of Power' (Review)', *Historische Zeitschrift* 188 (1959) 608-611, there 609.

16   Note that these opposites are not strictly contradictory, since in all titles Meinecke uses the word 'und'.

17   Sterling points out that Meinecke experienced the war on a political level as a personal tragedy, for the outcome was disastrous to the German nation, which was very dear to Meinecke. In addition to this, the outcome of the war had a corruptive effect on political life in Germany. Sterling, *Ethics in a World of Power*, 5-6.

18   Frederick C. Kreiling, 'Friedrich Meinecke and the Problems of Historicism', unpublished dissertation (New York 1959; microfilm 1967).

Meinecke's search for harmony. Kreiling states that harmony is in fact the one thing that is consistent in Meinecke's thought. Consequently, Kreiling also opposes Sterling's 'dualistic interpretation'. The views of Sterling and Kreiling are, however, reconcilable, because the longing for harmony presupposes dualism.[19]

Meinecke's striving for harmony is, in Kreiling's view, manifest on several levels. According to Kreiling, *Weltbürgertum und Nationalstaat* is, for example, an attempt to close the gap between Prussia and Germany; the unification of Germany was realised by means of a combination of the idea of national unity with consideration for the old German states, Prussia in particular.[20] And, in *Die Idee der Staatsräson*, according to Kreiling, Meinecke tries to connect Germany and Western Europe.[21] Likewise, on the level of theory, Meinecke tries to reconcile opposites. In this case 'Staatsräson' and 'historicism' are in Kreiling's opinion the two main overarching hypotheses.[22] He distinguishes not one fundamental (overarching) thought in the work of Meinecke, but different hypotheses – Kreiling's interpretation is in that sense pluralistic.

The German historian Carl Hinrichs (1900-1962) suggested yet another view on this 'continuity-thesis'.[23] Like Kreiling he was convinced that Meinecke's early studies already show a fundamental philosophy of history, which is visible throughout other works. Hinrichs, however, does not elaborate on concepts like harmony, monism, and dualism; instead he characterizes Meinecke's philosophy of history – his historicism – as a fundamental existential question (*Lebensproblem*). According to Hinrichs, Meinecke's philosophy of history essentially concentrates on the 'secret of personality'. That is, to Meinecke the fundamental problem is the 'unfathomability' and spontaneity of all individualities in history. In Hinrichs' view Meinecke denounced those views of history that emphasized causality and general laws. Instead, Meinecke suggested a historical view focussed on the individual, singular

---

19   Kreiling does not explicitly discuss Sterling or Hofer. However, he claims that after the First World War Meinecke substituted his conservatism for a more liberal stance. Though *Die Idee der Staatsräson* gives evidence of (according to Kreiling) a dualistic world view, Meinecke keeps searching for overarching concepts. Kreiling, 'Friedrich Meinecke', 129, 201. Stefan Meineke has shown that the First World War did not cause Meinecke to change his political views (from conservative to liberal): Meineke, *Friedrich Meinecke*, 295-313.

20   In the first chapter I will elaborate on this issue. Also: Kossmann, 'Friedrich Meinecke', 216.

21   Kreiling, 'Friedrich Meinecke', 204-205.

22   Ibidem, 22, 229, 275.

23   Hinrichs' article gives the initial impetus for Meinecke's fundamental thought. However, Hinrichs leaves many questions unanswered with regard to the principles of Meinecke's philosophy of history, the main subject of my research. Carl Hinrichs, 'Der Historismus als ein Lebensproblem Friedrich Meineckes' in: idem, *Preußen als historisches Problem. Gesammelte Abhandlungen*, ed. Gerhard Oestreich (Berlin 1964); First published in 1959 as an introduction to *Die Entstehung des Historismus*: Carl Hinrichs, 'Einleitung des Herausgebers' in: Friedrich Meinecke, *Die Entstehung des Historismus* (Munich 1965, first published 1936) vii-xlix.

and unique in history. In this sense, he remained true to the tradition of Humboldt, Ranke, and especially Droysen and Dilthey, who made him aware of the importance of the 'individual' in history in the first place.[24]

Meinecke's 'Lebensproblem' is in Hinrichs' view related to the problem of (ethical) relativism, which is a result of the principle of individuality (*Individualitäts-gedanke*). For, if values are singular, individual, and unique, can there really be any fixed or absolute values? According to Hinrichs, Meinecke agrees with Dilthey's conclusion that individuality leads to an 'anarchy of convictions'.[25] With regard to ethical relativism, Meinecke's notion of 'conscience' gave him, according to Hinrichs, something to hold on to. Moreover, Meinecke considers the individual conscience as fundamental in creating values in history. And within one's conscience the absolute and the individual intersect.[26] Hinrichs only reproduces this account of Meinecke's own description of conscience. He does not explain what this belief in conscience is based on, and how Meinecke managed to hold on to this belief when confronted with the many crises of his time, of which ethical relativism was only one.[27] Hinrichs only refers to a passage from Meinecke's autobiography in which his early interest in pantheistic poems is expressed; the early Meinecke believed in a pantheistic world view, which manifests itself in a 'World of ideals'.[28] Hinrichs fails to explain these philosophical and religious notions, which are vital for an understanding of Meinecke's thought. A reference to Goethe's influence on Meinecke's thought is also left undiscussed by Hinrichs – which is strange, since he devoted an entire study to Ranke and the 'Geschichtstheologie der Goethezeit'.[29] In a well-known article of 1925, also mentioned by Hinrichs, Meinecke hints at the complexities surrounding his notion of conscience: 'Culture and nature, or we could also call it God and nature, are a unity, be it a divided unity'.[30] This, as will become clear, refers to something that goes beyond a so-called God-related conscience, for this explains, on the one hand, the philosophical basis of Meinecke's

---

24   Hinrichs, 'Der Historismus als ein Lebensproblem', 364-371; Meinecke, 'Erlebtes', 52, 72; Friedrich Meinecke, 'Willensfreiheit und Geschichtswissenschaft' in: idem, *Zur Theorie und Philosophie der Geschichte* (Stuttgart 1965) 3-29, there 8.

25   Hinrichs, 'Der Historismus als ein Lebensproblem', 390.

26   Ibidem, 394-395.

27   Hofer gives the same account: Hofer, *Geschichtsschreibung und Weltanschauung*, 126-128, 224-231, 317.

28   Hinrichs, 'Der Historismus als ein Lebensproblem', 363; Meinecke, 'Erlebtes', 45.

29   Hinrichs, 'Der Historismus als ein Lebensproblem', 382, 392, 394; Carl Hinrichs, *Ranke und die Geschichtstheologie der Goethezeit* (Göttingen, Frankfurt and Berlin 1954).

30   Friedrich Meinecke, 'Kausalitäten und Werte' in: idem, *Zur Theorie und Philosophie der Geschichte* (Stuttgart 1965) 61-89, there 81; Hinrichs, 'Der Historismus als ein Lebensproblem', 367. 'Kultur und Natur, wir können auch sagen Gott und Natur, sind wohl eine Einheit, aber eine in sich gespaltene Einheit'.

notion of conscience and, on the other, it refers to a specific view of reality, which is always the point of departure of Meinecke's research. In sum, in this context Meinecke's panentheistic philosophy, in which monism and dualism coexist – they complement, and/or limit each other – are of vital importance. It is the analysis of this philosophy of history that is at the heart of my research.

Continuity or 'Umlernbereit'

During the sixties and seventies Meinecke's philosophy of history came under heavy criticism. The most important commentators in this respect are Hans Herzfeld (1892-1982) and Ernst Schulin (1929-2017).[31] Their interpretations reflect on, but also add to the positions of Hofer, Sterling and Kreiling. Like Sterling and Kreiling,

---

31   Ernst Schulin, 'Friedrich Meinecke', 39-57; Hans Herzfeld, 'Friedrich Meinecke: Der Geschichts-
denker' in: Richard Dietrich ed., *Historische Theorie und Geschichtsforschung der Gegenwart*
(Berlin 1964) 99-115; Hans Herzfeld, 'Friedrich Meinecke. Der Historiker, der Politiker und
der Mensch nach seinem Briefwechsel' in: idem, *Ausgewählte Aufsätze. Dargebracht als
Festgabe zum siebzigsten Geburtstage von seinen Freunden und Schülern* (Berlin 1962) 26-48; Hans
Herzfeld, 'Friedrich Meinecke. Zu seinem 90. Geburtstage', *Geschichte in Wissenschaft und
Unterricht* 3 (1952) 577-591. Also: Felix Gilbert, 'Friedrich Meinecke' in: idem, *History. Choice
and Commitment* (Cambridge Mass. and London 1977) 67-87; Far more critical are: Imanuel
Geiss, 'Kritischer Rückblick auf Friedrich Meinecke' in: idem, *Studien über Geschichte und
Geschichtswissenschaft* (Frankfurt am Main 1972) 89-107; Jörn Rüsen, 'Friedrich Meineckes
"Entstehung des Historismus". Eine kritische Betrachtung' in: Michael Erbe ed., *Friedrich
Meinecke heute. Bericht über ein Gedenk-Colloquium zu seinem 25. Todestag am 5. und 6. April
1979* (Berlin 1981) 76-100. For a more subtle analysis: Ernst Schulin, 'Das Problem der
Individualität'. During this period in Italy a relative historicization took place: Erbe, 'Zur
Meinecke-Rezeption im Ausland', 147-165; Herzfeld, 'Friedrich-Meinecke-Renaissance im
Ausland', 42-62; Fulvio Tessitore, *Friedrich Meinecke storico delle idee* (Florence 1969). Tessitore
prefers to understand (*verstehen*) Meinecke's work instead of criticising or even condemn
it. He is mainly interested in the philosophical influences of Hegel, Ranke, Droysen and
Dilthey. Tessitore is close to Herzfeld's interpretation, since Tessitore also emphasizes
continuity in Meinecke's thought. Far more critical is Sergio Pistone's political work: Sergio
Pistone, *Federico Meinecke e la crisi dello stato nazionale tedesco* (Turin 1969). He focussed on
the political dimension of Meinecke's studies, and he is particularly interested in Meinecke's
relationship with the national state. Pistone is very critical of Meinecke's idealistic pre-1914
studies. He expressed his appreciation for Meinecke's later work. He claims it possesses a
certain actuality. See: Herzfeld, 'Friedrich-Meinecke-Renaissance im Ausland', 56-59; Erbe,
'Zur Meinecke-Rezeption im Ausland', 152-153; Fulvio Tessitore, 'Friedrich Meinecke und die
Auflösung der onto-naturalistischen Hypothese' in: idem, *Kritischer Historismus. Gesammelte
Aufsätze* (Cologne, Weimar and Vienna 2005) 37-44, there 44; Fulvio Tessitore, 'Ernst
Troeltschs „Kompromiß" und Friedrich Meineckes „neuer Dualismus" in: idem, *Kritischer
Historismus. Gesammelte Aufsätze* (Cologne, Weimar and Vienna 2005) 45-50, there, 49.
Tessitore, 'Meinecke in Italien', 227-256.

Herzfeld also claims that continuity is fundamental to Meinecke's thought, albeit in a different way. This 'dualistic incentive to unity', says Herzfeld, was already apparent in his early works.[32] Schulin agrees with Hofer that Meinecke was able to adapt to the changing political and intellectual situations, which is why Schulin characterized Meinecke as an historian who is aware of tradition, but also willing to change or rethink (*Umlernbereit*).[33]

In Schulin's view (and also Hofer's), Meinecke's flexibility becomes particularly clear after the First World War, when he replaced his optimistic, monistic idealism with a pessimistic dualism. Instead of unity and harmony, Meinecke now seems to emphasize contrasts and disunity, which are expressed in (apparent) opposites like: *Kratos* and *Ethos*, 'causalities and values', enlightenment and historicism, individuality and development, absolute and relative.[34] Similar dualisms are, however, also apparent in Meinecke's earlier works, as Sterling already pointed out in his critique on Hofer. For example, Meinecke's first major study – as the title clearly shows – *Weltbürgertum und Nationalstaat*, already gives evidence of such a dualism. Meinecke's *Habilitationsschrift* on General Hermann von Boyen is also concerned with certain contrasts. Boyen, for example, was placed in a dilemma of either following Kantian ethics or military action. Thus, Boyen saw himself confronted with the relationship between 'personality' and the 'nation'.[35] Unlike Schulin, Herzfeld emphasized a continuity of opposites in Meinecke's work and he suggested not attaching too much importance to the rupture the First World War might have had on Meinecke's philosophy of history.[36]

Schulin and Herzfeld also have different views on Meinecke's concept of individuality. Inspired by Droysen and Dilthey, Herzfeld sees the 'unfathomable' indi-

---

32   Herzfeld, 'Friedrich Meinecke: Der Geschichtsdenker', 101; Herzfeld, 'Friedrich Meinecke. Der Historiker, der Politiker', 33, 35, 44; also: Herzfeld, 'Friedrich Meinecke. Zu seinem 90. Geburtstage', 577-579.

33   Schulin, 'Friedrich Meinecke', 39; Also: Schulin, 'Meineckes Leben und Werk', 117; Meineke, *Friedrich Meinecke. Persönlichkeit und politisches Denken*, 6-8.

34   Ernst Schulin, 'Friedrich Meinecke', 42-51.

35   Friedrich Meinecke, *Das Leben des Generalfeldmarschalls Hermann von Boyen* (Stuttgart 1896) Erster Band, 122; Friedrich Meinecke, *Das Leben des Generalfeldmarschalls Hermann von Boyen* (Stuttgart 1899) Zweiter Band, 411-412; Meinecke considered Boyen to be a transitory figure between Enlightenment, Idealism and Romanticism. He thus represented the transition of natural law to a historical way of thinking. I will come back to this issue in the following chapters. Also: Herzfeld, 'Friedrich Meinecke: Der Geschichtsdenker', 103; Hinrichs, 'Der Historismus als ein Lebensproblem', 373, 378. A thorough analysis of Boyen, however, won't be part of my research, for the Boyen biography is mainly concerned with concrete political history, not intellectual history. Also see my methodological principles at the end of this introduction.

36   Herzfeld, 'Friedrich Meinecke: Der Geschichtsdenker', 107; Herzfeld, 'Friedrich Meinecke. Zu seinem 90. Geburtstage', 577-578.

vidual as central to Meinecke's philosophy of history. This unfathomable character of the individual or the aforementioned 'secret of personality' alludes to what
Meinecke describes as an impenetrable element of spontaneity, which is unique
to every (historical) personality, and crucial within history.[37] This interest in the
individual, unique, spontaneous personality is, according to Herzfeld, the basis of
Meinecke's philosophy of history. This emphasis on individuality does not logically
lead to subjectivism, for Meinecke relates the personal to the general. Herzfeld
clarifies: 'Experience is for him [Meinecke, RK] to be concerned with our personal
existence as well as the major contemporary questions'.[38] So, to reflect on our
experience (*Erlebnis*) against the background of our own time is what is at stake
for Meinecke. However, the individual or personality in reality is, according to
Herzfeld, also an 'entity' which can strive for 'higher things'. Herzfeld concludes:
'Meinecke was, from beginning to end, convinced that historical practice in the end
– applied according to this fixation on the individual X: the uniqueness of personality and events – leads to a religious and metaphysical perspective'.[39] In Meinecke's
view, according to Herzfeld, the general and the absolute are simultaneously active within the concrete individual. Following Hinrichs, Herzfeld concludes: the
philosophy of history which characterises *Weltbürgertum und Nationalstaat* does not
differ all that much from what Meinecke later expounded upon in, for example,
Die Entstehung des Historismus.[40] After all, both studies revolve around notions like
'individuality', 'spontaneity', 'freedom', 'development', and 'the absolute'.

Schulin agrees with Herzfeld on Meinecke's concept of individuality and the
fact that it requires an acceptance of metaphysics. Contrary to Herzfeld, Schulin
considers this to be a weakness of Meinecke's philosophy of history.[41] In Schulin's
view Meinecke relates the individual with the ideal. This longing for an ideal, says
Schulin, is particularly manifest in Meinecke's post-World War I studies. According
to Schulin, both *Die Idee der Staatsräson* and *Die Entstehung de Historismus* rest on 'spiritualization'. The *Staatsräson*-study centers on a spiritualization of power politics.
And according to Meinecke, philosophers and historians like Hegel, Fichte, Ranke

37   Herzfeld, 'Friedrich Meinecke: Der Geschichtsdenker', 102-103; also: Meinecke, 'Willensfrei-
      heit und Geschichtswissenschaft', 3-29.
38   Herzfeld, 'Friedrich Meinecke: Der Geschichtsdenker', 104. 'Das eigene Erlebnis ist für ihn
      Auseinandersetzung mit den Problemen der ganz persönlichen Existenz wie mit den großen
      Fragen der eigenen Zeit'.
39   Ibidem, 111. 'Meinecke ist bis zu seinem Ende überzeugt geblieben, daß die so gehandhabte
      Geschichte von dem individuellen – X – des Einmaligen der Persönlichkeit und des
      Ereignisses letzten Endes den Weg in das Religiöse und Metaphysische zu weisen vermöge'.
      In my analysis of *Die Idee der Staatsräson* I will come back to Meinecke's X, which he borrowed
      from Droysen, see chapter 2.
40   Herzfeld, 'Friedrich Meinecke: Der Geschichtsdenker', 107.
41   Schulin, 'Das Problem der Individualität', 109.

and Treitschke were responsible for this spiritualization, which ultimately led to a 'coarsening of the idea of the power state'.[42] In *Die Idee der Staatsräson*, Schulin claims, Meinecke uncovered the dark realms of power politics, yet, in his study on historicism Meinecke is also guilty of a kind of spiritualizing, for he describes the emergence of historicism only until the so-called cultural ideal (*Kulturideal*) of the *Goethezeit*.[43] Meinecke discusses the emergence of historicism up to Goethe, because, according to Schulin, historicism at that time was not yet associated with the state, as was the case with later historicism.[44] Nonetheless, Schulin fails to see the relationship between politics (*Staatsräson*) and history (historicism), because the principle of *Staatsräson* – a statesman must know the history of his state to act prudently in the present – is, according to Meinecke, the precursor of historicism.[45]

Meinecke's 'ideal of Goethe' or the *Goethezeit* is, in Schulin's opinion, important, because it replaces the actual historical development. Put differently, the individual and the unique, which are related to the absolute, were now raised to an ideal level and thereby no longer related to history or the 'principle of development'. Thus, the 'horizontal' philosophy of history, in which the emphasis is on development is replaced by Meinecke's so-called 'vertical' philosophy of history, which is aimed at the eternal, the absolute.[46] As a result, according to Schulin, only a few historians could relate to Meinecke's historicism, for it was on such an exclusive and high level of abstraction.[47]

Despite Schulin's thorough analysis, he still relates Meinecke's philosophical turn to political realities like the First World War, which apparently was the cause of Meinecke's dualism. Furthermore, shortly before the Second World War Meinecke was, in Schulin's opinion, forced to abandon politics, and as a result his world view became bogged down in abstractions and ideals. Herzfeld, as mentioned earlier, assumed a fundamental principle in Meinecke's philosophy of history, which enabled him to intellectually withstand the changing environment, instead of adapting to the political circumstances. According to Herzfeld, this means that everything was

---

42   Ibidem, 108-109. I will come back to this issue in the chapters on Meinecke's *Die Idee der Staatsräson*; chapters 2 and 3.

43   Meinecke frequently makes use of the notion 'Goethe-era' (*Goethezeit*), however, it is a misleading term. Peter Matussek clarifies why it invites one to consider an: '(...) extrem unruhigen und in mehrfacher Hinsicht revolutionären Epoche durch seinen Namen eine einheitliche Identität zu unterstellen, während es sich in Wirklichkeit umgekehrt verhält: die Bedeutung Goethes für seine Zeit besteht nicht darin, daß er ihre Beben und Umbrüche kompensatorisch ausglich, sondern seismographisch erspürte und darauf mit einzigartiger Prägnanz reagierte'. Peter Matussek, *Goethe zur Einführung* (Hamburg 1998) 12.

44   Schulin, 'Das Problem der Individualität', 109.

45   Cf. the beginning of chapter 2.

46   Cf. chapters 4 and 5.

47   Schulin, 'Das Problem der Individualität', 109.

in the end united in personality, in the individual. Herzfeld claims Meinecke's metaphysics was indissolubly linked with the individual, the concrete, and personal. Herzfeld, however, concludes (as is clear from the aforecited quote) that Meinecke's philosophy of history ultimately ends in the realm of 'religion', which refers mainly to a (Rankean) Christian belief. Schulin, in contrast, considers it not so much a Christian, but rather a personally experienced harmony of *Anschauung*, *Denken* and *Gefühl* from which an inner acquiescent image of God (*Gottweltbild*) emerges.[48] Meinecke himself once called this a 'secularized religion'.[49] What this means and what purpose it served in his philosophy of history and his political theory will be central to my research.

Unity in diversity

With regard to the debate on Meinecke's dualistic or monistic '(dis)continuous' philosophy of history, the American historian Christian Russell Jensen was the first to suggest an alternative. According to Jensen – whose unpublished dissertation has until now been completely ignored – none of the aforementioned interpretations of Meinecke's philosophy of history are satisfactory.[50] According to Jensen all these studies explain Meinecke's argument in terms of 'either-or' instead of 'both-and'.[51] The so-called antinomies – contrasting concepts or judgements that can nevertheless both be true – are, in Jensen's view, at the heart of Meinecke's thought.[52] In his dissertation entitled 'Unity in Antinomy', Jensen argues that these 'contiguous polarities' enabled Meinecke to differentiate between polarities like *Weltbürgertum* and *Nationalstaat*, power and ethics, without tying himself down to either one of them. This made Meinecke more flexible than a monistic or dualistic position would allow for.[53] Hence, Jensen argues, Meinecke was able to maintain his balance in practice as well as in theory, that is: on the one hand he could

---

48   Ibidem, 101.

49   Friedrich Meinecke, 'Deutung eines Rankewortes' in: idem, *Zur Theorie und Philosophie der Geschichte* (Stuttgart 1965) 117-139, there 137.

50   Christian Russell Jensen, '"Unity in Antinomy". The Hermeneutics of Friedrich Meinecke' (Chicago 1974). unpublished dissertation.

51   Jensen, '"Unity in Antinomy"', 2. These both-and antinomies are also a typical feature of the Fin-de-Siècle – the age in which Meinecke lived. See: Michael Saler, 'Introduction', in: idem ed., *The Fin-De-Siècle World* (London and New York 2015) 1-8, there 4.

52   Ibidem, 4; Contrary to Jensen, Georg Masur argued that Meinecke's thought centers around contrasts instead of antinomies: 'Er dachte in Kontrasten, nicht in Antinomien, da, allen philosophischen Systemen zum Trotz, die historische Wirklichkeit ihm zeigte, daß Befruchtungen und Verschlingungen feindlicher Ideen möglich waren'. Masur seems to be referring to principles that are similar to Jensen's antinomies. Masur, 'Ethics in a World of Power', 609. Jensen does not mention Masur.

53   Jensen, '"Unity in Antinomy"', 4, 178.

for example criticize an uncontrolled power politics and on the other applaud a *Staatsräson* based on *Geist* and political power.[54]

The most important, what Jensen calls, 'polar relationship' in Meinecke's thought is that of *Geist* and the 'power-political state'. Although Meinecke ascribes different meanings to the notion of *Geist*, Jensen claims Meinecke mainly refers to the spiritual and intellectual life of the nation: the works of the major German poets and philosophers around 1800.[55] This interpretation of *Geist* is too strict and at the same time too general, for *Geist* in Meinecke's view is collective as well as individual.[56] Jensen admits Meinecke attributes different meanings and synonyms to the notion of *Geist*, yet he maintains it primarily refers to the products of the great poets and philosophers.[57] Furthermore, he argues that all other polarities basically fit the antinomy of *Geist* and *Staat*.[58] Nevertheless, not every work is compatible with this antinomy. For example, in *Die Entstehung des Historismus* politics are of secondary concern. The many different meanings Meinecke assigns to concepts like *Geist*, *Staat*, *Macht*, *Ethik* and *Natur*, cannot be reduced to an antinomy of only *Geist* and *Staat*, as Jensen wants us to believe.

It appears Jensen's antinomies are a variation on a dualistic theme, in which also Meinecke's alleged political and intellectual adaptability (as postulated by Hofer among others) is removed, because thanks to these antinomies Meinecke was flexible in his attitude towards a changing reality. Jensen's purpose was to examine the emergence and development of Meinecke's preference for antinomies and to deepen our understanding of Meinecke's thought. Nevertheless, it remains unexplained on what fundamental principles these antinomies are based. Jensen points at Meinecke's 'environment' in which he saw himself confronted with monistic and dualistic systems. And further, through Ranke, Droysen and Dilthey, Meinecke was able to orientate himself in the different political and intellectual crises which he encountered during his life. But in the end, it remains unclear how a 'unity in diversity' could also ensure political and moral stability.

Of course, Jensen does not shed any light on the principles of Meinecke's antinomies, since he thought these antinomies to be the very basis themselves. He concludes that to Meinecke antinomies were always crucial: 'his goal was to maintain the contiguous relationship between polarities he investigated'.[59] In my research it will become clear that Meinecke tried to reconcile or even transcend the many different contrasts and antinomies. Moreover, he strove for or even assumed a

---

54 Ibidem, 4-5. Cf. Chapters 2 en 3.
55 Jensen, '"Unity in Antinomy"', 5-6.
56 Meinecke's views on *Geist*, *Kultur*, *Natur* and *Macht* cf. chapter 2 and 5.
57 Jensen, '"Unity in Antinomy"', 59.
58 Ibidem, 6.
59 Ibidem, 238.

fundamental harmony. Jensen's antinomies can thus be reconciled by means of Meinecke's panentheistic philosophy of history.

Verstehen or criticising

During the sixties and seventies, a couple of studies on Meinecke were published, in which worship and admiration were replaced by severe critique. In 1965 the Dutch historian Maarten Brands (1933-2018) published a dissertation entitled 'Historicism as Ideology'. This study, in which among other things, Meinecke's historicism is analyzed, was a reaction to Walther Hofer's extensive work on Meinecke. In Brands' view, Hofer had failed to criticize Meinecke's historicism, which should be considered a conservative ideology.[60] Brands denounced what he characterized as Meinecke's 'uncritical' philosophy of history. 'Uncritical', in Brands' view, refers to conservative historicists like Meinecke, who 'only' want to explain and understand (*Verstehen*) the past instead of criticizing past ideologies that appear to be no longer viable.[61] Brands defines an ideology as follows: 'a mode of thought that is determined by a system, and yet, because of this determination it will, by definition, be superseded by historical development and therefore become alienated ever further

---

60   Maarten C. Brands, *Historisme als ideologie. Het 'onpolitieke' en 'anti-normatieve' element in der Duitse geschiedwetenschap* (Assen 1965) 100 note 1; Maarten Brands, 'Meinecke between "Macht" and "Innerlichkeit" in: Michael Erbe, ed., *Friedrich Meinecke heute. Bericht über ein Gedenk-Colloquium zu seinem 25. Todestag am 5. und 6. April 1979* (Berlin 1981) 176-185, there 176-177. Brands' argument also refers to his dispute with the Dutch historian Hermann von der Dunk (1928-2018). Cf. Herman Paul, *Het moeras van de geschiedenis. Nederlandse debatten over historisme* (Amsterdam 2012) 228-235. Also: H. W. von der Dunk, 'Friedrich Meinecke en het historisme; een blik terug', *Tijdschrift voor Geschiedenis* 79 (1966) 24-37. Recently Brands published a volume in which the same view on Meinecke is still articulated: Maarten Brands, *Het arsenaal van de geschiedenis. Over theorie en geschiedschrijving* (Amsterdam 2013) 247-262. Other critical works written in a similar vein are: Waldemar Besson, 'Friedrich Meinecke und die Weimarer Republik. Zum Verhältnis von Geschichtsschreibung und Politik', *Vierteljahreshefte für Zeitgeschichte* 7 (1959) 113-129; Gustav Schmidt, *Deutscher Historismus und der Übergang zur parlamentarischen Demokratie. Untersuchungen zu den politischen Gedanken von Meinecke / Troeltsch / Weber* (Lübeck 1964); Several critical accounts were also published in the US: Louis L. Snyder, *German Nationalism. The Tragedy of a People. Extremism contra Liberalism in Modern German History* (Harrisburg and Pennsylvania 1952) especially 255-283; and in the UK: Werner Stark's introduction to the English translation of Meinecke's *Die Idee der Staatsräson*: Friedrich Meinecke, *Machiavellism. The Doctrine of Raison d'état and its Place in Modern History*, transl. Douglas Scott (New Haven 1957) xi-xlvi; also published as: Werner Stark, 'Friedrich Meinecke' in: idem, *Social Theory and Christian Thought. A Study of Some Points of Contact. Collected Essays Around a Common Theme* (London 1958) 201-245. An exception in these decades is the subtle article by Schulin, 'Das Problem der Individualität'.

61   Brands, *Historisme als ideologie*, 3; Paul, *Het moeras van de geschiedenis*, 228-232.

from reality'.[62] According to Brands, Meinecke's historicism had become such an alienated mode of thought.

Brands claims Meinecke's historicism, with its emphasis on individuality and 'irrationality', should be understood as a conservative ideology, for every normative judgement is avoided in favour of a reproduction of a past reality.[63] Understanding the past was, therefore, more important to Meinecke than passing judgements or criticizing the past. Contrary to this view, Brands was convinced that judging the past should be at the heart of a critical view of history. That is why Brands prefers a 'progressive, critical view' of history to Meinecke's conservative 'irrational historicism'. For, says Brands: 'a future-oriented critique is the most important contribution history – as a social science – can make to the society to which it belongs'.[64]

To support the claim that Meinecke's historicism was a conservative political ideology, Brands points at the alleged connections between historicism, nihilism, and Nazism. With regard to Meinecke's historicism however, Brands is careful, although a remark like 'we have hesitated to relate the concept of nihilism to the etheric figure of Meinecke' seems to implicitly relate Meinecke's historicism to nihilism.[65] Brands' remarks on Meinecke's history of ideas are similar, for he char-

---

62   Brands, *Historisme als ideologie*, 7-8. '(...) een denkwijze of gedachtengoed, dat vastgesteld is in een *systeem* en vanwege dit feit per definitie door de historische ontwikkeling achterhaald wordt of reeds is en daarmee steeds verder van de werkelijkheid vervreemdt'. Brands', however, seems to be unaware of the fact that his own definition of ideology might also be outdated, or part of a 'superseded mode of thought'. In a later publication he admits his former critical view of history fell victim to the same blind spots: '(...) in spite of all its pretentions of being critical, the critical school has remained prone to the same kinds of limitations, of interpretations against which it was addressed'. Brands, 'Meinecke between "Macht"', 177.

63   Brands, *Historisme als ideologie*, 56, 80-81, 230-231, 249-250; Paul, *Het moeras van de geschiedenis*, 229.

64   Brands, *Historisme als ideologie*, 3. '(...) de belangrijkste bijdrage, die een sociale wetenschap als de geschiedenis behoort te leveren aan de maatschappij, waarbinnen zij bestaat, [is, RK] een op de toekomst georiënteerde kritiek (...)'. This point has been criticized by Von der Dunk, who states that Brands' view presupposes an objective viewpoint which is derived from its own time and serves as a basis on which the past can be submitted to critical analysis in order to decide which values are valid for the present and the future. H.W. von der Dunk, 'Friedrich Meinecke en het historisme', 28; Also: E.E.G. Vermeulen, *Waarden en geschiedwetenschap. Een vergelijking van de standpunten ingenomen door H.W. von der Dunk, A.G. Weiler, M.C. Brands. Met notities over die van J.M. Romein, G. Harmsen, J.H.J. van der Pot* (Assen 1978); on presentism (to which Brands' view leans): Ed Jonker, 'Hedendaags historisme en anachronisme', *Theoretische Geschiedenis* 21 (1995) 1-15, there 2-3.

65   Brands, *Historisme als ideologie*, 230. 'Wij hebben (...) geschroomd het begrip nihilisme uit te spreken ten aanzien van de etherische figuur van Meinecke'.

acterizes it as: 'a naïve trust in values (...), while it is, and has become abundantly clear that these values could also result in something brutish [*naturhaftes*]'.[66]

Relating Meinecke's historicism to nihilism and implicitly to Nazism – aside from the fact that both are anything but conservative ideologies – sounds like finalism.[67] It does, however, clarify why Brands wants to break with Meinecke's historicism, for according to Brands, it leads to a value relativism, which created the preconditions for Nazism. Yet, this is a fundamental misunderstanding, since Nazism was far from relativistic; in fact it advocated a series of extreme values and ideals. Hence, value relativism could not have led to Nazism. Moreover, historicism is about the individual is every respect, Nazism was mainly focussed on the collective, in particular the masses.[68] Brands, as just mentioned, was in fact referring to the lack of *resistance* to Nazism which was, in his opinion, endorsed by an 'anti-normative' view like Meinecke's philosophy of history. Instead, Brands suggests a future-oriented critique, which helps to (politically) orientate a society. Another argument quoted in evidence by Brands to support his thesis is Meinecke's alleged 'state-worship', which is, according to Brands, the result of Meinecke's state-centered view of history.[69] And this 'state-worship' contributed to the 'derailment' of Germany. Hence, when taking to heart Brands' critical view of history, historians should distance themselves from this nineteenth-century 'historical irrationalism'.[70]

---

66 Ibidem, 148. 'een naïef vertrouwen in waarden (...), waarvan toch maar al te zeer duidelijk werd en is geworden, hoe ook deze [waarden, RK] op hun beurt tot iets 'naturhaftes' kunnen worden'.

67 Cf. Hermann von der Dunk in his review of Brands' work: 'Ik kan echter niet akkoord gaan met de doctrinair-moralistische toon waarop hier Meinecke verweten wordt niet daar te staan waar een Nederlands historicus in 1965 staat' and: 'De causaliteit Historisme-Nihilisme-Nazisme, die hij [Brands, RK] toch duidelijk suggereert, lijkt mij dan ook onhoudbaar'. Von der Dunk, 'Friedrich Meinecke en het historisme', 28, 36. Cf. Waldemar Besson, 'Friedrich Meinecke und die Weimarer Republik', 113, 128-129; Iggers, *The German Conception of History*, 277. Stefan Meineke gives an overview of the politically-oriented secondary literature, although he only pays attention to German historiography. Brands, Iggers and the Italian reception are left out. Stefan Meineke, 'Parteien und Parlamentarismus im Urteil von Friedrich Meinecke' in: Gisela Bock and Daniel Schönpflug ed., *Friedrich Meinecke in seiner Zeit. Studien zu Leben und Werk* (Stuttgart 2006) 51-93, there 55-63; Meineke, *Friedrich Meinecke*, 1-59.

68 I will come back to these issues in the last chapter.

69 Brands, *Historisme als ideologie*, 111.

70 Ernst Schulin described such a form of breaking with traditions as a principle of what he called *Traditionskritik* and *Rekonstruktionsversuch*. According to Schulin, these ruptures in traditions (like for example the French Revolution or the world wars) lead to reconstructions of traditions, be it in a negative or affirmative manner. Ernst Schulin, *Traditionskritik und Rekonstruktionsversuch. Studien zur Entwicklung von Geschichtswissenschaft und historischem Denken* (Göttingen 1979) 16-23.

The German-American historian Georg Iggers (1926-2017) agrees with Brands on almost every level. Iggers fully endorsed Brands' association of historicism with nihilism and Nazism. For example, in 1967 Iggers claims: '[Meinecke, RK] did not recognize the responsibility of the German idealistic tradition for preparing the intellectual road to Nazism'.[71] A year later Iggers' major study on the tradition of the German conception of history was published, and again, in this study he states: '(...) it is difficult to escape the thought that the political ethics of historicism in its recognition of the rights of the state and its denial of minimal universal norms of political behavior contributed in a significant way to breaking down the barriers against political nihilism and totalitarianism in German[y]'.[72] Iggers did not agree with Brands on equating historicism with conservatism, for he stressed German historical thought was far more complex than Brands suggested. Iggers argues: 'For histori[ci]sm as an ideology played as central a role in liberal and democratic theory in Germany from the Wars of Liberation to the Weimar Republic as it did in conservative thought'.[73]

Both Brands and Iggers concentrate only on the political aspects of Meinecke's work.[74] Meinecke's view of history is disregarded as a reactionary, conservative ideology. Moreover, Brands and Iggers consider Meinecke's thought to be solely about politics. According to them, even the 'unpolitical' and 'withdrawn' study *Die Entstehung des Historismus*, which was published in 1936, was highly political precisely because of the timing.[75] With regard to Meinecke's historicism, Brands' and

---

71   Georg G. Iggers, 'The Decline of the Classical National Tradition of German Historiography', *History and Theory* 6 (1967) 382-412, there 395.

72   Georg G. Iggers, *The German Conception of History. The National Tradition of Historical Thought from Herder to the Present* (Middletown 1983, first published 1968) 277, on Meinecke and Nazism: 223-228. A critique on this study: H. Butterfield, 'Review', *The English Historical Review* 86 (1971) 337-342, in particular: 338-339, 342. Robert Pois also relates historicism to Nazism. His study, however, is full of misrepresentations, and excessively critical. For example, he states – without any explanation – that Meinecke was condemned to 'that unhappy tendency in German political thinking to consider the state as a unit'. Or: Meinecke allegedly was not capable of 'understanding an ideology'. Pois' argument becomes ridiculous when he starts questioning Meinecke's mental condition: 'in the inner most reaches of his [Meinecke, RK] rather confused psyche (...)'. And, finally, Pois claims that Meinecke in the end abandoned both politics and history! Pois, *Friedrich Meinecke and German Politics*, 104, 114, 79, 146. Also: Kossmann, 'Friedrich Meinecke (1862-1954)', 224; Ernst Schulin, 'Robert A. Pois, Friedrich Meinecke and German Politics in the Twentieth Century', *Historische Zeitschrift* 217 (1973) 454-456; Klaus Böhme, 'Meinecke als politischer Historiker', *Neue Politische Literatur* (1975) 110-114. In the last chapter I will come back to Meinecke's attitude towards Nazism.

73   Georg G. Iggers, 'Review', *History and Theory* 6 (1967) 112-117, there 116-117.

74   Schulin, 'Meineckes Leben und Werk', 131.

75   Brands, *Historisme als ideologie*, 111, 220-221; Iggers, *The German Conception of History*, 218-222.

Iggers' views are very limited and strict, for Meinecke's historicism is far too complex to reduce it to a conservative ideology.[76] Of course, politics was an important factor in Meinecke's ideas and life; nevertheless, his notion of historicism contains far more than Iggers and Brands suggest, as will become clear in my research.

Umdefinition or new paradigm

The value relativism debate, which was also associated with Meinecke's historicism, was revived during the eighties and nineties of the twentieth century.[77] Initially – and contrary to Brands and Iggers – criticism was not aimed specifically at the political, ethical, or ideological aspects, but concentrated solely on Meinecke's historicism in relation to ethical relativism. The German historian Otto Gerhard Oexle (1939-2016) and his pupil Annette Wittkau are the key instigators in this debate. They claim Meinecke's interpretation of historicism marks a 'wrong development' (Fehlentwicklung). Below I will only discuss Oexle's argument, since Wittkau wholly agrees with him.[78]

Oexle's criticism consists of five points, which can be summarized as follows. First, at the time when historicism was under great pressure Meinecke, according to Oexle, attributes a 'positive' content to historicism by reducing historicism to the nineteenth century. And by doing so, Meinecke, in Oexle's view, eliminated the problem of ethical relativism, which was of great concern at the time when Meinecke wrote his book on historicism. Oexle asserts ethical relativism is inherent

---

76 Even though Brands admits Meinecke's historicism consists of more than one dimension, he still emphasizes only the political one.

77 This revival of a desire for (Neo-Kantian) fixed values is probably the result of certain characteristics historicism has in common with postmodernism. On this issue of similarities and differences between these concepts: F. R. Ankersmit, *History and Tropology. The Rise and Fall of Metaphor* (Berkeley 1994) 182-238.

78 Wittkau's point of view regarding Meinecke is often over-simplified and tendentious: 'Die Ausgrenzung der Wertfrage und die nationalistische Umdeutung gaben dem Historismus eine neue Dimension, die ihn mit den politischen Grundtendenzen jener Jahre konform erscheinen ließen'. According to Wittkau Meinecke's interpretation of historicism as 'pure' German fits the political climate of the thirties, the period *Die Entstehung des Historismus* was published. Annette Wittkau, *Historismus. Zur Geschichte des Begriffs und des Problems* (Göttingen 1994) 195. Wittkau's argument has come under heavy criticism from Muhlack and Bahners: Ulrich Muhlack, 'Gibt es ein "Zeitalter" des Historismus? Zur Tauglichkeit eines wissenschaftsgeschichtlichen Epochenbegriffs' in: Otto Gerhard Oexle and Jörn Rüsen eds., *Historismus in den Kulturwissenschaften. Geschichtskonzepte, historische Einschätzungen, Grundlagenprobleme* (Cologne, Weimar and Vienna 1996) 201-219, there 211-219; Patrick Bahners, 'Literaturkenntnis schützt vor Denkmalsturz. Völlig losgelöst: der Historismus nach Annette Wittkau', *Frankfurter Allgemeine Zeitung* (17 May 1993); Also: Ernst Schulin, 'Neue Diskussionen über Historismus', *Storia della storiografia* 33 (1998) 109-118, there 111.

to historicism in the sense Ernst Troeltsch once stated: historicism entails 'an all-embracing historicization of the world and thought'.[79] Oexle's second, third, fourth and fifth points of criticism are closely connected. He claims (2) Meinecke's interpretation of historicism does not assume a 'historical dependent socio-cultural process', but a 'conception of ideas' generated by individuals. Subsequently, these individuals are considered the spiritual fathers of historicism. Meinecke, however, restricts these founding fathers, according to Oexle, to the Germans; historicism, therefore, should be regarded as a (3) particularly German phenomenon. Moreover, Oexle claims it is reduced to (4) the idealism of Humboldt and Ranke, and that is why (5) Meinecke's historicism should be considered as closed, for it is limited to the nineteenth century.[80]

Meinecke did indeed interpret historicism positively, but he did not ignore the problem of ethical relativism at all. He was an outstanding example of an historian who was conscious of the process of historicization, comparable to what Burckhardt and Troeltsch experienced. His focus on conscience as a 'fixed agency' to counter ethical relativism is also proof of that. This 'conscience' solution is characterized by Oexle as a turn or conversion (*Umorientierung*) to a philosophy of life comparable with Dilthey: 'This kind of triumph of historicism to Meinecke was subjective, individual, and ultimately based on a religious conviction'.[81] Oexle, however, fails to explain what, in this context, the concept 'religion' entails, and how it is tied to the 'individual' and the 'subjective' – this will be analyzed in detail in my research.

Oexle's distinction between a process of historicization and a 'conception of ideas' generated by individuals was not an issue for Meinecke. Instead, he regarded both as closely related. Oexle's accusation regarding Meinecke's alleged strict notion

---

79   Otto Gerhard Oexle, '"Historismus". Überlegungen zur Geschichte des Phänomens und des Begriffs' (1986), in: idem, *Geschichtswissenschaft im Zeichen des Historismus. Studien zu Problemgeschichte der Moderne* (Göttingen 1996) 41-72, there 64. 'einer umfassenden Historisierung der Welt und des Denkens'. Otto Gerhard Oexle, 'Meineckes Historismus. Über Kontext und Folgen einer Definition' in: idem, *Geschichtswissenschaft im Zeichen des Historismus. Studien zu Problemgeschichte der Moderne* (Göttingen 1996) 95-136, there 118; Otto Gerhard Oexle, 'Krise des Historismus – Krise der Wirklichkeit. Eine Problemgeschichte der Moderne' in: idem ed., *Krise des Historismus – Krise der Wirklichkeit. Wissenschaft, Kunst und Literatur 1880-1932* (Göttingen 2007) 11-116, there 95-96; Wittkau, *Historismus*, 190-196.

80   Oexle, '"Historismus". Überlegungen zur Geschichte', 64-65; Oexle, 'Meineckes Historismus', 118-119. Wittkau states: 'Mit dem Problemkreis, der bis in die 20er Jahre hinein unter dem Schlagwort des "Historismus" diskutiert worden war, dem Problem des Wertrelativismus, der Frage nach dem Verhältnis von Historie und Leben sowie auch mit der Frage nach den Möglichkeiten und Grenzen der empirischen Geschichtswissenschaft, hat auch der Historismus im Sinne Meineckes nichts mehr zu tun'. Wittkau, *Historismus*, 191.

81   Oexle, 'Meineckes Historismus', 120. 'Diese Art der Überwindung des Historismus war für Meinecke individuell, subjektiv und letztlich religiös begründet'.

of historicism and it being specifically German is also not convincing, for Meinecke considered historicism, apart from Germany's important role, as a European phenomenon.[82] Moreover, Meinecke's historicism is not restricted to the historical discipline but includes the 'whole of life'.[83] It therefore surpasses the boundaries of the historical sciences. And finally, Oexle's last point of critique fails to convince, for Meinecke's historicism is not 'closed' in the sense of being limited to the nineteenth century, because he searched for answers, historical insight and solutions to current problems; ethical relativism was one of them. In sum, Oexle's criticism is one-sided and reveals a distorted interpretation (*Umdeutung*) of Meinecke.[84]

During the eighties and nineties, the distinguished German historian Jörn Rüsen, Oexle's rival, considered historicism as a scientific paradigm and warned against Oexle's strict interpretation of historicism, which renders the twentieth century discussion absolute.[85] Instead, Rüsen suggests viewing historicism as a mode of thought between Enlightenment and modernity – thus, it should roughly be associated with the nineteenth century. This distinction is, according to Rüsen, not strict, because the Enlightenment already anticipated the different aspects of historicism.[86] In that sense, historicism was not a clean break, but the crowning glory of the Enlightenment.[87] Both Enlightenment and historicism were, in Rüsen's opinion, a phase in modernity, which nevertheless faded at the beginning of the twentieth century and was overcome by other, newer modes of thought.[88] Given the fact that Rüsen considered historicism as a new paradigm, it should not surprise

---

82   *Die Entstehung des Historismus* as well as *Die Idee der Staatsräson* discuss English, French, Italian, Scottish and Irish philosophers, politicians and historians.

83   I will elaborate on this issue in chapter 4.

84   Cf. Chapter 4. Schulin also criticizes Oexle for being one-sided in his analysis of Meinecke: Ernst Schulin, 'Neue Diskussionen über Historismus', 111.

85   Jörn Rüsen, 'Historismus als Wissenschaftsparadigma. Leistung und Grenzen eines strukturgeschichtlichen Ansatzes der Historiographiegeschichte' in: Otto Gerhard Oexle and Jörn Rüsen eds., *Historismus in den Kulturwissenschaften. Geschichtskonzepte, historische Einschätzungen, Grundlagenprobleme* (Cologne, Weimar and Vienna 1996) 119-137, there 136; Friedrich Jaeger and Jörn Rüsen, *Geschichte des Historismus. Eine Einführung* (Munich 1992) 8. Also: Schulin, 'Neue Diskussionen über Historismus', 113.

86   Jaeger and Rüsen, *Geschichte des Historismus*, 8-10; Jörn Rüsen, 'Historismus als Erkenntnisprinzip und Wissensform – einige Gesichtspunkte', in: idem, *Konfigurationen des Historismus. Studien zur deutschen Wissenschaftskultur* (Frankfurt am Main 1993) 17-28, there 18.

87   Jaeger and Rüsen, *Geschichte des Historismus*, 10; This has already been argued by Gadamer: Hans-Georg Gadamer, *Wahrheit und Methode. Grundzüge einer philosophischen Hermeneutik* (Tübingen 1960) especially the second part.

88   That is Rüsen's and Jaeger's argument in: *Geschichte des Historismus*; also: Georg G. Iggers, 'Historismus – Geschichte und Bedeutung eines Begriffs. Eine kritische Übersicht der neuesten Literatur' in: Gunter Scholtz ed., *Historismus am Ende des 20. Jahrhunderts. Eine internationale Diskussion* (Berlin 1997) 102-126, there 117, 119.

us that he disagrees with Meinecke's broad interpretation of historicism, that is, historicism as world view (*Weltanschauung*) or existential question (*Lebensproblem*). Rüsen considers such an interpretation as far too subjective.[89]

Ulrich Muhlack also explicitly turns against Oexle, and like Rüsen he considers historicism to be a paradigm, but, unlike Rüsen, this paradigm is not connected to a certain period. Historicism, in Muhlack's view, is a lasting accomplishment, which is still viable.[90] Muhlack essentially agrees with Meinecke's notion of historicism, in the sense that its inception is to be situated in the late eighteenth and beginning of the nineteenth century.[91] Muhlack situates the *professionalization* of history, however, three centuries earlier than Meinecke: in the sixteenth century. Muhlack does not consider this sixteenth-century historicism comparable with nineteenth-century historicism.[92] Further, he attacks Oexle's and Wittkau's claim that Meinecke was responsible for a 'wrong development' (*Fehlentwicklung*). In this context Muhlack demonstrates in detail Meinecke's interference in the debate on ethical relativism, and shows that Meinecke came up with a solution to these problems.[93] Muhlack, like Hinrichs, also points at Meinecke's notion of 'conscience' as a fundamental principle. Muhlack, however, also fails to explain the fundamental principle, the (metaphysical) ground, on which this conscience is based.

Continuity and historicization

During the mid-nineties, in reaction to Brands', Iggers', Oexle's, and Rüsen's criticism, a shift to a more understanding (*verstehen*) view of Meinecke came to the fore. In 1995 the German historian Stefan Meineke published a biography of the development of Meinecke's political ideas up to the end of the First World War.[94] *Verstehen* is at the heart of this biography, for, according to Stefan Meineke, previous criticism had pushed this aspect into the background. He aims to 're-examine the

---

89   For an elaborate analysis on this issue, see chapter 4. In his recent studies Rüsen became an advocate of a 'new' humanism. For example: Jörn Rüsen and Henner Laass eds., *Humanism in Intercultural Perspective: Experiences and Expectations* (Bielefeld 2009).

90   Muhlack, 'Gibt es ein "Zeitalter"', 201-219. Also: Horst Walter Blanke, 'Aufklärungshistorie und "Historismus" im Denken Friedrich Meineckes' in: W. Bialas and G. Raulet ed., *Die Historismusdebatte in der Weimarer Republik* (Frankfurt am Main 1996) 142-160.

91   It is remarkable that Muhlack also agrees with Benedetto Croce's definition of historicism, which is a completely different version of historicism. Cf. section 4.6. Muhlack, *Geschichtswissenschaft im Humanismus und in der Aufklärung. Die Vorgeschichte des Historismus* (Munich 1991) 19-22.

92   Muhlack, *Geschichtswissenschaft*, 27, 42, 412; Also: Iggers, 'Historismus – Geschichte und Bedeutung eines Begriffs', 118; Schulin, 'Neue Diskussionen über Historismus', 113.

93   Muhlack, 'Gibt es ein "Zeitalter" des Historismus?', 212-219.

94   Meineke, *Friedrich Meinecke*, 2.

relationship between Meinecke's world view and his historical-political thought'.[95] His main objective is to demonstrate that the First World War was not a cause for a deep crisis in Meinecke's thought, for already in his youth Meinecke adopted a 'classical liberal-humanistic' conviction, which gave him the tools to withstand the political crises Germany encountered. 'Only', says Stefan Meineke, 'because he held on to his world view and the ideals of classical liberalism, was Meinecke able to commit himself to the Weimar Republic and fight against the rise of National Socialism'.[96] Stefan Meineke, in short, refutes Hofer's dominant and persistent idea of a rupture in Meinecke's thought. Meinecke, on the contrary, was not a flexible mind or an historian willing to change (*Umlernbereit*) in times of crisis, but a man of 'consistent political principles', and, according to Stefan Meineke: 'Since Meinecke did not identify his values with certain political systems or institutions, he could remain open to historical change'.[97] Stefan Meineke concludes that the so-called 'turn-thesis' (*Wandlung-These*) – Meinecke supposedly underwent a crisis caused by the First World War – has little to do with Meinecke's political thought, but instead, it is an indication of the condition of German historiography around 1945 and after.[98]

Stefan Meineke concentrates on Meinecke's political articles, his studies up to and including *Weltbürgertum und Nationalstaat*, his autobiographical works, and his correspondence. What is striking about this political biography is the use of a so-called 'developmental psychology' to analyze Meinecke's political commitment and his political 'fundamental conception' (*Grundvorstellung*).[99] This psychological method assumes that convictions and 'philosophical orientations' of a person are developed between the age of fifteen and twenty-five.[100] Stefan Meineke explains how different conflicts in Meinecke's personal life – in particular his childhood – contributed to moulding his political convictions.[101] In this context, it supposedly was the poet Wilhelm Raabe (1831-1910) who was of great importance for Meinecke's oeuvre, because this poet's work apparently articulated Meinecke's 'youth issues'. The shaping of Meinecke's identity was particularly complicated by the pressure

---

95   Ibidem, 43. 'den Zusammenhang zwischen Meineckes weltanschaulichem und historisch-politischem Denken erneut nachzugehen'.

96   Ibidem, 307. 'Nur weil er in seinem weltanschaulichen Denken an den Wertidealen des klassischen Liberalismus unverbrüchlich festgehalten hat, war es Meinecke möglich, sich für die Weimarer Republik und gegen den aufkommenden Nationalsozialismus zu engagieren'.

97   Ibidem, 56, 306. 'gleichbleibender politischer Grundvorstellungen'. 'Da Meinecke seine Wertideale nicht mit bestimmten politischen Systemen oder Institutionen identifizierte, konnte er für historische Wandlungen offen sein'.

98   Ibidem, 312.

99   Ibidem, 44.

100 Ibidem, 48. 'weltanschaulichen Orientierungen'.

101 Ibidem, 60-89.

– the obligations concerning the Church (*Kirchenzwang*) – of his strict pietistic father.[102] Meinecke eventually broke with his father's pietism to replace it with literature – Stefan Meineke claims Meinecke's interest in literature became a substitution for religion (*Ersatzreligion*).[103] This struggle with his father, the humanism and idealism he found in poets like Raabe, Mörike, and later on in Humboldt had thus a tremendous effect on Meinecke's political ideas.[104]

Literature was indeed very important for shaping Meinecke's world view. Stefan Meineke, however, exaggerates Raabe's influence on Meinecke's (political) thought. In fact, Raabe – who is only mentioned in Meinecke's letters and autobiography – is of minor importance regarding Meinecke's studies. Furthermore, considering the many references to, and articles on Goethe and Schiller, it is striking that Stefan Meineke fails to mention either of these poets, who were in fact extremely important to Meinecke's personal and intellectual development; in particular his philosophy of history and political theory are shaped by the thoughts of these poets.[105] With regard to Meinecke's break with pietism, I also disagree with Stefan Meineke. Even though Meinecke distanced himself from pietism, in which by the way the legacies of Neoplatonism and panentheism are interwoven, it did leave an indelible mark on his thought, which is clearly articulated in his studies – in this research I will elaborate on this issue at length.

Stefan Meineke's main sources to support his thesis consist of Meinecke's political writings and his autobiography.[106] He demonstrates that both these sources show that the First World War did not cause an intellectual crisis. The political writings indeed give evidence of this, however, to use Meinecke's autobiography and correspondence as evidence for this statement is not very convincing. What is worse, many different quotes from Meinecke's autobiography and letters serve as a guide for the thesis of Stefan Meineke's biography. This poses a problem, because thinkers, in this case Meinecke, are not always the best interpreters of their own work and intellectual development (*Werdegang*).

To justify his choice to use Meinecke's personal life as a basis for understanding his political views, Stefan Meineke points at Meinecke's own statements on 'life' as a foundation for his world view.[107] He assumes Meinecke equates 'life' with his own life, or life as such, while 'life' in Meinecke's explanation refers to life as a continuous 'process of becoming' and thereby it points at a view (of history) in which collective and individual history merge. Meinecke is not first and foremost interested in

---

102   Ibidem, 68, 74.
103   Ibidem, 75.
104   Ibidem, 73-77.
105   Meinecke, 'Erlebtes', 32, 42.
106   Meineke, *Friedrich Meinecke*, 43, 298
107   Ibidem, 58.

understanding the personal convictions of individuals, but wants to fathom (ideas of) history; *that* is the fundamental or existential question (*Lebensproblem*) Meinecke saw himself confronted with.

I do agree with Stefan Meineke that Meinecke had a 'fundamental conviction', however, it was not limited to 'classical liberalism' and it did not originate in his childhood. Meinecke's world view developed in close interaction with his time and especially with the tradition in which he positioned himself. And, in that sense, the 'consistent political principles', which Stefan Meineke referred to, can be based on Meinecke's panentheistic conviction: the leitmotiv of my research.

From the year 2000 onwards Meinecke studies still hold understanding (*verstehen*) to be the right way of analyzing his life and work. Instead of judging Meinecke's oeuvre historicizing his life and thought is more and more accepted as the standard.[108] What strikes one most is the general acceptance of the 'continuity-thesis'. Meinecke is no longer portrayed as an historian who got carried away with political and intellectual revolutions, but as an historian with a firm conviction, who in particular tried to explain the present against the background of the past. The main study in which Meinecke is portrayed in this sense, is the volume *Friedrich Meinecke in seiner Zeit. Studien zu Leben und Werk* (2006) edited by Gisela Bock and Daniel Schönpflug.[109] This volume was published on the occasion of the fiftieth anniversary of Meinecke's death and the symposium 'Friedrich Meinecke: Erinnern, Gedenken, Historisieren', which was organized in 2004 in remembrance of Meinecke.[110] In their preface Bock and Schönpflug state that their volume, in many respects, concurs with the conference volume of 1981, which was published on the occasion of the twenty-fifth anniversary of Meinecke's death.[111] This volume was the result of a colloquium on Meinecke, which was also organized with the aim

---

108   Laube is the exception. He is critical in the tradition of Oexle: Reinhard Laube, *Karl Mannheim und die Krise des Historismus. Historismus als wissenssoziologischer Perspektivismus* (Göttingen 2004) in particular chapter 2: 'Von Heussis "Krisis des Historismus" über Meineckes "Entstehung des Historismus" zu Wehlers "Kritischer Wissenschaft" – und zurück', 87-197.

109   Bock and Schönpflug, *Friedrich Meinecke in seiner Zeit*; at the same time a study was published on Meinecke and his students: Friedrich Meinecke, *Akademischer Lehrer und emigrierte Schüler*. The tenth volume of Meinecke's collected works was edited by Ritter and Bock (in cooperation with Stefan Meineke, among others): Meinecke, *Neue Briefe und Dokumente* (Munich 2012). Also: Ernst Schulin, 'Treitschke und Max Weber, Meinecke und Gerhard Ritter. Politik und Geschichte in Freiburg zwischen 1863 und 1967' in: Achim Aurnhammer en Hans-Jochen Schwier ed., *Poeten und Professoren. Eine Literaturgeschichte Freiburgs in Porträts* (Freiburg 2009) 193-220; Carla du Pree, 'Intellectueel in tijden van crisis. Friedrich Meinecke en het publieke debat in Duitsland van 1914 tot 1954', *Streven: cultureel maatschappelijk maandblad* 79 (2012) 226-236.

110   Bock, 'Vorwort', 7.

111   Michael Erbe ed., *Friedrich Meinecke heute. Bericht über ein Gedenk-Colloquium zu seinem 25. Todestag am 5. und 6. April 1979* (Berlin 1981).

to historicize Meinecke's work. The Dutch historian Ernst Kossmann (1922-2003), who was present at this colloquium, describes the critical and slightly awkward atmosphere of this gathering in 1979:

> In general, the discussion was critical. The organizers thought it even necessary to somewhat apologise for the whole venture. In Germany at that time it was not considered a matter of course or even worthwhile to study and discuss his [Meinecke's, RK] oeuvre; and when it was considered useful one had to be careful to avoid the impression of too much praise.[112]

A similar description is offered by another Dutch historian, Offringa, in his review of the volume of 1981. He concludes: '(...) we have, for the time being, finished discussing Meinecke. We have to await a new generation, which will rediscover him in the light of different experiences'.[113]

Evidently, that new generation is represented in the volume of 2006. A major difference between the articles of the 1981 volume and the 2006 volume is the positive tone, but more importantly in 2006 the controversies surrounding Meinecke's life and especially his work are discussed with historical distance. In that sense these controversies have become themselves subjects of historical research. This corresponds with the quote of Bock and Schönpflug that I mentioned at the beginning of this introduction: the emphasis is no longer on 'praise and reproach', but on acquiring and deepening insight into Meinecke's work. What is most striking regarding the 2006 volume is the slight rehabilitation of Meinecke. He is portrayed as an historian who was a 'strong opponent of National Socialism'.[114] This trend,

---

112   Kossmann, 'Friedrich Meinecke', 210. 'De toon van de discussie was over het algemeen kritisch. De organisatoren vonden het zelfs noodzakelijk de onderneming enigszins te verontschuldigen. Het sprak toen in Duitsland blijkbaar niet vanzelf dat zijn [Meineckes, RK] oeuvre de moeite van studie en bespreking waard was en men moest, toen men dat wel meende, zeer zorgvuldig de indruk vermijden als zou men bezig zijn bloemenkransen op zijn graf te leggen'.

113   C. Offringa, 'Friedrich Meinecke herdacht', *Theoretische Geschiedenis* 14 (1987) 220-223, there 223. '(...) wij zijn voorlopig over Meinecke uitgesproken. Het wachten is op een generatie die hem vanuit andere ervaringen zal herontdekken'.

114   According to the back cover of the 2006-volume: Gisela Bock and Daniel Schönpflug ed.,*Friedrich Meinecke in seiner Zeit*. 'als entschiedener Gegner des Nationalsozialismus äußerte'. Ritter's introduction also emphasizes Meinecke's aversion to National Socialism: Meinecke, *Akademischer Lehrer und emigrierte Schüler*, 24-27; In a review on the 2006 volume and Ritter's study, Georg, Iggers attacks this issue: '(...) Meinecke's position in the Third Reich was by no means as clear-cut as it appears from the correspondence in the Ritter volume, or in the essays in the Bock and Schönpflug book (...)' And: 'He [Meinecke, RK] kept his distance from the Nazi regime and was aware of its negative sides, but he did not spell out what they were'. Georg G. Iggers, 'Book reviews', *Central European History* 41 (2008) 151-155, there 154. In chapter 6 I will come back to this issue.

as it turns out, is also manifest in the tenth volume of Meinecke's collected works, published in 2012. Gisela Bock, one of the editors of this volume, claims in the introduction: 'His letters (like his studies of this age) reveal a deep reservation towards the Nazi regime, and simultaneously show evidence of a desire to openly grapple with the politics of National Socialism instead of turning away from it'.[115] A 'rehabilitating historicization' of Meinecke's life and work are, in short, the central themes in this recent publication.[116]

Two other recent studies on Meinecke are worth mentioning in relation to my research.[117] The first one by Wolfgang Kämmerer discusses Meinecke in the context of his time and is primarily concerned with the 'problem of historicism'.[118] His main aim is to determine why, how, to what extent and with what arguments Meinecke participated in the debate on the problems of historicism.[119] His argument is reminiscent of the positions taken previously by Hofer, Jensen, Muhlack and Stefan Meineke. His argument against Oexle and Wittkau is in the same vein as Muhlack: Meinecke did indeed position himself in the debate on the crisis of historicism.[120] Kämmerer is close to Hofer in the sense that he primarily explains Meinecke's attitude by describing his concept of historicism against the background of Meinecke's time. Kämmerer's view on Meinecke's notion of conscience is in the end close to Jensen's explanation, which is in fact a description of what Meinecke himself

---

115 Gisela Bock, 'Friedrich Meinecke und seine Briefe. Eine Einführung' in: Friedrich Meinecke, *Neue Briefe und Dokumente* (Munich 2012) 1-23, there 10. 'Seine Briefe (ebenso wie seine Schriften dieser Zeit) zeugen von tiefer Distanz zum NS-Regime, aber auch von seinem Willen, nicht einfach die Augen zu verschließen, sondern sich mit der nationalsozialistischen Politik gesprächsweise auseinanderzusetzen'. Recently a re-issue of Meinecke's *Die deutsche Katastrophe* was published; this edition has an extensive part on the reception of this book until 2018. I will come back to *Die deutsche Katastrophe* in the last chapter of this book. The re-issue: Bernd Sösemann, ed., *Friedrich Meinecke, Die deutsche Katastrophe. Betrachtungen und Erinnerungen. Edition und internationale Rezeption* (Berlin 2018).

116 122 The famous Dutch historian Hermann von der Dunk also briefly discusses Meinecke in his study: H.W. von der Dunk, *De glimlachende sfinx. Kernvragen in de geschiedenis* (Amsterdam 2011) 45-47. Frits Boterman, *Duitse dichters en denkers. Het belang van cultuur in de moderne Duitse geschiedenis* (Amsterdam and Antwerp 2008) 209-212. On a theoretical level criticism has become less severe. Pyper describes an historic and philosophical discussion (and correspondence) between Meinecke and Croce. Pyper troubles over Hegel's influence on Croce and Meinecke. Pyper, by the way, claims Meinecke's post-war (World War I) philosophy of history to be dualistic. I my research I will have a more subtle approach to this issue. Pyper, 'Meinecke, Croce, and the Individual', 68.

117 I will discuss Nicholas Berg's argument in chapter six, for his study is not entirely on Meinecke and only discusses Meinecke's last book, which I will elaborate on in my last chapter.

118 Wolfgang Kämmerer, *Friedrich Meinecke und das Problem des Historismus* (Frankfurt am Main 2014). This is the published version of Kämmerer's dissertation of 2000.

119 Kämmerer, *Friedrich Meinecke und das Problem des Historismus*, 24.

120 Ibidem, 194-195.

said about conscience, not an analysis of how Meinecke philosophically grounds this idea of conscience. Kämmerer thus fails to address the tradition in which Meinecke was rooted, that is: German Idealism, Romanticism, and also the tradition of Neoplatonism. These traditions are important when discussing Meinecke's fundamental mode of thought. So, like Jensen, Kämmerer fails to answer the question of what Meinecke's idea of conscience was actually based on. Like Jensen, Kämmerer only describes what Meinecke claims about the notion of conscience.[121] However, Kämmerer does pose the relevant questions regarding Meinecke's notion of conscience, for he asks himself what Meinecke thought conscience meant, since Meinecke at first glance failed to really explicate what conscience is.[122] To answer such a question, one needs to know what the foundation of such a notion entails. Kämmerer, like many commentators before him, considers it to be a religious foundation, a religous belief.[123] Since Meinecke broke with the idea of a 'personal God' and replaced it – at a young age – with an abstract notion, a philosophical concept, this explanation does not hold ground. So it must be something else. Kämmerer suggests something which he calls 'Dazwischen', an in-between or go-between, a median between humans and a religious sphere, but it remains vague.[124] In the present research, I will try to give a satisfactory answer to the question of what conscience is founded on in Meinecke's philosophy of history.

Krämmerer admits that there is a gap in his research with regard to a discussion of Neoplatonism.[125] It is in fact regrettable, since a discussion of this philosophy is pre-eminently important to come to a balanced interpretation of Meinecke's philosophy of history. Moreover, Kämmerer states that he will only discuss Meinecke's analysis of thinkers like Herder, Goethe and Ranke, but he does not discuss whether Meinecke's interpretations are accurate or not.[126] In the same vein as Stefan Meineke, Kämmerer is more interested in Meinecke's interpretations, and even more so in his intentions and motives.[127] So, my comment on Stefan Meineke also applies to Kämmerer, for the question is: are thinkers or historians, in this case Meinecke, the best interpreters of their own work? Instead, a research on Meinecke also requires a thorough analysis of the thinkers he discusses, or one should at least understand what these thinkers were interested in. Then questions like why and how Meinecke selects their ideas become important. That way, it will become clear that Meinecke constructed a tradition of historicist thinkers. My

---

121   Ibidem, 237-239.
122   Ibidem, 285.
123   Ibidem, 275.
124   Ibidem, 279.
125   Ibidem, 315.
126   Ibidem, 104-105.
127   Ibidem, 16, 24. Since Gadamer and postmodern hermeneutics it is almost a provocative enterprise to claim, as Kämmerer does, to search for the author's intentions and motives.

research situates Meinecke in the tradition of thinkers he discusses, and thereby not only gives evidence of how this tradition was still viable during his age, but also how Meinecke positions himself within this tradition, while simultaneously constructing this tradition. The result of omitting a discussion of the philosophical tradition Meinecke was part of, leads in the end to Kämmerer claiming that Meinecke became more and more uncertain and in doubt during his life – that is almost Hofer's argument of monism turning into dualism and even pessimism.[128] Instead, I claim Meinecke was never really in doubt, since his philosophy gave him the certainty to withstand crises.

Speaking of a feeling of crisis, the second recent study on Meinecke is focussed entirely on Meinecke's experience of the many crises he witnessed.[129] The main argument of Ute von Lüpke's study is on how historians describe, interpret and evaluate crises they encounter during their lives.[130] As a case study Meinecke functions as the representative historian of the twentieth century. What is at stake in Lüpke's study is how Meinecke and his contemporaries perceived their present, and what historiographical, theoretical, methodical and political consequences they drew in ordering and coping with times they experienced as revolutionary.[131] Lüpke claims her study is not a continuation of Stefan Meineke's research.[132] Thus for Lüpke, it is not a (political) biography but an intellectual biography.[133] Essentially, Lüpke's research can be characterized as an historicist account of Meinecke's reaction to the many crises he witnessed.

Similar to Kämmerer, Lüpke discusses Meinecke and his context, and thus she also leaves out the tradition of thinkers Meinecke discusses, and thereby omits how this tradition was still viable during his age, and how it helped Meinecke position himself in several crises. Such an inquiry – researching the tradition Meinecke constructed and in which he positioned himself – would have lead to Meinecke's fundamental mode of thought and would lead to an explanation for Meinecke's notion of conscience for example. Lüpke refers to Meinecke's notion of conscience, but remains on the same level as Kämmerer and Jensen, that is: no explanation is given for the foundation of Meinecke's conscience, only vague references to 'religion'.[134] Which is regrettable, for it is pre-eminently his philosophy of history on

---

128   Kämmerer, *Friedrich Meinecke*, 133.

129   Ute von Lüpke, *Zäsuren – Katastrophen – Neuanfänge. Friedrich Meinecke und die Umbrüche der deutschen Geschichte im 20. Jahrhundert* (Hamburg 2015). In the author's preface it is mentioned that this book was originally a thesis defended in 2013/2014. This published version is slightly revised, but the *Forschungsstand* corresponds with the year 2013.

130   Lüpke, *Zäsuren – Katastrophen – Neuanfänge*, 18-24.

131   Ibidem, 18.

132   Ibidem, 29.

133   Ibidem, 13.

134   Ibidem, 340-346.

which Meinecke's notion of conscience is based and which gave him the tools to withstand the many crisis or at least cope with sudden ruptures during his intellectual life. It is, therefore, no surprise that Lüpke, like Jensen, considers Meinecke to be a self-confident historian who simultaneously had doubts when confronted with (intellectual) crises, but he in the end was – as Schulin stated earlier – *Umlernbereit*, willing to change.[135] Hence, Lüpke fails to explain why Meinecke hovers between these positions; this dovetails with her research for it is aimed at how historians experience crises, how they describe, interpret, and evaluate them. It does not explain why Meinecke takes this stand in the first place. An explanation for Meinecke's seemingly ambivalent position should address his – to him self-evident – philosophy of history, and should discuss his tradition and how he relates to this tradition. This is all crucial for an account of Meinecke's notion of conscience, his whole *Weltanschauung*, and his reaction to and attitude towards crises.

Before ending this detailed overview of Meinecke studies, some last remarks on recent publications and articles which do not focus exclusively on Meinecke but take his ideas as a reason to dismiss his conception of history or as a departure to correct or add to them and use them for particular puposes, for example: to legitimize one's own research. The historian D. Timothy Goering in his introduction to the volume *Ideengeschichte heute* more or less claims Meinecke is history and at the same time considers him the founder of the study of the history of ideas, which is the main subject of Goering's volume.[136] He only discusses Meinecke very briefly and only his first major work, which is considered the foundation of the *politische Ideengeschichte*. Such a discussion is of course not fair to Meinecke's whole development and misrepresents Meinecke – as Goering does – as an historian who was solely interested in great thinkers and their ideas, independent of the problems and issues of his own time.[137] In other words, Goering fails to discuss Meineckes *Historik*, historicism and *Weltanschauung*. All three could have contributed not only

---

135   Ibidem, 457, 483-484. Schulin, 'Friedrich Meinecke' 39. For an appreciative review of Lüpke's book: Hans-Christof Kraus, 'Ute von Lüpke, Zäsuren – Katastrophen – Neuanfänge. Friedrich Meinecke und die Umbrüche der deutschen Geschichte im 20. Jahrhundert (Review)', *Historische Zeitschrift* 306 (2018) 936-937. On the ambivalent attitude of Meinecke with regard to 'Umlernen': Hans-Christof Kraus, 'Friedrich Meinecke als Korrespondent. Zu den neuen Briefeditionen', *Historische Zeitschrift* 298 (2014) 89-100, there 91. With regard to Meinecke's alleged 'will to change' in relation to the Second World War: Franka Maubach, '"Wie es dazu kommen konnte". 1933 als Fluchtpunkt deutsch-deutscher Ursachensuche im frühen Kalten Krieg' in: Franka Maubach and Christina Morina ed., *Das 20. Jahrhundert erzählen. Zeiterfahrung und Zeitforschung im geteilten Deutschland* (Göttingen 2016) 143-189, there 161-162.

136   D. Timothy Goering, 'Einleitung. Ideen- und Geistesgeschichte in Deutschland – eine Standortbestimmung' in: idem ed., *Ideengeschichte heute. Traditionen und Perspektiven* (Bielefeld 2017) 7-53, there 7, 13.

137   Goering, 'Einleitung', 14, 49.

to a better understanding of Meinecke's notion of the history of ideas, but also would have given a fair discussion of his philosophy of history. Goering is of course interested in the tradition of *Ideengeschichte* and how it developed after the Second World War and whether certain older views are still valid. However, such a discussion does not justify a one-sided account of Meinecke's philosophy of history, which, by the way, was always grounded in reality, that is, Meinecke starts every research from contemporary problems and tries to find (in a dialogue with his past) a solution, and hopes to gain insight into and maybe an answer to current problems. It is all the more astounding that Goering discusses the relation between ideas and practice. Meinecke is not mentioned, neither is Croce (history as thought and action), or Collingwood; historians of action par excellence.[138] Goering's most unwarranted claim is at the end of his introduction where he states that a revaluation of Meinecke's work is long overdue.[139] Considering the above overview and recent publications on Meinecke, this statement is clearly evidence of a lack of insight into the most up-to-date Meinecke studies. Nonetheless, I will take this request to heart and hope my study will constribute to the lacuna in the revaluation of Meinecke's work.

Finally, a word on a study that is also not directly on Meinecke, but on historicism; however, the immediate motivation for this study was Meinecke's conception of historicism. And moreover, it is a study aimed at introducing an Anglophone audience to German historicism, and is therefore important to discuss with regard to its interpretation on Meinecke. First, the American philosopher Frederick Beiser's *The German Historicist Tradition* is a welcome and much-needed follow-up on Iggers' study on the German conception of history, but it falls short of a thorough discussion of Meinecke. Beiser takes Meinecke's *Die Entstehung des Historismus* as a starting point (a 'model' and 'precedent', as Beiser states) for his own research, which he considers to be an addition and correction to Meinecke's book on historicism.[140]

---

138   Rik Peters, *History as Thought and Action. The Philosophies of Croce, Gentile, de Ruggiero, and Collingwood* (Exeter and La Vergne 2013).

139   Goering, 'Einleitung', 49. Recently Borowksi made an attempt. Apart from the many inaccurate references in her chapter, her argument as a whole – Meinecke unintentionally contributed to the defeat of German historicism – is a step back to the views of Iggers, Pois, G. Strauss, Oexle and Wittkau. Further, she fails to address the foundations of Meinecke's notion of conscience (which she equates with consciousness). And when she discusses Meinecke's and Goethe's idealism, she fails to acknowledge that Goethe in fact was also an empiricist – one of the reasons Meinecke made use of his ideas – which draws her to the conclusion that Meinecke and Goethe are only interested in the 'mystical'. Audrey Borowski, 'Friedrich Meinecke's *Historism* or the Defeat of German Historicism' in Herman Paul and Adriaan van Veldhuizen eds., *Historicism: A Travelling Concept* (London 2020) 165–185. Cf. Chapters 4, 5, 6.

140   Frederick C. Beiser, *The German Historicist Tradition* (Oxford 2011) vii.

In his preface he states three points on which he dismisses Meinecke's work on historicism. First, Meinecke supposedly ignores crucial figures in his discussion of historicism. Second, Beiser considers his interpretations anachronistic. And third, his conceptual scheme is misleading and simplistic.[141] This is quite severe criticism, all the more since Beiser does not give an in-depth analysis of Meinecke's historicism, nor does he examine any of Meinecke's other works. With regard to Beiser's claim of Meinecke's alleged anachronism and his simplistic and misleading conceptual scheme, I will come back to those issues in the next chapters. Beiser's first point of criticism can be dealt with here.

Beiser reproaches Meinecke for ignoring certain thinkers, that is, thinkers who, according to Beiser, belong within the tradition of historicism. Meinecke's study is, however, not 'the' history of the rise of historicism, but an account of Meinecke's *own* view of historicism. The same applies to Beiser's study, in which he mainly considers historicism from the perspective of a scientific tradition. His definition of historicism – as the Dutch philosopher of history and historian Herman Paul clearly states in his review – is quite narrow: Beiser's historicism is a nineteenth-century German philosophical tradition, which is aimed at justifying the scientific position of the practice of history.[142] So, it is about an epistemology of the historical sciences, not a world view as Meinecke thought it to be.[143] All thinkers that do not fit Beiser's definition of historicism are cast aside, which ultimately results in a Whig history of historicism.[144]

Moreover, it is striking that Beiser, who has written several penetrating studies on German Idealism and Romanticism, did not use that knowledge to analyse Meinecke's historicism, in which these traditions are well-attested. In short, Beiser fails to elaborate on Meinecke's study on historicism, or give a fair account of it against the background of the philosophical traditions, and he overlooks the complexity of Meinecke's work in favour of his own conception of historicism, which is in the end a scientistic view of historicism.[145]

This overview of Meinecke's effective history (*Wirkungsgeschichte*) shows a wave-like movement in which admiration, disparagement, differentiation and revalua-

---

141   Beiser, *The German Historicist Tradition*, viii, also: 28-29, 101.

142   Herman Paul, 'Historisme op een procrustusbed', *Tijdschrift voor Geschiedenis* 126.1 (2013) 134-136.

143   Paul, 'Historisme op een procrustusbed', 135; Beiser, *The German Historicist Tradition*, 8.

144   Paul, 'Historisme op een procrustusbed', 135.

145   Idem. Next to Beiser, Jeremy Popkin mentions Meinecke in a few pages in *From Herodotus to H-Net. The Story of Historiography* (New York and Oxford 2016) 69, 76, 105. His brief account on Meinecke tells the general story, which also includes the idea that Meinecke changed his position with regard to historicism due to the Nazi era; I will come back to these issues in the last chapter. Moreover, Popkin claims Ranke is the central figure in Meinecke's *Die Entstehung des Historismus*, whereas Goethe is in fact the key thinker in this magnum opus.

tion of Meinecke's oeuvre took turns in quick succession. In close connection to this, Meinecke's work and its development are characterized variously as monistic, dualistic, harmonious, politically conservative, ideological, ethically relativistic, and estranged from reality and therefore bogged down in (one-sided) philosophical abstractions. As far as praise and reproach are focussed on aspects of his work, the different characterizations of Meinecke's studies are related to the argument on the (dis)continuity of (the development of) his work. In sum, it is the shift in the Meinecke studies and also the character of Meinecke's oeuvre, which are the causes of the quick succession of contrasting and more subtle and nuanced views. To address the (dis)continuities in Meinecke's oeuvre, I therefore suggest the consideration of a panentheistic view of history, that is, *Meinecke's* panentheistic philosophy of history. It is this philosophy or worldview that gave him the tools to withstand the many political and intellectual crises of his age. Moreover, it is this philosophy which is key to understanding his philosophy of history and life.

Methodological principles

This research is in line with the recent historicizing approaches, albeit without an (explicit) rehabilitation of Meinecke. This book is *not* a biography of Meinecke, but an intellectual history. Being a specific form of cultural history, intellectual history concentrates roughly on the coherence and/or development of ideas, views and theories of past (groups of) individuals. In general, it is about understanding political, scientific and philosophical ideas, that came into being in interaction with the historical context (*Umwelt*) and the individuals that came up with these ideas.[146]

This is a book on Meinecke's intellectual development (*Werdegang*). The emphasis is on his ideas and not so much on his life and the socio-cultural context.[147] These personal and socio-cultural perspectives will of course pass in review – after all Meinecke was not isolated from his own time – but the emphasis will be on

---

146   Frank Ankersmit, *Meaning, Truth, and Reference in Historical Representation* (Ithaca 2012) 130. There is an interaction between the social-cultural contexts and the (personal) ideas. For example, a study, in this case Meinecke's study, is an expression of ideas of an individual (Meinecke), but can also be the reflection of the political, cultural-intellectual climate. Donald Kelly has aptly formulated this overlap: 'cultural history is the outside of intellectual history, and intellectual history is the inside of cultural history'. Donald R. Kelly, 'Intellectual History and Cultural History: the Inside and the Outside', *History of the Human Sciences* 15 (2002) 1-19, there 12.

147   There are enough (extensive) biographical sketches on Meinecke's life available. Stefan Meineke's biography gives a good, albeit mainly politically oriented, description of Meinecke's life up to and including 1918. A complete biography of Meinecke's undeniably interesting and long life would be very welcome.

(the development of) his views. Furthermore, the dialogue with his contemporaries will also play a minor part in this research. For the characteristics of Meinecke's ideas in comparison to his contemporaries are of little importance compared to the significance of Meinecke's panentheism and the role it plays in his oeuvre. This, therefore, is not a comparative research. Judging by his intensive correspondence with colleagues, Meinecke was in close contact with his contemporaries; however, he was primarily in discussion with the tradition in which he positioned himself. In other words, in his oeuvre he tried to interpret the current political and cultural-intellectual events by analyzing the *history* of those events.[148] In history he thus searched for answers and meaning: Meinecke tried to understand Germany's past and present, and his own position in history and historiography. The best way to uncover his attempts at understanding – for this will lead to Meinecke's fundamental principle of history – is a chronological discussion of his trilogy *Weltbürgertum und Nationalstaat, Die Idee der Staatsräson* and *Die Entstehung des Historismus*. It is important to situate this trilogy – as well as his theoretical articles and short studies after the Second World War – in the complex tradition in which he positioned himself. It is a tradition in which Platonism, Aristotle's philosophy, Neoplatonism, Leibniz, Spinozism, Pietism, 'Goetheanism', Romanticism and German Idealism merge into one. Meinecke was deeply rooted in, and trained by, these traditions, which enabled him to respond to present events. That is why I emphasize Meinecke's nineteenth-century intellectual roots.[149] With regard to Meinecke's thought, this is also the first time the meaning of that century, and in particular the importance of Goethe, is analyzed in detail.[150]

At first glance, the complex tradition in which Meinecke positioned himself, and the *Wirkungsgeschichte* of his oeuvre, does not point to an unambiguous harmonious philosophy of history. The famous American philosopher and intellectual historian Arthur O. Lovejoy (1873-1962) would applaud an exposition of an author's views emphasizing the 'inner tensions – fluctuations or hesitancies between opposing ideas or moods', for he claims that more often than not we are inclined to

---

148   Kossmann, 'Friedrich Meinecke', 210.

149   In his memoirs Meinecke describes how both his grandparents were born and raised around 1770: the time of Frederick the Great. Due to several *Spätgeburten* (late births) the generations of his family are stretched-out over decades, more than is usually the case. For example, Meinecke's father witnessed the Wars of Liberation as a child in 1813. Meinecke thought himself to be rooted in the early nineteenth century: 'Und ich gebe mich gern dem Gedanken hin, daß ich blutmäßig eigentlich im frühen und nicht im späten 19. Jahrhundert wurzele (...)'. Meinecke, 'Erlebtes', 6.

150   Nicolas Berg points at Goethe's role in Meinecke's *Die deutsche Katastrophe*, but leaves the rest of Meinecke's oeuvre undiscussed. Given the subject of his study that is of course understandable: Nicolas Berg, *Der Holocaust und die westdeutschen Historiker. Erforschung und Erinnerung* (Göttingen 2003) 69-76. I will elaborate on Berg's argument in Chapter 6.

'over-unify' the views of an author so as to present his thinking 'all-of-a-piece'.[151] I agree with Lovejoy, but I am also inclined to add to this view what the American philosopher and art critic Arthur C. Danto (1924-2013) claimed with regard to the systematic nature of philosophy: 'The problems of philosophy are so interconnected that the philosopher cannot solve, or start to solve, one of them without implicitly committing himself to solutions for all the rest. (...) philosophy as such is architectonic, and imposes an external regimen upon its least systematic practitioners, so philosophers are systematic through the nature of their enterprise'.[152] A second argument Danto suggests is the rather common idea that historians reconstruct afterwards the life or development of the thinkers in question – that is, historians, for example, look back on earlier writings with the later ones in mind and thereby are able to see a unity in the thought of the thinkers in question.[153] Both these views apply to Meinecke, being an historian as well as a philosopher of history. I therefore claim a unity and continuity can be distilled from the development of his philosophy of history, and it is this philosophy that I will characterize as the leading principle of his oeuvre.

As stated before, in this research I will focus on Meinecke's (contrasting) notions, ideas, views, theories, as well as the political and intellectual events, which form the context of this discussion. The main question concerns how these different contradicting notions relate to each other in his major *theoretical* studies. This means I will not discuss Meinecke's political-historical studies. For example, a detailed discussion of Meinecke's biographies on Boyen and Radowitz will be left out, for instead of a philosophical or theoretical analysis, these studies are primarily concerned with the development of the individuals in question and their historical context.[154] *Das Zeitalter der deutschen Erhebung* (1906) is also left undiscussed, since it is a preparatory study; the same subject is discussed in detail in Meinecke's first major systematic study *Weltbürgertum und Nationalstaat*. And, to conclude, Meinecke's *Geschichte des deutsch-englischen Bündnisproblems 1890-1901* will also be left out, for the underlying theory of this concrete political history on the *practical* problems of *Realpolitik* is analysed in *Die Idee der Staatsräson*. In sum, the emphasis is on Meinecke's theoretical and philosophical views.

151   Arthur O. Lovejoy, *Essays in the History of Ideas* (Baltimore 1948) xiv-xv.

152   Arthur C. Danto, *Nietzsche as Philosopher* (New York and London 1965) 24.

153   Danto, *Nietzsche as Philosopher*, 25.

154   Friedrich Meinecke, *Das Leben des Generalfeldmarschalls Hermann von Boyen* (Stuttgart I 1896, II 1899); Friedrich Meinecke, *Radowitz und die deutsche Revolution* (Berlin 1913).

Structure of the book

To analyze Meinecke's development I will chronologically discuss his studies. In the first chapter I will discuss Meinecke's first major study *Weltbürgertum und National-staat*, in which he claims Germany's unification of 1871 caused, or was caused by, a synthesis of power (*Macht*) and culture (*Kultur*). But was this synthesis actually realised? And what was the underlying thought with regard to Meinecke's desire for this synthesis? In chapters two and three I will discuss *Die Idee der Staatsräson* – the study known for the alleged rupture in Meinecke's thought. As mentioned, Walther Hofer was the first to postulate this theory of a rupture. With regard to Meinecke's reconciliatory view I will correct this misrepresentation. An analysis of several thinkers put forward by Meinecke will be the core of chapter two and serves as a guide to Meinecke's Interbellum thought. In addition, in chapter three, I will clarify the ultimate consequences of Meinecke's argument in *Die Idee der Staatsräson* by means of his conception of conscience, which is related to his panentheistic philosophy of history. In chapters four and five the focus will be on the last major study: *Die Entstehung des Historismus*. I will first explain Meinecke's construction or even creation of the tradition of historicism (chapter four). Then, in chapter five, I will give a detailed account of Goethe's importance to Meinecke's historicism. Chapter six discusses the synthesis of Meinecke's philosophy of history. Themes which will be analyzed are: Meinecke's struggle with the destruction caused by the Second World War, his last works, the publication of his memoirs, and above all, the final synthesis of different contradicting views by means of his panentheistic philosophy of history. In the conclusion I will evaluate Meinecke's panentheistic philosophy, discuss the importance of his thought with regard to the tradition of historicism, elaborate on what is left of all this, and what we can still learn from his example.

1. The Ideal of a unity

'Wo das gelehrte beginnt,
hört das politische auf'[1]

Introduction

Friedrich Meinecke was born in a small Prussian town called Salzwedel, which lies
between Hamburg and Berlin. His father was a postmaster in this town until 1871
when he was transferred to Berlin. This transition from the idyllic Salzwedel to
the urban and industrializing city of Berlin – in which the social relationships
were not as harmonious as they had been in Meinecke's native village – had a
huge impact on the eight-year-old Meinecke.[2] In his memoirs he describes that
the contrast between the two had 'romantically influenced his disposition in life'.[3]
Meinecke's admiration for the German romantics – the poets and the philosophers
– is a striking example of this romantic disposition (*Lebensstimmung*), which is also
articulated in his work. Of course, his romantic disposition was not only the result
of the move to Berlin. His education and his first steps of his scientific career were
also of great importance.

Meinecke studied in Berlin and Bonn; he was trained by Johann Gustav Droysen, Heinrich von Sybel and Heinrich von Treitschke. In 1893, on recommendation
of Sybel, Meinecke began an archival research on the life of General Hermann
von Boyen (1771-1848). The first volume of this biography was published in 1896.[4]
The year prior to this publication, Meinecke's teacher Sybel had died – since 1893
Meinecke had co-edited the *Historische Zeitschrift* with him. After Sybel's death;
Meinecke shared the editorial activities, for a short while, with Treitschke, who
suddenly died in 1896. Meinecke then became editor-in-chief of the *Historische
Zeitschrift*. In this role he positioned himself in the so-called *Methodenstreit*, which
had been unleashed a few years earlier with the publication of the first volume of

---

1   Friedrich Schiller and Johann Wolfgang von Goethe, 'Xenien: 'Das Deutsche Reich'' in:
    Friedrich Schiller, *Sämtliche Werke in fünf Bänden* (Munich 2004) Volume I, 267.
2   Meinecke, 'Erlebtes', 25; also: Meineke, *Friedrich Meinecke*, 60-65.
3   Meinecke, 'Erlebtes', 25. 'ganze Lebensstimmung beeinflußt und romantisch disponiert hat'.
4   Friedrich Meinecke, *Das Leben des Generalfeldmarschalls Hermann von Boyen* (Stuttgart 1896).

Karl Lamprecht's *Deutsche Geschichte* (1891). This debate, in which Georg von Below, Otto Hintze and Hermann Oncken participated among others, centered in the first place on the contradiction between Lamprecht's holistic philosophy of history – a collective view of history focussed on causal and social connections – and a political history of ideas and individualizing philosophy of history, which was especially advocated by Meinecke. It would lead too far afield to discuss in detail Lamprecht's position, the content and development of the debate, but with regard to Meinecke it is important to recognize his influence on this debate as editor-in-chief of the *Historische Zeitschrift*; for in several articles he reprimanded Lamprecht, ignored him and finally he silenced him by publishing a very critical article by Below.[5] Meinecke considered Lamprecht a danger to what Meinecke regarded as the 'right' view of history; a view which should be aimed at the individual and unique in history, instead of Lamprecht's general laws.

By the end of the *Methodenstreit* in 1899 Meinecke finished his second volume of his Boyen biography, which secured him a professorship in Strasbourg, where he stayed from 1901 to 1906. This, according to Meinecke, was the most fruitful period of his life. For it was supposedly in Strasbourg that he saw his whole lifework in his mind's eye.[6] The short study *Das Zeitalter der deutschen Erhebung* was the first result of this fruitful period. This study gives a brief exposition of the theme also discussed in the Boyen biography. Moreover, it is a prelude to Meinecke's first major *geistesgeschichtliches* work (intellectual history): *Weltbürgertum und Nationalstaat* – which is the subject of this chapter.[7]

In early 1908 *Weltbürgertum und Nationalstaat* was published, and it made Meinecke instantly famous. For in this study he made a connection between 'power' and 'culture'. This was a tremendous accomplishment, because this dichotomy was heavily discussed in 'Germany' for roughly over a century.[8] Goethe and Schiller

---

5   For a detailed discussion of this debate see: Roger Chickering, *Karl Lamprecht. A German Academic Life (1856-1915)* (Atlantic Highlands 1993) 212-245. Also: Krol, 'De "scheppende spiegel" van Friedrich Meinecke', 75-84; Jaeger and Rüsen, *Geschichte des Historismus*, 141-146; Jo Tollebeek, 'Het Duitse debat. Geschiedenis rond 1900' in: Herman Beliën and Gert Jan van Setten ed., *Geschiedschrijving in de twintigste eeuw. Discussie zonder eind* (Amsterdam 1991) 15-40, there 29-33; also: Manuel Stoffers, *Het nerveuze tijdperk en zijn historici: de opkomst van de mentaliteitsgeschiedenis in Duitsland 1889-1915* (Maastricht; dissertation 2007) 23-45.

6   'Die Wurzeln der drei geistesgeschichtlichen Werke, die ich in den drei Jahrzehnten von 1907 bis 1936 veröffentlichen konnte, liegen in dem, was mir hier [Strasbourg, RK] durch den Kopf ging, teils schon als fest ergriffener Leitgedanke, teils als Wendung des Interesses auf neue lockende, mir bisher fernliegende Erscheinungen des geschichtlichen Lebens'. Meinecke 'Erlebtes', 159.

7   Meineke, *Friedrich Meinecke*, 92; Friedrich Meinecke, *Das Zeitalter der deutschen Erhebung (1795-1815)* (Göttingen 1906); Friedrich Meinecke, *Weltbürgertum und Nationalstaat* (Munich 1969) 25.

8   Schulin, 'Friedrich Meinecke', 40.

had already asked themselves the question in one of their famous *Xenien* of 1796: 'Germany? But where is it?' On the one hand this quote points at the political disunity of the German Reich, which consisted of a multitude of small states and principalities.[9] On the other hand this 'Where is it?' refers to different cultural criteria, which might form the basis of a national unity, but remained isolated form German politics.[10] In *Weltbürgertum und Nationalstaat* Meinecke now claimed to have reconciled the age-old dichotomy of politics and culture.

When Meinecke was writing *Weltbürgertum und Nationalstaat*, the relationship between power and culture, or *Macht* and *Kultur* was still very much under discussion. In particular the role that both had played in the establishment of the German national state in 1871 was key to the debate.[11] During his research Meinecke realised that the overarching Spirit (*Geist*), culture (*Kultur*), or what he referred to as 'cosmopolitan ideals', were far more closely connected with national-political ideas than had been acknowledged thus far.[12] 'Cosmopolitan', in Meinecke's view, should not be interpreted as a 'concrete' cosmopolitanism, in the sense of 'the ambition to adapt to the character of all nations', rather one should understand it as a 'universal' ideal, which could vary between nations.[13] By means of several nineteenth-century 'cosmopolitan' and 'individually' oriented poets, philosophers and statesmen Meinecke argues in *Weltbürgertum und Nationalstaat* that the German idea of the national state had developed in close interaction between cosmopolitan ideals and national ideas, and between *Kultur* and *Macht*, between literature and politics, between the individual and the state.[14] It was not the development from cosmopolitan *to* national ideas which Meinecke expounded, but the dynamic

---

9   Frits Boterman, *Moderne geschiedenis van Duitsland 1800-heden* (Amsterdam and Antwerp 2005) 30, 34.

10 Boterman, *Moderne geschiedenis*, 11, 34. Cultural criteria like language, literary creations by poets and philosophers, ethnicity, and a shared (collective) past.

11 Schulin, 'Friedrich Meinecke', 40.

12 Friedrich Meinecke, 'Straßburg – Freiburg – Berlin. Erinnerungen' in: idem, *Autobiographische Schriften* (Stuttgart 1969) 135-320, there 160.

13 This is obviously a simplification, because there are numerous different views in the eighteenth and nineteenth century with regard to the notion of cosmopolitanism. On this issue: Andrea Albrecht, *Kosmopolitismus. Weltbürgerdiskurse in Literatur, Philosophie und Publizistik um 1800* (Berlin and New York 2005) 31. In his discussion of different thinkers, Meinecke also points at these different views; however, in the end he assumes one distinct movement.

14 Schulin, 'Friedrich Meinecke', 40. According to Schmidt, Meinecke constructed a Kulturnation that did not do justice to the situation around 1800; moreover he claims Meinecke's idea of the nation and state had deeper and different origins: Georg Schmidt, 'Friedrich Meineckes Kulturnation. Zum historischen Kontext nationaler Ideen in Weimar-Jena um 1800', *Historische Zeitschrift* 284 (2007) 597-621.

between both notions – not without reason the title of Meinecke's study is Welt-bürgertum *und* Nationalstaat.[15] This 'und', according to Meinecke refers to the relationship between cosmopolitan and national ideas which arose during the prelude to 1871. That means Meinecke thought he had united the classical-romantic Spirit with the German unification. He considered it a union of Goethe and Bismarck.[16]

Weltbürgertum und Nationalstaat also attracted attention for a different reason. With this study Meinecke introduced a whole new kind of political history or a specialization within the history of ideas: the *politische Ideengeschichte*. Meinecke is considered the founding father of this sub-discipline, which was (not only in Germany) influential well into the twentieth century.[17] In general, the history of ideas is concerned with the historical development of a particular idea, or how ideas change under the influence of different historical circumstances.[18] Meinecke's political variation on the history of ideas concentrates predominantly on the (apparent) opposites between two notions. As already stated, in *Weltbürgertum und Nationalstaat* Meinecke analyses the interaction between 'cosmopolitan' ideals and 'national' ideas in German thought. By means of several great thinkers Meinecke follows the development of these notions, for 'The examination of political ideas', says Meinecke, 'can never be separated from great personalities, creative thinkers'.[19] Meinecke, by the way, concentrates only on those 'creative thinkers' whose political development at a certain moment in time had contributed to the inception of the German national idea. So, when discussing the realization of the German national state, it was important to find its origin in the ideas of different thinkers, who had constructed their world view under the influence of, and in interaction with, the prevailing spirit of the time.[20] Ideas, according to Meinecke, were not only to be

---

15   Eberhard Kessel, 'Friedrich Meinecke in eigener Sicht' in: Michael Erbe ed., *Friedrich Meinecke heute. Bericht über ein Gedenk-Colloquium zu seinem 25. Todestag am 5. und 6. April 1979* (Berlin 1981) 186-195, there 190.

16   Meinecke, 'Straßburg – Freiburg – Berlin. Erinnerungen', 318.

17   Ernst Schulin, 'Friedrich Meinecke und seine Stellung', there 27. Arthur Lovejoy's *The Great Chain of Being* (1936) also fits this tradition, although Lovejoy concentrated mainly on the history of one idea, for Meinecke was interested in the contrast between different notions. Also: Jacob Talmon (1916-1980): Jacob L. Talmon, *The Origins of Totalitarian Democracy* (London 1952); Jacob L. Talmon, *The Myth of the Nation and the vision of Revolution. The Origins of Ideological Polarisation in the Twentieth Century* (London 1980).

18   Cf. my Introduction.

19   Meinecke, *Weltbürgertum*, 24-25: 'Die Untersuchung politischer Gedanken darf niemals losgelöst werden von den großen Persönlichkeiten, den schöpferischen Denkern (...)'. Friedrich Meinecke, *Cosmopolitanism and the National State* trans. Robert Kimber (Princeton 1970) 22.

20   Gilbert, 'Friedrich Meinecke', 71.

followed back to concepts and notions, but also to 'life' and 'personality' itself.[21] When looking back on writing *Weltbürgertum und Nationalstaat* Meinecke described it as follows: 'What I really wanted to describe was the gradual transformation and regeneration of these ideas, and the connection with life and personality as such'.[22]

Meinecke's interest in great personalities was and still is very much under discussion. It is held against him that his analysis of great thinkers does not tally with his conception of historicism. For example, Iggers claims Meinecke 'seriously violated the "historical sense" (*historischen Sinn*) which he demanded'[23], which means: Meinecke allegedly isolates the thinkers and therefore he did not represent them as individuals for their own sake. According to Iggers: 'All of them seemed to be less ends in themselves than stages in the development of an objective process'.[24] Meinecke's philosophy, according to Iggers, is permeated with a certain amount of Hegelian dialectics – Hegel considered individualities as transitory modes of expression of the self-realization of Spirit.[25] To put it briefly, the individual as such is trivial in this kind of philosophy (of history), Iggers concludes.[26]

Iggers' view seems to be at odds with Meinecke's conception of historicism since, Meinecke of all people starts from the unique individual. Moreover, the interaction between the individual and his time was Meinecke's main concern.[27] The American historian Felix Gilbert – a pupil of Meinecke – claims Meinecke did

---

21   Meinecke, *Weltbürgertum*, 25. In Chapter 3 I will elaborate on Meinecke's interpretation of notions like 'life' and 'personality'.

22   Friedrich Meinecke, 'Zur Beurteilung Rankes' in: idem, *Zur Geschichte der Geschichtsschreibung* (Munich 1968) 50-65, there 53. 'Was ich in Wahrheit darstellen wollte, ist die allmähliche Wandlung und Erneuerung des Lebensblutes in diesen Gedanken, überhaupt ihr Zusammenhang mit Leben und Persönlichkeit'. First published: Friedrich Meinecke, 'Zur Beurteilung Rankes', *Historische Zeitschrift* 111 (1913) 582-600.

23   Iggers, *The German Conception*, 202.

24   Idem.

25   Rüsen and Jaeger, *Geschichte des Historismus*, 37.

26   This is an outdated characterization of Hegel's philosophy. I will come back to this in Chapter 4, section 6.

27   Meinecke's interest in great thinkers is in that sense comparable to Goethe's interest in biographies, for he claimed history is essentially the history of the sciences, and since this history is told by certain *gebildeten* individuals, history consists of biographies: 'Der Konflikt des Individuums mit der unmittelbaren Erfahrung und der mittelbaren Überlieferung ist eigentlich die Geschichte der Wissenschaften: denn was in und von ganzen Massen geschieht, bezieht sich doch nur zuletzt auf ein tüchtigeres Individuum, das alles sammeln, sondern, redigieren und vereinigen soll (...)', thus Goethe in his *Farbenlehre*. Johann Wolfgang von Goethe, 'Materialien zur Geschichte der Farbenlehre' in: idem, *Goethes Werke. Hamburger Ausgabe in 14 Bänden*, ed. Erich Trunz (Munich 1982) Volume XIV, 7-269, there 51; Also: Nicholas Boyle, 'Geschichtsschreibung und Autobiographik bei Goethe (1810-1817)', *Goethe Jahrbuch* 110 (1993) 163-172, there 164-165. In Chapter 5 I will give a detailed account of Goethe's view of history and its relevance to Meinecke.

indeed discuss the individual thinkers for their own sake, because 'For him *man* was the medium through which ideas worked in history; they established a common basis for thought and action among men but they also gave each individual the opportunity for developing a distinctive personality'.[28]

Instead of criticizing Meinecke's discussion of great poets and philosophers, it is far more interesting to analyze his choice of these specific personalities. What is behind Meinecke's selection of these 'great individuals'? The answer to this question will give us a clear understanding of Meinecke's political theory and philosophy of history in this period. Another important question is: what did Meinecke have in mind with *Weltbürgertum und Nationalstaat*, and the union of *Macht* and *Kultur*? Besides the question of *whether* this union indeed had been realized, there is also the question how Meinecke managed to unite these apparently mutually exclusive notions, and furthermore why he emphasized this harmony between *Macht* and *Kultur* with regard to the unification of Germany? Put differently, what is the underlying principle, and what is the importance of such a synthesis to Meinecke and his time?

In this chapter on *Weltbürgertum und Nationalstaat* I will demonstrate that Meinecke created a tradition of thinkers, which was, according to him, relevant to the union of *Macht* and *Kultur*. Moreover, his choice of thinkers reflects his conception of historicism, which concentrates on the individual and unique in history. All thinkers discussed by Meinecke concentrate on 'cosmopolitan ideals' and to a greater or lesser extent the focus is on the concrete (individual) national state. Politics and history are, therefore, closely connected in Meinecke's historicism. This is clearly expressed in his account of Leopold von Ranke, who was, according to Meinecke, the first to bridge the gap between the actual individual state and the cosmopolitan ideals. That is why he is the key figure in Meinecke's study.

Meinecke's discussion of Ranke, Humboldt, Schlegel, and Fichte is not only an expression of his conception of historicism, but also of his ethical-political convictions. I will demonstrate that Meinecke's moral point of view is aimed at rec-

---

28   Gilbert, 'Friedrich Meinecke', 71. My italics. Gilbert characterizes this as a 'psychological empathic method'. According to Gilbert, Meinecke borrowed this idea from Dilthey, who used it for individual biographies, whereas Meinecke applied it to a group of thinkers. See: Gilbert, 'Friedrich Meinecke', 73. In this context Herder also had a considerable influence on Meinecke; on this issue: Chapter 4. Brands basically agrees with Gilbert: 'This [the idea in the sense of a trend, RK] does not devaluate the originality of the personality. This way the character and the extent to which the originality of the individual struggles with historical forces are clearer than is possible in a regular biography': Brands, *Historisme als ideologie*, 146-147. 'Deze [the idea in the sense of a trend, RK] doet echter geenszins afbreuk aan de originaliteit van de persoonlijkheid. Eerder worden op deze wijze (...) het karakter en de mate van oorspronkelijkheid van het individu in zijn strijd met de historische krachten duidelijker naar voren gehaald dan in de gewone biografie gebeurt'.

onciliation and based on the partially panentheistic world views of the thinkers which he discussed. In addition to this, Meinecke's synthesis of *Macht* and *Kultur* is related to a political ideal, which is, as far as I know, never been put forward. I am convinced that Meinecke's thesis in *Weltbürgertum und Nationalstaat* is closely related to a social-political synthesis, which was advocated by the politician and clergyman Friedrich Naumann a few years before Meinecke published his book. In particular the failure of Naumann's synthesis urged Meinecke to fall back on the more comprehensive synthesis of 'Goethe and Bismarck'.[29]

Before analyzing Meinecke's constructed canon, Ranke's synthesis, and Naumann's suggestion, I will first distinguish several different forms of nations. This is necessary for a better understanding of Meinecke's explanation of the relationship between the national state and 'cosmopolitanism'.

1.1   Nations, the national state and cosmopolitanism

In *Weltbürgertum und Nationalstaat* Meinecke draws a distinction between two forms of nations: cultural nations (*Kulturnation*) and political nations (*Staatsnation*).[30] The 'political nation', according to Meinecke, is based on a political unity, a shared political history and a constitution. The 'cultural nation' is founded on a unity of language, literature and religion. In Meinecke's view, the lines between cultural and political nations are, both on the 'inside' and 'outside', difficult to identify.[31] Cultural nations can be part of a political nation, and vice versa: a cultural nation can include different political nations. Switzerland, according to Meinecke, is a good example of a political nation with several cultural nations. Nineteenth-century Germany with its multitude of political entities, on the other hand, is a good

---

29 Robert Pois touched upon the relationship between Naumann and Meinecke; however, he mainly discussed the subject of the 'mass'. Pois failed to discuss Meinecke's disappointment with Naumann's unsuccessful synthesis and Meinecke's ideal of *Weltbürgertum und Nationalstaat*. Pois, *Friedrich Meinecke*, 4-25, 144.

30 Meinecke, *Weltbürgertum*, 10. Meinecke, *Cosmopolitanism*, 10. For a critique and correction on Meinecke's description of the *Kulturnation*: Schmidt, 'Friedrich Meineckes Kulturnation. Zum historischen Kontext nationaler Ideen in Weimar-Jena um 1800', 597-621.

31 This is characteristic for Meinecke's thought, for throughout his oeuvre he tried to transcend several different contrasting notions. It is also clearly articulated in a quote from the poet Friedrich Rückert (1788-1866) which serves as a motto for *Weltbürgertum und Nationalstaat*: 'Grenzpfähle steckest du, um ein Gebiet zu messen; Doch daß du sie nur steckst, das sollst du nicht vergessen. Der grade Gegensatz setzt grad' die Wahrheit schief, Weil stets in Wahrheit eins ins andre sich verlief'. Here the theme of his second major study, *Die Idee der Staatsräson*, resounds: the (inter)relation between *Ethos* and *Kratos*. Moreover, Rückert's quote can be interpreted as an implicit form of panentheism, which must be the underlying thought to counterbalance the relativism in which Rückert's view results.

example of a cultural nation, for apparently Germany had an overarching culture.[32] Both examples clearly demonstrate that a political nation and a cultural nation can develop into a national state.[33]

Meinecke's notion of the 'cultural nation' is derived from Herder's notion of 'national consciousness'. Herder's ideas are in many respects a starting point for Meinecke. He considers Herder an 'instigator of a new historical sense'.[34] This alludes primarily to Herder's discovery of the so-called 'principle of individuality', a mode of thought which is central in Meinecke's view of history. In *Weltbürgertum und Nationalstaat*, for example, Meinecke considers the state to be an individuality, that is, the state is not only an institution created by a group of people for the common good, but also an individuality with its own principle of life.[35] So, Meinecke's notion of 'individuality' refers not only to individuals and individualities, but also to collective institutions.[36]

Considering the state as an individuality also implies, in Meinecke's view, that the nature of a state changes in a way similar to individuals: in mutual contact. Meinecke claims we should consider these contacts as a form of 'community life' between different states and nations, which are in the end all united in a universal or cosmopolitan community. Since the boundaries between 'community life' within different states and nations are difficult to identify, Meinecke suggests we should consider the 'national' and universal' in (world) history as 'a massive interweaving and crossing of national and universal developments'.[37]

Meinecke claims this 'interweaving' offers the historian a perspective on the characteristics or individuality of a national state. Next, the historian can examine the individual or individuality in its historical development, because: 'for the entire subject matter of historical research: society, culture, state, nation and all of mankind exist only in and through the ideas, feelings and actions of individuals. They are the basic cells of historical life', according to Meinecke.[38] He searched for

---

32   Meinecke, *Weltbürgertum*, 12, 20.

33   Ibidem, 20.

34   Friedrich Meinecke, 'Aphorismen' in: idem, *Zur Theorie und Philosophie der Geschichte* (Stuttgart 1965) 215-263, there 216. 'Erwecker eines neuen historischen Sinnes'. Originally published as part of: Friedrich Meinecke, *Aphorismen und Skizzen zur Geschichte* (Leipzig 1941) 7-68, there 8. In *Die Entstehung des Historismus* Meinecke regards Herder as the father of historicism. Cf. Chapter 4.

35   Kossmann, 'Friedrich Meinecke', 216.

36   With this view, Meinecke built on Ranke's tradition; Ranke also considered the state a personality. E.H. Kossmann, 'Een kennismaking met Ranke (1795-1886)' in: idem, *Vergankelijkheid en continuïteit. Opstellen over geschiedenis* (Amsterdam 1995) 160-191, there 183.

37   Meinecke, *Cosmopolitanism*, 19. 'eine gewaltige Verflechtung und Durchkreuzung nationaler und universaler Entwicklung'. Meinecke, *Weltbürgertum*, 22.

38   Meinecke, *Cosmopolitanism*, 20. '(...) alle die übrigen Objekte historischer Forschung: Gesellschaft, Kultur, Staat, Nation und Menschheit existieren nur in und durch die

the 'intersection' of the national and universal, and accordingly the inner self of the great historical personalities is relevant, for it is here that the dynamics between both notions are apparent.

According to Meinecke, the cosmopolitan, like the national, 'arose' (simultaneously) from the interaction between different states: '"It was only from the conflict of two nationalities that the concept and the word [cosmopolitanism, RK] could grow"'.[39] To realise that national states differ from each other gives us insight into what unites both: humanity.[40] According to Meinecke the cosmopolitan idea gives the initial impetus to 'limit the state's claims on the individual, to distinguish between nationality and humanity'.[41] 'Cosmopolitanism' in Meinecke's view, as I stated before, refers to universal ideals, and can vary between nations. For example, in France after the French Revolution a national state arose based on universal ideals.[42] According to Meinecke, under the influence of the French Revolution, 'Cosmopolitanism' and 'the national state' became (instead of limiting each other) evermore mutually supporting concepts.[43] Nationality and universality are, in that sense, two sides of the same cosmopolitan coin. In Meinecke's words: 'At that time, too, an undertaking beyond national interests helped ignite the national idea in the hearts of men'.[44] Consequently, Meinecke asked himself in *Weltbürgertum und Nationalstaat* how in Germany the idea of the individual national state was reconciled with the 'cosmopolitan' ideals.

1.2   The universal and the national

Meinecke claims the German national idea was first introduced by thinkers such as Herder, Wilhelm von Humboldt, Novalis, Friedrich Schlegel, Fichte, Hegel and Ranke. It is striking that Meinecke fails to give an account of the major figures of

---

Vorstellungen, Empfindungen und Handlungen der Individuen, und diese seien die Urzellen geschichtlichen Lebens'. Meinecke, *Weltbürgertum*, 22.

39   Meinecke, *Cosmopolitanism*, 20. Meinecke cites Richard Reitzenstein (1861-1931): 'Erst der Konflikt zweier Nationalitäten hat den Begriff und das Wort geschaffen'. Meinecke, *Weltbürgertum*, 23.

40   F.R. Ankersmit, 'What is Wrong with World History from a Cosmopolitical Point of View?'. Conference paper, Shanghai November 2008.

41   Meinecke, *Cosmopolitanism*, 20. 'das Recht des Staates auf das Individuum abzugrenzen, zwischen Nationalität und Menschentum zu scheiden'. Meinecke, *Weltbürgertum*, 23.

42   Meinecke, *Weltbürgertum*, 23.

43   Ibidem, 35.

44   Meinecke, *Cosmopolitanism*, 21. 'Eine universale Aufgabe hat damals auch im Inneren der Menschen die nationale Idee mit entzünden helfen'. Meinecke, *Weltbürgertum*, 35.

the German Enlightenment.[45] Lessing for example, although he is mentioned in one sentence, is brushed aside by Meinecke, for Lessing apparently falls outside of his scope.[46] This strict distinction between Enlightenment and historicism characterizes all of Meinecke's major studies, as will become clear in the next chapters.

Although Meinecke discusses a liberal like Humboldt, for the greater part, he ignores the liberal movement in *Weltbürgertum und Nationalstaat*. He is only interested in the romantic-conservative movement, for this movement is, according to him, less known than the frequently discussed and praised liberal movement.[47]

On the other hand, Meinecke also does not discuss all romantics. Considering his interest in the relationship between 'unity' and 'diversity', it is remarkable that he does not even mention Friedrich Hölderlin (1770-1843), an outstanding example of a romantic who was interested in unity and contrasts.[48] Perhaps Hölderlin was, in Meinecke's view, too vague concerning German politics and the formation of the national state; nevertheless as an exponent of the 'cultural nation' he qualifies for Meinecke's study. Even more striking is the absence of Goethe. Besides mentioning him only once in the context of Fredrick the Great – Goethe was 'Fritzisch gesinnt', 'favouring Frederick'[49] – Meinecke does not discuss his views. This is strange, because Goethe is generally viewed as the personification, and the epitome

---

45   Whereas at the beginning of the eighteenth century cosmopolitanism became a: 'politisch, rechtlich, pädagogisch, kulturell, ökonomisch oder moralisch kodierten Programmbegriff der europäischen Aufklärung'. Albrecht, *Kosmopolitismus*, 31. Also: T.J Reed, *Light in Germany. Scenes from an Unknown Enlightenment* (Chicago and London 2015) 2-3.

46   Meinecke, *Weltbürgertum*, 33. With regard to the tensions between cosmopolitanism and the national state, it is also very strange that Meinecke ignores a thinker like Christoph Martin Wieland (1733-1813).

47   Meinecke, *Weltbürgertum*, 25.

48   The *George Kreis* and Rilke rediscovered Hölderlin at the beginning of the twentieth century. This may be an explanation for the absence of Hölderlin in Meinecke's early works. Also, with regard to Meinecke's education: he was probably not familiar with Hölderlin, for Prussian grammer schools did not discuss him. Many thanks to Professor H.W. von der Dunk for making me aware of this. However, it remains strange that Meinecke, throughout his oeuvre, never really discussed Hölderlin, for contemporary writers like Thomas Mann and Stefan Zweig indeed did. With regard to Hölderlin's interest in unity and diversity, Terry Pinkard gives a clear example of how Hölderlin saw both of them coalesce with regard to 'love'. In the words of Pinkard: 'Love existentially solves the problem of how to unite spontaneity and responsiveness in that in it there is awareness and recognition of both unity and difference, a recognition of each other as uniquely existing individuals in a unity with each other; indeed, love can exist only where there is a full responsiveness to the independent and full reality of the other which is at the same time a liberation, a feeling of complete autonomy'. Terry Pinkard, *German Philosophy 1760-1860. The Legacy of Idealism* (Cambridge 2002) 143.

49   Cited in Meinecke, *Weltbürgertum*, 407.

of the 'cultural nation'.[50] Nonetheless, throughout *Weltbürgertum und Nationalstaat* Goethe's 'spirit' is indeed present – this will become clear in the following chapters.

In the present section Meinecke's analysis of Humboldt, Schlegel and Fichte is central. His discussion of these thinkers, who in many ways tried to reconcile the contrasts between the 'universal' and the 'national', will clearly reveal the principles of Meinecke's own philosophy of history and political theory.

Bildung

At the end of the eighteenth century a group – later to be called the German 'early Romantics' (*Frühromantiker*) – gathered around Jena. Among these early Romantics, who were deeply influenced by Herder, are August and Friedrich Schlegel, the theologian Friedrich Schleiermacher, the writer Ludwig Tieck, the philosopher Friedrich Schelling, and the poet-philosopher Novalis. It was a circle which manifested itself philosophically, literary, but also anti-philosophically and poetically, aiming to call into question the boundaries of philosophical systems.[51] The most important issue which occupied these early Romantics was the need to express and develop their individuality. This *feeling* for one's own individuality came into conflict with the rationalist Kantian and Fichtean idea of subjectivity, which were, according to these Romantics, terribly dry, formal and aimed too much at capturing everything in a frame. All this, in the view of the early Romantics, did not tally with the chaos and unpredictability of human life.[52]

The early Romantics lived in a politically divided 'Germany', which consisted of numerous political communities and principalities. A feeling of solidarity during this period was found in the idea of belonging to a 'cultural nation'. Wilhelm von Humboldt, often associated with the early Romantics, was, according to Meinecke, the first to identify this feeling of belonging. Like the early Romantics, Humboldt was interested in the development of personal individuality, and he managed to connect this with the universal or cosmopolitan, which is mainly why Meinecke was interested in his views.

Humboldt's central notion is the so-called *Bildungsideal* (this is a term for which there is no sufficient English translation). This ideal of *Bildung*, in Humboldt's view, means to educate or cultivate the individual personality, and enable an unlimited development of all (spiritual) abilities for the purpose of a harmonious individual

50   Meinecke, *Weltbürgertum*, 52, 74; Boterman, *Moderne geschiedenis*, 35.
51   Pinkard, *German Philosophy*, 132-136; Karl Ameriks, 'Introduction: "Interpreting German Idealism"' in: idem ed., *The Cambridge Companion to German Idealism* (Cambridge 2000) 1-17, there 11.
52   Pinkard, *German Philosophy*, 136.

personality.[53] A personality is not developed in isolation, but in interaction with 'higher things'. In other words, *Bildung*, as an active process of character-building, is aimed at the spiritual, which is closely connected with the absolute ideals of the Good, True and Beautiful.[54] Humboldt expressed it as follows: 'When we speak of 'Bildung', we refer to something both higher and inward, namely, the disposition, which, on the basis of our consciousness and feeling, harmoniously affects our state of mind and character, both spiritually and morally'.[55] On a moral-intellectual level, in Humboldt's view, there seems to be an unfathomable relationship or harmony between 'higher' ideals and the individual. With regard to the 'cultural nation' (*Kulturnation*) it is about a unification of an individual and a collective identity, which means: an interweavement of the absolute and the individual within the (collective) personality.[56]

---

53   Aleida Assmann, *Arbeit am nationalen Gedächtnis. Eine kurze Geschichte der deutschen Bildungsidee* (Frankfurt, New York and Paris 1993) 34; Isaiah Berlin, *Political Ideas in the Romantic Age. Their Rise and Influence on Modern Thought*, ed. Henry Hardy (Princeton and Oxford 2006) 175.

54   Arnold Labrie, *'Bildung' en politiek, 1770-1830. De "Bildungsphilosophie" van Wilhelm von Humboldt bezien in haar politieke en sociale context* (Amsterdam 1986) 30-33. Initially, Bildung had a theological connotation. It referred to the idea of 'developing oneself in God's image', and aspire to the Kingdom of God on earth. When the theological meaning was replaced with a moral and ethical meaning (by which it lost its religious content), the emphasis shifted to the ideal world of the arts and culture, which still kept some sort of sacred aura. See: Willem Erauw, 'De relatie tussen cultuur en politiek tegen de achtergrond van de Duitse natievorming: een inleiding', *De bijdragen tot de eigentijdse geschiedenis* 11 (2003) 7-19, there 10. Also: Reinhart Koselleck, 'Einleitung – Zur anthropologischen und semantischen Struktur der Bildung' in: idem ed., *Bildungsbürgertum im 19. Jahrhundert. Teil II Bildungsgüter und Bildungswissen* (Stuttgart 1989) 11-46; Kristin Gjesdal, 'Bildung' in: Michael N. Forster and Kristin Gjesdal ed., *The Oxford Handbook of German Philosophy in the Nineteenth Century* (2015) DOI: 10.1093/oxfordhb/9780199696543.013.0035.

55   Cited in: Assmann, *Arbeit am nationalen Gedächtnis*, 25. 'Wenn wir aber in unserer Sprache "Bildung" sagen, so meinen wir damit etwas zugleich Höheres und mehr Innerliches, nämlich die Sinnesart, die sich aus der Erkenntnis und dem Gefühle des gesamten geistigen und sittlichen Strebens harmonisch auf die Empfindung und den Charakter ergießt'. Also see: Labrie, *'Bildung' en politiek*, 32.

56   Assmann, Arbeit am nationalen Gedächtnis, 10-25. Also: Frederick C. Beiser, *Enlightenment, Revolution, and Romanticism. The Genesis of Modern German Political Thought, 1790-1800* (Cambridge, Mass. and London 1992) 113: '(...) Bildung, the all-round development of the personality; an insistence on the value of individuality and the right of the individual to resist the pressures of social conformity; a recognition of the role of art in educating the whole personality, (...), and an ardent devotion to beauty and truth'. With regard to nineteenth-century Germany the notion of Bildung is often used in comparison with the classical Greeks. The realization of a harmonious individual personality meant, in practice, a reborn, classical, preferably a 'Greek' man. The late eighteenth-century and early nineteenth-century Romantics considered the politically divided German nation to be related to classical Greece,

Besides a cultural notion, *Bildung* also refers to politics. In fact, it falls within the scope of (cultural) utopia and political program.[57] This is also expressed by Humboldt; to him the state should provide conditions like freedom and security which enables the individual to dedicate himself to *Bildung*.[58] In *Ideen zu einem Versuch, die Grenzen der Wirksamkeit des Staates zu bestimmen* (1792) Humboldt claims: 'The State is to abstain from all solicitude for the positive welfare of the citizens, and not to proceed a step further than is necessary for their mutual security and protection against foreign enemies; for with no other object should it impose restrictions on freedom'.[59] Humboldt was aware that this view also aimed at an ideal, because as a statesman he knew that it was not a given to acquire and secure freedom in (a political) reality.[60] Humboldt's view on the state's actions are a result of his ideal of *Bildung*, because the individual and the nation at large can only develop when (relative) freedom and security are provided by the state.

Humboldt considers nations (particularly the German nation) to be separate individualities with their own spirit. On the basis of his *Ideenlehre* (theory of (historical) ideas) he claims nations are the embodiment of an idea.[61] This requires some explanation. For Humboldt, and also for Ranke, historical reality is a product of ideas, that is to say: ideas are active *in* historical reality, and therefore the driving force of history. These ideas cannot be reduced to a certain part of history; in fact, they govern the whole of history. In the idea both the general and the individual come together. In other words, the ideas are potentialities which enable the development of certain characteristic cultural, intellectual, political and social entities.[62] Next, it is the historian's responsibility to uncover these ideas. According to Humboldt, this is possible since the ideas and the historian are of the same

---

beause like classical Greece, Germany also consisted of numerous different states, but at the same time it was culturally united. I will come back to this issue in my analysis of Meinecke's discussion of Schlegel.

57   Assmann, *Arbeit am nationalen Gedächtnis*, 9.

58   Beiser, *Enlightenment, Revolution, and Romanticism*, 111-137; Labrie, *'Bildung' en politiek*, 92.

59   Wilhelm von Humboldt, *The Limits of State Action* trans. J.W. Burrow (Cambridge 1969) 37. 'Der Staat enthalte sich aller Sorgfalt für den persönlichen Wohlstand der Bürger und gehe keinen Schritt weiter, als zu ihrer Sicherstellung gegen sich selbst und gegen auswärtige Feinde notwendig ist; zu keinen anderen Endzwecke beschränke er die Freiheit'. Wilhelm von Humboldt, 'Ideen zu einem Versuch, die Gränzen der Wirksamkeit des Staats zu bestimmen' in: idem, *Werke in fünf Bänden*, ed. Andreas Flitner and Klaus Giel (Stuttgart 1960-1981) Volume I, 56-223, there 90.

60   Labrie, *'Bildung' en politiek*, 92.

61   Ibidem, 90.

62   This resembles the notion of 'entelechy'. This Aristotelian concept entails the notion that certain things change and grow according to a principle already present within. For example, an acorn grows into an oak. Ankersmit, *Denken over geschiedenis*, 176; Ankersmit, *Meaning, Truth, and Reference*, 11.

'Spiritual nature' (*Geistnatur*), that means: subject and object have the same 'ground' or origin.[63] This Spinozist thought, in which subject and object essentially coincide, is clearly expressed by Humboldt in *Über die Aufgabe des Geschichtsschreibers*:

> All understanding presupposes in the person who understands, as a condition of its possibility, an analogue of that which will actually be understood later: an original, antecedent congruity between subject and object. Understanding is not merely an extension of the subject, nor is it merely a borrowing from the object; it is, rather, both simultaneously. Understanding always is the application of a pre-existent general idea to something new and specific. When two beings are completely separated by a chasm, there is no bridge of communication between them; and in order to understand each other, they must, in some other sense, have already understood each other.[64]

The shared spiritual nature (*Geistnatur*) of both historian and historical reality enables an understanding (*verstehen*) of history and thus an understanding of historical ideas.[65] This philosophy is extremely important to Meinecke, and it is crucial in reconciling cosmopolitan ideals and the national state. Moreover, in his later studies, Meinecke further developed the idea of a common ground or *Geistnatur* – a thought which is similar to the principle of panentheism.

Humboldt claimed that every nation is unique and embodies a specific idea; as a result, a *weltbürgerliche Verschmelzung* (cosmopolitan fusion) is impossible. Instead, only a national differentiation remains. Humboldt, in Meinecke's interpretation, thought it possible to infer certain universal characteristics by studying different 'cultural nations'; in history he searched for a confirmation of the 'idea of humanity'. According to Meinecke, he tried to extract 'the highest human values

---

63   Ankersmit, *Denken over geschiedenis*, 175-176; Jaeger and Rüsen, *Geschichte des Historismus*, 39-40.

64   Wilhelm von Humboldt, 'On the Historian's Task', *History and Theory* 6 (1967) 57-71, there 65. 'Jedes Begreifen einer Sache setzt, als Bedingung seiner Möglichkeit, in dem Begreifenden schon ein Analogon des nachher wirklich Begriffenen voraus, eine vorhergängige, ursprüngliche Uebereinstimmung zwischen dem Subject und Object. Das Begreifen ist keineswegs ein blosses Entwickeln aus dem ersteren, aber auch kein blosses Entnehmen vom letzteren, sondern beides zugleich. Denn es besteht allemal in der Anwendung eines früher vorhandenen Allgemeinen auf ein neues Besonderes. Wo zwei Wesen durch gänzliche Kluft getrennt sind, führt keine Brücke der Verständigung von einem zum andren, und um sich zu verstehen, muß man sich in einem andren Sinn schon verstanden haben'. Wilhelm von Humboldt, 'Über die Aufgabe des Geschichtsschreibers' in: idem, *Wilhelm von Humboldt Werke in fünf Bänden*, ed. Andreas Flitner and Klaus Giel (Stuttgart 1960-1981) Volume I, 585-609, there 596-597.

65   Labrie, *'Bildung' en politiek*, 88-90. Humboldt's 'Idea', not like Plato's philosophy of ideas, is limited to our representation of the eternal, whereas Plato's Idea is equal to the eternal. Labrie, *'Bildung' en politiek*, 95.

from history'.[66] Humboldt thought he could reconcile universalism with individualism, or better: cosmopolitanism with the national state. Humboldt's recognition
of a variety of individualities is in that sense also an acceptance of universalism –
this is, once more, the key of the theory of ideas (*Ideenlehre*).

A notion closely related to *Bildung*, which also clearly demonstrates the connection between the cosmopolitan and the national, is that of *Menschheitsnation*
(nation of all mankind). At first sight this seems to be a contradiction in terms,
for a nation cannot simultaneously contain mankind. However, against the background of Humboldt's theory of ideas (*Ideenlehre*), in which everything has a common spiritual origin, it makes sense. Besides considering the German nation to be
representative of all mankind, Humboldt and the Romantics went a step further:
they also believed in the universal calling of the German nation; that is it should be
the cultural educator of all nations.[67] Hence, Humboldt's *Bildungsideal*, which was
initially restricted to the individual and the collective (the 'cultural nation'), was
now raised to the transnational, the global, to humanity, the universal.

---

66   Meinecke, *Cosmopolitanism*, 43. 'höchst menschliche Werte aus der Geschichte'. Meinecke,
      Weltbürgertum, 50. In a similar fashion at the beginning of the twentieth century
      Ernst Troeltsch tried to extract the highest human values in history, be it from a
      different epistemological and ethical-philosophical perspective. Cf. Troeltsch' ideas and his
      appreciation of Humboldt: Ernst Troeltsch, *Der Historismus und seine Probleme. Erstes Buch:
      Das logische Problem der Geschichtsphilosophie* (Tübingen 1922) 208-213. I will come back to
      Troeltsch in Chapter 5.
67   Meinecke, *Weltbürgertum*, 66. Like Humboldt, Novalis (1772-1801) was also convinced of
      the universal calling of the German nation. However, Novalis had a view coloured by
      Romantic ideas, which is expressed in his definition of Romanticism: 'Die Welt muß
      romantisiert werden (...) Indem ich dem Gemeinen einen hohen Sinn, dem Gewöhnlichen
      ein geheimnisvolles Ansehen, dem Bekannten die Würde des Unbekannten, dem Endlichen
      einen unendlichen Schein gebe, so romantisiere ich es'. Novalis, *Werke, Tagebücher und Briefe*
      (Munich 1978) Volume 2, 334. Novalis advocated a new, open-minded view on the world
      surrounding us. With regard to the contrasts between the national and the universal,
      he states in line with his definition of Romanticism: 'Alles Nationale, Temporelle, Lokale,
      Individuelle läßt sich universalisieren (...) dieses individuelle Kolorit des Universellen ist
      sein romantisierendes Element'. Cited in: Meinecke, *Weltbürgertum*, 66. This way Novalis
      managed to reconcile the individual with the universal, and consider them as one. However,
      according to Meinecke, this reconciliation caused Novalis to situate himself at such a
      high and abstract level above history that 'sky and earth' fused together, or as Novalis
      puts it: 'Menschheit und Weltall zusammenströmen'. Meinecke, *Weltbürgertum*, 66, 68. The
      similarities with Schelling's philosophy will be discussed in Chapter 5. Hans Jörg Sandkühler
      ed., *Handbuch Deutscher Idealismus* (Stuttgart 2005) 331, 335-336; Charles Larmore, 'Hölderlin
      and Novalis' in: Karl Ameriks ed., *The Cambridge Companion to German Idealism* (Cambridge
      2000) 141-160, there 157; Frederick Beiser, *German Idealism. The Struggle against Subjectivism,
      1781-1801* (Cambridge Mass. 2002) 13, 373, 428.

The ideal state

Like Humboldt, Friedrich Schlegel (1772-1829) – one of the founders of the Jena-circle – and the philosopher Johann Gottlieb Fichte (1762-1814) also start from cosmopolitan ideals. Meinecke considers their views as 'transitional ideas'. Their ideas, for the first time, demonstrate a change with regard to the notions of the universal and the national, which means they were the first to clearly articulate the idea of the national state.[68]

Initially, Schlegel was very much interested in politics. For example, in a letter from 1796 to his brother, August Schlegel (1767-1845), he writes: 'I shall be happy when I can finally revel in politics'.[69] According to Meinecke, Schlegel was not interested in the political state as such, but in the state as it should be.[70] Hence, Schlegel starts from an ideal. Meinecke even claims it is a rationalistic and ahistorical view. According to him, this view is also expressed in Schlegel's article 'Versuch über den Begriff des Republikanismus' (1796) – written in response to Kant's *Zum ewigen Frieden* (1795) – which is on the question what the best form of government should be. Initially, Schlegel thought a republic the best form of government; so, he was not opposed to the ideals of the French Revolution. He was convinced that a democratic republic was feasible, and eventually a so-called *Völkerstaat* (international state) or even a *Weltrepublik* (world republic) was a possibility. Kant, nevertheless, thought a world republic to be unrealistic. Both Schlegel and Kant, however, started from the assumption that different 'peoples' (*Volk*) should be politically independent and autonomous. To Schlegel and Kant, it was not the autonomy of a concrete, historical state (*Staatspersönlichkeit*), as Meinecke clearly shows, but the autonomy or sovereignty of a people, based on natural law and thus based on general human postulates of reason (*Vernunftpostulaten*), which, in the words of Meinecke 'sees individual states only as subdivisions of humanity'.[71] In Schlegel's political thought the 'general' was, therefore, far more important than the 'individual'. With regard to the autonomy of the actual, historical state, Meinecke advocated the exact opposite, which is why he considers Schlegel's thought to be still deeply rooted in a world very different from Romanticism.[72]

Contrary to Schlegel's politically un-Romantic attitude, his poetic and aesthetic romanticism had indeed developed early on. He, so to speak, heard or had already

---

68 For a critique on this issue: Schmidt, 'Friedrich Meineckes Kulturnation. Zum historischen Kontext nationaler Ideen in Weimar-Jena um 1800', 610, 619.

69 Cited in Meinecke, *Cosmopolitanism*, 59. 'Ich werde glücklich sein, wenn ich erst in der Politik schwelgen kann'. Meinecke, *Weltbürgertum*, 70.

70 Meinecke, *Weltbürgertum*, 70.

71 Meinecke, *Cosmopolitanism*, 60; '(...) aus den einzelnen Staaten nur Unterabteilungen der Menschheit machte'. Meinecke, *Weltbürgertum*, 71.

72 Meinecke, *Weltbürgertum*, 71.

known 'the soul and the voice of the nation' for a long time.[73] Schlegel, thus attached far more meaning to the cultural nation (*Kulturnation*) than to the political nation (*Staatsnation*). He found the ideal ingredients for the cultural nation in the Hellenic culture and classical Greek poetry, which, in his view, originated from the Greek nation. His interest in the classical Greek nation also stimulated his interest in the state.[74] He viewed the classical Greek political ideal as: 'the image of a great, independent, and unique cultural nation that underlay a number of independent, unique, but still related and connected state structures', as Meinecke explains it.[75] The classical Greek ideal symbolized nineteenth-century Germany, which likewise consisted of a multitude of small states with an overarching cultural nation. In this context, to Schlegel, as well as to Humboldt, *Bildung* remains the most important aspect of their thought regarding the nation. 'In this respect', Meinecke concludes, 'the early Romantics were still the genuine sons of the generation that had created the humanistic ideal, the generation of Herder, Goethe, Schiller, and Kant'.[76]

Considering the above, the question presents itself why, in Meinecke's view, Schlegel should be considered as a key figure in this debate on the universal and the national. After all, for Schlegel the cultural nation remains the most important. The answer to this question has everything to do with a new phase in Schlegel's thought. According to Meinecke, Schlegel's intellectual life can be divided into two phases. The first I just discussed. The second began around 1800 when Schlegel converted to Catholicism. In this second phase, according to Meinecke, he substituted a free individualistic Romanticism for a Romanticism bound to the Church and politics.[77]

In comparison with his first phase, Schlegel now expressed his views on the national state more explicitly. His most radical idea, in Meinecke's opinion, was that language should be the basis of the ordering of states. Schlegel considers language not only to be spiritually binding, but also the proof of a common origin.[78] As Schlegel puts it: 'The older, purer, and less interbred the race, the more customs there will be; and the more customs there are and the more true persistence in them and attachment to them, then all the more will the group having them be

---

73 Cited in Meinecke, *Cosmopolitanism*, 60. 'Seele und Stimme der Nation'. Meinecke, *Weltbürgertum*, 71.

74 Meinecke, *Weltbürgertum*, 72-73.

75 Meinecke, *Cosmopolitanism*, 61. 'das Bild einer großen, selbständigen, eigenartigen Kultur-nation, deren Boden eine Mehrzahl von selbständigen, eigenartigen, aber untereinander verwandten und verbrüderten Staatsgebilden trug' Meinecke, *Weltbürgertum*, 73.

76 Meinecke, *Cosmopolitanism*, 62. Meinecke, *Weltbürgertum*, 74.

77 Meinecke, *Cosmopolitanism*, 64; Meinecke, *Weltbürgertum*, 76.

78 Of course, Herder already demonstrated this in *Über den Ursprung der Sprache* (1772). Cf. Beiser, *The German Historicist Tradition*, 120-127.

a nation'.[79] The purity of a race is thus central to Schlegel's thoughts. Meinecke considers this idea the result of a misunderstanding, for nations are not solely based on a unified language and kinship. Moreover, Schlegel was apparently not aware, thus Meinecke, of the inconsistency of this claim with his own ideas on freedom and the uniqueness of life in every nation: 'Freedom was reduced to the nativistic doctrine of opposing all foreign influence, and character was equivalent to the strictest maintenance of tradition, to stagnation and archaism in the characters of nations', writes Meinecke.[80] Schlegel was therefore, the first to express a conservative interpretation of the principle of nationality in Germany. When Schlegel realised that his political ideal was never to be realised, he became more conservative – even reactionary – for he now preferred a (medieval) aristocratic-monarchic state.[81]

Though Schlegel concentrated on politics, his ideas on the national state still remained based on the cultural nation, for in the end, he also wanted a state that provided freedom and security and fostered an individual *Bildung*. Moreover, in his conservative phase Schlegel developed a fear of the masses. He now began to oppose the ideals of the French Revolution.[82] In the end he even rejected universal despotism (*Universaldespotie*), but, according to Meinecke, he did this in a way which could also be understood as 'oriented towards the universal', because, like Humboldt, Schlegel defended the idea of the national state by means of universal ideals.[83]

His study of the Romantics made Meinecke aware of the ethical and religious connotations of ideas like 'cosmopolitanism' and 'universalism'.[84] Especially his discussion of Schlegel's view on power politics is proof of this. According to Schlegel, a state is sound when peace is guaranteed by the Church, because the lust for power (*Herrschsucht*) disrupted the ties between the different Christian states and also disturbed the inner national feudal systems (*Ständeverfassungen*).[85] Schlegel and other

---

79   Cited in Meinecke, *Cosmopolitanism*, 65. '(...) je älter, reiner und unvermischter der Stamm, desto mehr Sitten, und je mehr Sitten und wahre Beharrlichkeit und Anhänglichkeit an diese, desto mehr wird es eine Nation sein'. Meinecke, *Weltbürgertum*, 77-78.

80   Meinecke, *Cosmopolitanism*, 66. 'Die Freiheit wurde nativistisch vergröbert zur Fernhaltung alles fremden Blutes, die Eigentümlichkeit, lief hier hinaus auf möglichste Erhaltung des Überlieferten, auf Stagnation und Altertümlichkeit der Nationalcharactere'. Meinecke, *Weltbürgertum*, 77-78.

81   Meinecke, *Weltbürgertum*, 80-81; Beiser, *Enlightenment, Revolution, and Romanticism*, 261-262.

82   Beiser, *Enlightenment, Revolution, and Romanticism*, 252.

83   Meinecke, *Weltbürgertum*, 81-82.

84   Ibidem, 83.

85   Meinecke, *Weltbürgertum*, 82; Dieter Sturma, 'Politics and the New Mythology: the Turn to Late Romanticism' in: Karl Ameriks ed., *The Cambridge Companion to German Idealism* (Cambridge 2000) 219-238, there 232.

Romantics considered power politics to be unethical.[86] That is why the Romantics stick to their universal ideals: 'They [the Romantics, RK] moralized from without instead of understanding the nature of the state from within. It was not clear to them that ethics have an individually determined aspect along with their universal aspect and that the apparent immorality of a power-hungry political egoism can be justified from this individually determined side', according to Meinecke.[87] Consequently, he claims the individual state has in essence a characteristic individual moral nature. From the outside the effects of power politics could be interpreted as immoral, but, according to Meinecke, seen from the inside they are a necessary result of self-determination and self-preservation, and for that reason they are by definition not unethical. A crucial sentence in *Weltbürgertum und Nationalstaat* reveals Meinecke's view on ethics and power politics: '(...) nothing that springs from the deepest individual nature of a being can be immoral'.[88] This moral optimism with regard to the power state and power politics is characteristic for Meinecke's argument in *Weltbürgertum und Nationalstaat*, because, in spite of his criticism of the Romantics and their universal view, Meinecke also adopts a Humboldt-like ethical *Geistnatur*.[89] That is, 'the deepest individual nature' of a power state *emanates* from the eternal or universal 'good', which is situated in the commonly shared *Geistnatur*. For, only this justifies Meinecke's claim that political actions of an individual state are always ethically just. The crucial difference between Meinecke and the early Romantics is that Meinecke attributes freedom of development to the individual or individualities, whereas the Romantics restrict the individual to the universal, the absolute. In my discussion of Schelling's philosophy of nature in Chapter five, I will elaborate on this difference and the relationship between the individual and the absolute.

Next to Schlegel's views, Meinecke discusses Fichte's views on universal ideals and the German idea of the national state. According to Meinecke, Fichte was

---

86   The ultimate consequences of power politics are discussed in Meinecke's *Die Idee der Staatsräson*.

87   Meinecke, *Cosmopolitanism*, 70. 'Sie [the Romantics, RK] moralisierten von außen her, statt das Wesen des Staates von innen heraus sich verständlich zu machen; sie machten es nicht klar, daß das Sittliche überhaupt neben seiner universalen auch eine individuell bestimmte Seite hat und daß von dieser Seite her auch die scheinbare Unmoral des staatlichen Machtegoismus sittlich gerechtfertigt werden kann'. Meinecke, *Weltbürgertum*, 83.

88   Meinecke, *Cosmopolitanism*, 70. '(...) unsittlich kann nicht sein, was aus der tiefsten individuellen Natur eines Wesens stammt'. Meinecke, *Weltbürgertum*, 83. This quote also alludes to Goethe's notion of *Natur*, which I will elaborate on in Chapter 5.

89   On Meinecke's ethical optimism Georg Iggers comments: 'The catastrophes of war and totalitarianism in the twentieth century have revealed the tragic shallowness of this optimism'. Iggers, 'The Decline of the Classical National Tradition', 412. For my discussion of Iggers see the historiographical overview in the introduction.

initially against any form of nationalism, or better: patriotism. Like most of the early Romantics, Fichte was more interested in the universal than the national. It was, therefore, striking when in 1807 Fichte published his *Reden an die deutsche Nation*. Initially it was considered a radical turn in Fichte's thought, but, according to Meinecke, this rests on a misunderstanding. Fichte neither abolished his universal ideals, nor did he completely dive into the 'national'. Meinecke claims that Fichte aimed at a reconciliation of the national and the universal in the educated (*gebildeten*) man. In Meinecke's words: '(...) love of fatherland is realised in such a man's deeds; cosmopolitanism is realised in his thoughts. The first is the manifestation of the spirit; the second, the spirit itself "the invisible in the visible"'.[90] With 'cosmopolitanism as thought' Meinecke refers to the universal ideals of the cultural nation (*Kulturnation*) and with 'love of the fatherland as deed' he refers to the creation of a power state, which is shaped in the political nation (*Staatsnation*). According to Meinecke, both are united in Fichte's thought.

'Cosmopolitanism' and the 'national idea' go together in Fichte's thought, as clearly expressed in statements like: '"cosmopolitanism cannot really exist at all and [that] in reality it must necessarily become patriotism"'.[91] In other words, a cosmopolitan is simultaneously an individual that is connected to a nation. This means that universal ideals, which manifest themselves in the *Geist* (Spirit), only have effect in the reality of the nation, for: 'a will bent on these objectives [the universal, RK] must have and can have an immediate effect only on its immediate environment, i.e., its sphere of influence is the nation', Meinecke writes.[92] In the end Fichte's goal is the education (*Bildung*) of mankind as such and that can only be realised by means of nation building. That is why Meinecke claims Fichte was both a cosmopolitan and a patriot.[93]

Fichte was the first to recognize that the nation was no longer a creation or modification (*Modifikation*) of a higher universal, but instead the other way around:

---

90   Cited in Meinecke, *Cosmopolitanism*, 73. 'Vaterlandsliebe ist seine Tat, Weltbürgertum ist sein Gedanke; die erstere die Erscheinung, der zweite der innere Geist dieser Erscheinung, "das Unsichtbare in dem Sichtbaren"'. Meinecke, *Weltbürgertum*, 86-87. Apart from this synthesis between cosmopolitanism and the national idea, Fichte also tried to construct a more abstract synthesis. Kant's dualism of phenomenal (*Erscheinung*) and noumenal (*innere Geist*) reality should be, according to Fichte, reconciled within a harmonious, monistic system. Rolf-Peter Horstmann, 'The early philosophy of Fichte and Schelling' in: Karl Ameriks ed., *The Cambridge Companion to German Idealism* (Cambridge 2000) 117-140, there 117.

91   Cited in Meinecke, *Cosmopolitanism*, 74. 'daß es gar keinen Kosmopolitismus überhaupt wirklich geben könne, sondern daß in der Wirklichkeit der Kosmopolitismus notwendig Patriotismus werden müsse'. Meinecke, *Weltbürgertum*, 88.

92   Meinecke, *Weltbürgertum*, 88.

93   Idem. Habermas' essay on Europe is strikingly close to Fichte's discussion of Germany: Jürgen Habermas, *Zur Verfassung Europas – Ein Essay* (Berlin 2011). Habermas does not mention Fichte.

'the primal and individual impulses [*Triebe*, RK] of nations appear to be the force that creates the general and supra-national', according to Meinecke.[94] Thus, Fichte acknowledges the individuality of a nation. This shift in reflecting on the national state – to consider it as an individual – is of great importance to Meinecke's philosophy of history, for his historicism concentrated on the individual in history.[95] To Meinecke Fichte's acknowledgment of the individuality of nations is reason enough to connect Fichte's philosophy of history with Ranke's *Geschichtsauffassung* – which I will discuss below.[96] The affinity between Fichte and Ranke, and also Schlegel, is demonstrated by the fact that the power of a state should be interpreted as a force with its own natural impulse (*Lebenstrieb*), which is in harmony with a moral world view (*Weltanschauung*). As Meinecke puts it: 'If the state was not ruled merely by the will of a prince alone nor merely by interest in its own self-preservation, but was maintained by a living national community, and if this national community became valuable for mankind precisely because of its own particular character, then the voracity of the state, too, was ennobled and elevated'.[97] This reminds us of Meinecke's statement that every political act of a state is essentially morally just, since the individual nature of the state has emanated from the (ethical) *Geistnatur*. In short, with regard to the power state we can characterize Meinecke's attitude as 'optimistically Rankean' – that is one of the main hallmarks of *Weltbürgertum und Nationalstaat*.

Despite Fichte's interest in the individuality of the nation, in reality he considered the state to be an 'idea' or an ideal. His ideal is a 'perfect state', a state which is beyond history, in some sort of 'realm of reason' (*Vernunftreich*).[98] In fact, according to Meinecke, Fichte's nation is an 'intellectual nation' (*Vernunftnation*).[99] Fichte was

---

94 Meinecke, *Cosmopolitanism*, 74. 'die ursprünglichen und individuellen Triebe der Nationen als die Kraft, die das Allgemeine und Übernationale hervorbringt'. Meinecke, *Weltbürgertum*, 93-94.

95 Meinecke claims the revolution of historicism resulted in a substitution of a generalising view of history by an individualising view. This more or less is Meinecke's well-known definition of historicism, as formulated in *Die Entstehung des Historismus*. Cf. Chapter 4.

96 On the moral aspects of Ranke's philosophy of history: chapter 2 and 3.

97 Meinecke, *Cosmopolitanism*, 79-80. 'Wenn der Staat nicht bloß durch den Willen des Fürsten, auch nicht bloß durch kalte Interesse seiner eigenen Selbsterhaltung getrieben, sondern von einer lebendigen Volksgemeinschaft getragen wurde, und wenn diese Volksgemeinschaft gerade durch ihre Eigentümlichkeit wertvoll für die Menschheit wurde, dann wurde auch die Pleonexie [greed, RK] des Staates geadelt und versittlicht'. Meinecke, *Weltbürgertum*, 94.

98 Hegel's Reason and its self-realization in history, differs from Fichte's *Vernunftreich*, because Fichte's Reason is beyond history, Hegel's realization takes place within history. Friedrich Meinecke, *Die Idee der Staatsräson in der neueren Geschichte* (Munich 1963, first published 1924) 440.

99 Meinecke, *Weltbürgertum*, 96. According to Meinecke, this it not strange at all, for Fichte only wanted to promote his philosophy under the pretext of the idea of the national state: '(...) alle

deeply influenced by cosmopolitan thinking, which was predominantly focussed on an ideal. In that sense, Fichte's *Reden an die deutsche Nation* was not inspired by nationalism, but idealism.[100] Nevertheless, Meinecke considered his ideas a source of inspiration for the conception of the idea of the German national state.

All Romantics clung to the 'universal'. Time and again the national idea was inspired by universal ideals and in the end disappeared into the transcendental. It remains to be seen how Meinecke thought both could be reconciled, especially when these universal ideals continuously managed to break free from the national idea. After all, there appears to be no real unification or reconciliation between the two notions, only a brief contact. The next section will discuss how Meinecke thought it indeed possible to unite *Macht* and *Kultur*.

1.3   The reconciliation of cosmopolitanism and the national state?

At the end of the first part of *Weltbürgertum und Nationalstaat* Meinecke discusses 'the three great liberators of the state': Hegel, Ranke and Bismarck.[101] Meinecke labels them 'liberators' because they apparently liberated political thought from the 'non-political universalistic ideas' and thereby paved the way for the national state.[102] His use of terms like 'liberators' or 'saviours' is striking in this context, since it implies there was no such thing as a synthesis of *Macht* and *Kultur*, rather it seems *Kultur* was replaced by *Macht*. Whether or not this is true will be the subject of this section.

In his discussion of the three liberators of the state Meinecke says it might seem daring to name Hegel – the great thinker who perfected and synthesized the speculative-idealistic movement – in the same breath with the 'two great empiricists' Ranke and Bismarck.[103] Yet, since Hegel always unified the opposite, a discussion of his philosophy is, in Meinecke's view, essential for an analysis of contrasting concepts. Moreover, Hegel's ideas regarding the state have been extremely influential. In the words of Meinecke: 'He stands in the foremost ranks of the great thinkers

---

seine oft so wunderlich berührenden Wandlungen im Verhältnis zu Staat und Nation waren nichts anderes als wechselnde Erwägungen über die tauglichsten Mittel zur Erreichung seines idealen Endzwecks, die Menschheit vom Banne der Sinnlichkeit zu erlösen und zu der höheren Welt der Freiheit, zu ihrem göttlichen Ursprunge hinaufzuheben'. Meinecke, *Weltbürgertum*, 90.

100   Gilbert, 'Friedrich Meinecke', 72.

101   Meinecke, *Cosmopolitanism*, 197. '(...) die drei großen Staatsbefreier'. Meinecke, *Weltbürgertum*, 236.

102   Idem. 'unpolitisch-universalen Ideen'. Meinecke, *Weltbürgertum*, 236; also see: Brands, *Historisme als ideologie*, 165.

103   Meinecke, *Cosmopolitanism*, 197; Meinecke, *Weltbürgertum*, 236.

of the nineteenth century who promulgated a favourable attitude toward the state and a conviction of its necessity, greatness, and ethical dignity'.[104]

Meinecke's justification of his discussion of Hegel reveals his attitude toward him. Meinecke strongly disapproves of Hegel's philosophy of history, and in particular of his idea of the 'self-realization of the spirit in history'. According to Meinecke, Hegel's conviction that everything in history serves the realization of the spirit leads to a distortion of history. Otherwise, all individualities in history are robbed of their singular and unique character. Moreover, Hegel's theory aims at a universalistic view which is so far removed from historical reality that it completely obscures empirical insight, Meinecke writes.[105] In that sense, Hegel is very close to the early-Romantic interweaving of the cosmopolitan and national ideas. Likewise, Hegel's confirmation of the early-Romantic notion of the *Menschheitsnation* also is a proof of this. Contrary to the early-Romantic's interpretation of this notion – the universal calling of the German nation – Meinecke claims Hegel is convinced that: '(…) in every epoch of world history [there is, RK] a "nation of world historical consequence" that acts as the bearer of the universal spirit in its current state of development'.[106] In short, Meinecke concludes Hegel's view (*Anschauung*) inevitably leads to a reduction of all historical individualities to mere instruments and functionaries of the world spirit.[107]

Meinecke claims that even when Hegel discusses the Herderian notion of the unique national spirit (*Volksgeist*) of the nation or state he considers it a means to an end, instead of something unique and of value in itself, as the Romantics, and especially Herder considered it to be.[108] The only distinctive characteristic Hegel attributes to a state, according to Meinecke, is the 'national principle': the spiritual legacy of a common past, which is related to the present and future needs of the state, that is: the right to autonomy.[109] On the basis of its own political interest, the state has, according to Hegel, the right to maintain and to develop itself. That

---

104   Meinecke, *Cosmopolitanism*, 198. 'Unter den großen Denkern des 19. Jahrhunderts, die überhaupt Staatsgesinnung, Überzeugung von der Notwendigkeit, der Größe und sittlichen Würde des Staates verbreitet haben, steht er in der vordersten Reihe'. Meinecke, *Weltbürgertum*, 237.

105   Meinecke, *Weltbürgertum*, 240.

106   Meinecke, *Cosmopolitanism*, 201. '(…) in jeder Epoche der Weltgeschichte ein "weltgeschichtliches Volk" gäbe als Träger der jeweiligen Entwicklungsstufe des allgemeinen Geistes'. Meinecke, *Weltbürgertum*, 240-241.

107   Meinecke, *Cosmopolitanism*,201. Meinecke, *Weltbürgertum*, 241. Again, this is an outdated interpretation, which I will correct in Chapter 4 section 6.

108   Frederick C. Beiser, *The Fate of Reason. German Philosophy from Kant to Fichte* (Cambridge Mass. and London 1987) 145.

109   Meinecke, *Weltbürgertum*, 241.

means war is justified or even inevitable, for how is perpetual peace possible when all states have their own sovereign will that guides their individual state interest?

Meinecke concludes that Hegel's transcendent point of view caused the concrete, the uniqueness of individualities to fade into the background. Hence, Meinecke sums up his discussion of Hegel with the following statement: 'For it was universalism in its most extreme form that motivated Hegel and let the world spirit advance by means of its unconsciously functioning instruments'.[110] To be sure, Meinecke is not against universal principles, he only wants to draw clearer boundaries between the individual and the universal, because: '(...) it was possible to go still further in recognizing the proper claims of historical individualities; it was possible to embrace them more warmly, yet at the same time to direct the spiritual vision upwards to the eternal constellations', Meinecke says.[111] In Meinecke's view, the most successful historian who situated himself at the intersection of the absolute and the individual, was and will forever be Ranke. But how did Ranke manage to avoid the Hegelian (speculative) 'errors'?

Meinecke claims the universalistic characteristics in Ranke's view of history are the result of the influence of Romanticism. Initially, Ranke's national sense is not political but of a cultural-intellectual or even of a more spiritual nature. Meinecke characterizes this feeling as a 'pantheistic relationship between his spirit and the spirit of the nation'.[112] According to Meinecke, a crucial difference between Ranke and Hegel (and the Romantics) is that Ranke's 'own German national feelings and his awareness of the universal existed side by side'.[113] For example, Ranke no longer adopts the notion of the universal calling of the German nation (*Menschheitsnation*), for 'His consciousness and the consciousness of his time were too realistic and concrete for that'.[114] Ranke does consider the real (*Reale*) always together with the spiritual (*Geistige* or *Gott*).[115] In this context, Ranke's notion of the 'moral energy'

---

110   Meinecke, *Cosmopolitanism*, 202. 'Denn Universalismus in der höchsten Steigerung war es, was Hegel trieb und den Weltgeist treiben ließ durch seine bewußtlosen Werkzeuge'. Meinecke, *Weltbürgertum*, 242.

111   Meinecke, *Cosmopolitanism*, 202. 'Es war (...) möglich, noch weiter zu gehen in der Anerkennung des Eigenrechts der geschichtlichen Individualitäten, sie noch herzhafter zu umfassen und doch dabei das geistige Auge auch immer nach oben zu den ewigen Gestirnen gerichtet zu erhalten'. Meinecke *Weltbürgertum*, 243.

112   Meinecke, *Cosmopolitanism*, 205. '(...) pantheistischen Verhältnisses zwischen seinem Geiste und dem Geiste der Nation'. Meinecke *Weltbürgertum*, 246.

113   Meinecke, *Cosmopolitanism*, 207. 'seine eigene deutsche Nationalempfindung und sein Sinn für das Universale an ihren Grenzen zusammenstoßen'. Meinecke, *Weltbürgertum*, 249.

114   Meinecke, *Cosmopolitanism*, 207. 'Dazu war sein Sinn und der Sinn seiner Zeit schon zu real und konkret geworden (...)'. Meinecke, *Weltbürgertum*, 249.

115   In that sense Ranke is very close to Goethe, who also assumed a connection between the real and the spiritual. In Goethe's correspondence with Schiller this is clearly expressed. Confer: Chapter 5.

(*moralische Energie*) is important, for it refers to the 'politicization' of the individual. The moral energy, according to Ranke, demands of the state's citizen 'to elevate moral duty to self-reliance, and commands to freedom'.[116] Because of this notion of 'moral energy' Ranke has been accused of 'spiritualizing' power, and thus of losing sight of the true nature of power: (unintentional) 'evil'.[117] In an essay in 1913 Meinecke claims this interpretation of Ranke is false. He explains that: 'Every expert on Ranke knows that the 'moral energy', which he saw in history, not only encompasses higher spiritual values, but also all human effort'.[118] As a result, in Ranke's thought the real and the spiritual always go together. Meinecke praises Ranke for this achievement because: 'No historian has ever managed to treat things both so realistic and so transcendent'.[119]

To Meinecke Ranke is the 'synthesis' of the real and the spiritual, for he managed to unite both. That is, Ranke reconciled the universal-cultural spirit with the real national state and therefore connects idea and reality. Consequently, Meinecke considers Ranke the bridge between Humboldt and Bismarck: 'This connection between the three men (...) shows that truly idealistic and truly realistic thinking are always bound to come together again'.[120] Finally, Meinecke concludes: 'the forces of the German cultural nation and of the individual state (...) created the new German political nation'.[121] Meinecke is convinced that it was Bismarck who accomplished the last step in this 'creation' towards the formation of the national state.

With regard to politics and culture, the advent of Bismarck marks a new era in German history. The final reconciliation of *Kultur* and *Macht* culminated, according to Meinecke, in Bismarck's realization of the German national state; Bismarck combined the *Kulturnation* with the *Staatsnation* in Germany's unification. In addition to the unification of Germany Bismarck also managed to maintain the respect for the old German states, Prussia in particular – which is the subject of the second

---

116   Cited in Meinecke, 'Zur Beurteilung Rankes', 59. 'die Zwangspflicht sich zur Selbsttätigkeit, das Gebot zur Freiheit erhebe'.

117   Kossmann, 'Een kennismaking met Ranke', 189; Meinecke, 'Zur Beurteilung Rankes', 59.

118   Meinecke, 'Zur Beurteilung Rankes', 59. 'Jeder Kenner Rankes weiß, daß die "moralischen Energien", die er in der Geschichte aufsucht und darstellt, nicht nur höhere geistige Werte, sondern auch Anspannung aller menschlichen Kräfte für sie umfassen'. After the First World War Meinecke changes his mind. See Chapters 2 and 3.

119   Friedrich Meinecke, 'Rankes "Große Mächte"' in: idem, *Zur Geschichte der Geschichtsschreibung* (Munich 1968) 66-71, there 70. 'Keinem Historiker der Welt ist es je gelungen, zugleich so realistisch und so transzendent die Dinge zu behandeln'.

120   Meinecke, *Cosmopolitanism*, 217. 'Dieser (...) symphonische Zusammenhang zwischen den dreien ist ein Zeugnis, daß das wahre ideelle und das wahre reale Denken sich immer wieder finden müssen'. Meinecke, *Weltbürgertum*, 261.

121   Meinecke, *Cosmopolitanism*, 218. 'Einmal ist die neue deutsche Staatsnation ja eben geschaffen worden durch ein Zusammenwirken der Kräfte der deutschen Kulturnation und des Einzelstaates'. Meinecke, *Weltbürgertum*, 263.

part of *Weltbürgertum und Nationalstaat*.[122] At first glance, this political unification under the leadership of Prussia was successful; however, it appears that due to this political unification of Germany, the *Kulturnation* (the cultural nation) ceased to exist. Meinecke even claims Bismarck did not have any affinity with the '(...) conception (...) and ideas of a cultural and universal nation of the spirit that Fichte, the early Romantics, and the Classical Idealist had held'.[123] Meinecke postulates Bismarck's indifference can be explained as follows: 'All these delicate and profound ideas could become vivid experiences for a mind inclined to contemplation but not for one, like Bismarck's, inclined to action'.[124]

Bismarck's poor understanding of German cultural-intellectual life is also clearly expressed in his Olmütz address of 1850. In this lecture he explicitly points out that: 'The only sound foundation for a major state – and in this it is fundamentally different from a minor state – is political egoism and not Romanticism. It is not worthy of a great state to fight for a cause that does not touch on its own interests'.[125] Later on Bismarck characterized this as 'Prussian national policy'.[126] Especially this reference to 'Prussia' reveals that Bismarck only created a German Empire on paper. In reality 'the Germans' were far from united, for, apart from the fact that it was a unification under Prussian leadership, the contrast between, for example, Protestants and Catholics did not diminish at all;

---

122 Kossmann, 'Friedrich Meinecke', 216.

123 Meinecke, *Cosmopolitanism*, 221. 'Vorstellungen von der geistigen Kultur- und Universalnation, die in Fichte, den Frühromantikern und den klassischen Idealisten lebten'. Meinecke, *Weltbürgertum*, 265-266.

124 Meinecke, *Cosmopolitanism*, 221. 'Alle diese feinen und tiefsinnigen Ideen konnten inneres Erlebnis eines zum Betrachten, aber nicht eines so zum Handeln gestimmten Geistes werden, wie er es war'. Meinecke, *Weltbürgertum*, 266. In one of his *Maximen* Goethe states: 'Der Handelnde ist immer gewissenlos; es hat niemand Gewissen als der Betrachtende'. Johann Wolfgang von Goethe, 'Maximen und Reflexionen' (posth. 1833) in: *Goethes Werke. Hamburger Ausgabe in 14 Bänden*, ed. Erich Trunz (Munich 1982) Volume XII, 399; cf. Paul Dauner, 'Das Gewissen' (Stuttgart 2008) 233. unpublished dissertation. To reflect on one's own past, one's own thinking, and will, one has to interrupt one's actions. Goethe's statement is in that sense very close to Bismarck's political standpoint. For Bismarck fails to account for the spiritual (moral) past of his own country on which he could have based his politics. Instead he starts from the (Prussian) political act. In order to preserve the state, his decisions are based on power and prudence. In the next chapter I will elaborate on the notion of 'prudence' when I analyse Meinecke's *Die Idee der Staatsräson*.

125 Cited in: Meinecke, *Cosmopolitanism*, 225. 'Die einzige gesunde Grundlage eines großen Staates, und dadurch unterscheidet er sich wesentlich von einem kleinen Staate, ist der staatliche Egoismus und nicht die Romantik, und es ist eines großen Staates nicht würdig, für eine Sache zu streiten, die nicht seinem eigenen Interesse angehört'. Meinecke, *Weltbürgertum*, 270.

126 Meinecke, *Cosmopolitanism*, 225. 'nationale preussische Politik'. Meinecke, *Weltbürgertum*, 270.

in fact it got more intense. Moreover, under the pretext of nationalism, the 'real' Germans within the empire turned against minorities like Jews and the non-German speaking population.[127] One can conclude that Bismarck's unification was a realization of Prussian political history.[128] Ranke's reconciliation of spirit and reality was replaced by Bismarck's emphasis on politics and political action.[129]

Nevertheless, Meinecke considers Bismarck to be the last phase in the 'liberation of the state', for he liberated the state from the 'universal idea(l)s'. Initially the Romantic cosmopolitan ideals were of great value to political practice, but gradually they became impractical and eventually they had 'fateful effects on practical politics and on the power of the state'.[130] And Meinecke, therefore, concludes the universal 'poison' had to be 'cast off if the body was to function naturally again'.[131] The 'doctor' who could extract this poison was Bismarck.[132]

The above, of course, is still consistent with Meinecke's main theme in *Weltbürgertum und Nationalstaat*: an analysis of the political triumph of the German national state.[133] Yet, the disappearance of the universal ideals in the creation of the German national state indicates that a synthesis between the political nation and the cultural nation did not really occur. Meinecke maintains a synthesis had in fact been realised. Nonetheless, in the end he had to admit it was more of a 'substitution' than an actual synthesis:

> Universal and spiritualized national ideas invaded the state simultaneously and in the closest contact with each other. The national ideas gave the universal ones strength and warmth and helped them to enter the state. Later, the universal ideas had to be put aside to permit the Prussian national state to develop into the German national state.[134]

---

127 Boterman, *Moderne geschiedenis*, 136, 158-159.

128 Ibidem, 137

129 In this context the American historian Sterling remarks: 'If Ranke had introduced reality into political thought, Bismarck had reintroduced it into political action'. Sterling, *Ethics in a World of Power*, 73.

130 Meinecke, *Cosmopolitanism*, 229. 'verhängnisvolle Wirkungen auf die praktische Politik, auf die Machtstellung des Staates'. Meinecke, *Weltbürgertum*, 275-276.

131 Meinecke, *Cosmopolitanism*, 229. 'das der Körper wieder ausscheiden mußte, wenn er wieder natürlich funktionieren sollte'. Meinecke, *Weltbürgertum*, 275-276.

132 Meinecke, *Weltbürgertum*, 276. Also see: Sterling, *Ethics in a World of Power*, 71: 'This was the crux of Meinecke's criticism of the cosmopolitan idea. It provided no basis for practical and purposeful political association between the rulers and the ruled. It produced instead anarchy, revolution, and the tyranny of imperialism because it ignored the individual peculiarities of the states and the historic development of societies in which group integration was always achieved by means of group differentiation'.

133 Sterling, *Ethics in a World of Power*, 36.

134 Meinecke, *Cosmopolitanism*, 230. '(...) die Invasion der universalen und der vergeistigt nationalen Ideen in den Staat erfolgte gleichzeitig und in engster Verbindung miteinander.

Meinecke considers the cultural nation to be a driving force in the German (polit-ical) unification, but when this unification was realised by Bismarck, the cultural nation disappeared, or it no longer served any purpose. In that sense, Bismarck did not realise a synthesis between *Macht* and *Kultur*, for the result of the German unification was a decrease of cosmopolitan ideals.[135] In an essay in 1913 Meinecke claims that Ranke and Bismarck freed themselves from the 'political Romantic web of the period of the Restoration and confirmed an autonomous Realpolitik applied by modern national states'.[136] The synthesis of *Macht* and *Kultur* was thus only realised in the *ideas* of Ranke; yet Meinecke was convinced that the intellectual aspects of the nation (the *Kulturnation*) had not lost its value in the newly created Bismarckian political climate: 'The idealistically true and the realistically vital el-ements they contain remain preserved for us (...)'.[137] Nevertheless, it seems the *Kulturnation* was played out with Bismarck's unification of Germany.

In this context the views of many of Meinecke's contemporaries are illuminat-ing. For example, the group of poets and writers close to Stefan George (1868-1933) also thought that the cultural nation had disappeared with the unification of Germany. Consequently, they rejected Bismarck's culturally empty politics.[138] Like the early Romantics a century before, members of the *George-Circle* were also predominantly interested in the cultural nation. The early Romantics considered the cultural nation a more or less political instrument to form a (mainly cultural) unity, whereas George *cum suis* were not interested in politics at all.[139]

Die einen gaben den anderen Kraft und Wärme und halfen ihnen, in den Staat einzudringen. Später mußten die einen wieder ausgestoßen werden, um den preußischen Nationalstaat zum deutschen Nationalstaat weiter entwickeln zu können (...)'. Meinecke, *Weltbürgertum*, 275-276.

135   On the other hand, to Meinecke (and many other Prussians at the time) Macht was also associated with order and discipline and therefore with civilization and thus *Kultur*. Many thanks to Wessel Krul for making me aware of this issue. Also: Henry Pachter, 'Masters of Cultural History II: Friedrich Meinecke and the Tragedy of German Liberalism', *Salmagundi* No. 43 (Winter 1979) 12-42, there 25.

136   'Gespinst der politischen Romantik, der Restaurationszeit durchbrachen und die autonome Realpolitik des modernen Nationalstaates zur Geltung brachten (...)'. Meinecke, 'Zur Beurteilung Rankes', 52.

137   Meinecke, *Cosmopolitanism*, 217. 'Das ideell Wahre wie das real Lebendige, was sie enthielten, bleibt uns unverloren (...)'. Meinecke, *Weltbürgertum*, 261.

138   Boterman, *Duitse dichters en denkers*, 56-57.

139   In *Enlightenment, Revolution, and Romanticism* Beiser shows that the early Romantics and the whole of German Idealism were in fact politically engaged movements. Beiser corrects the one-sided image that these movements were allegedly apolitical. Beiser's argument is in that sense very close to Meinecke's *Weltbürgertum und Nationalstaat*, since Meinecke also shows that the poets and thinkers were involved with politics.

Another advocate of the cultural nation next to George was Oswald Spengler (1880-1936). He preferred the 'power of the spirit' and 'life' – the 'philosophy of life' and notions like 'life' were fashionable during this period.[140] Friedrich Nietzsche (1844-1900) was also associated with this philosophy of life, and he was a major influence on both George and Spengler. Meinecke denounced Nietzsche's philosophy and especially his nihilism: 'Such an anarchy of subjectivism I could not bear', Meinecke writes in his memoirs.[141] Meinecke also found it difficult to accept Spengler's views. Meinecke disagreed with his view of history, because it reminded him too much of the 'system of a previous philosophy of history', which had become outdated.[142] Meinecke did not really elaborate on George, but when he mentions him, he is quite reserved in his opinion. Meinecke considered George's focus on the eternal and timeless as one-sided and too subjectivistic, too mystical and therefore too disconnected from the reality of the past. Meinecke thought George and his Circle were too focussed on an imagined past, instead of analysing the 'real' past.[143]

In the process of the German unification the cultural nation disappeared. It would be better to change the motto of this chapter to: 'where the political begins, the learned ends'.[144] Meinecke, who probably sensed this, emphasized Ranke's 'actual' synthesis of *Macht* and *Kultur*. The question remains why Meinecke, in contrast to some of his contemporaries, wanted to bridge politics and culture, the national and the universal. This will be the subject of the next section.

1.4   The tradition of the ideal

When Bismarck unified Germany in 1871 Meinecke was only eight years old. Even though this event made a deep impression on the young Meinecke, it is not very likely that he already considered the unification or synthesis as a triumph of the German national state.[145] By means of his education and his study and in particular his familiarity with the writings of Ranke and the Prussian School – primarily Droysen, Sybel and Treitschke – his knowledge of German (political) history increased,

---

140   Boterman, *Duitse dichters en denkers*, 79.

141   'Eine solche Anarchie des Subjektivismus konnte ich nicht ertragen'. Meinecke, 'Erlebtes', 105.

142   'überwundenen Systemgeist der früheren Geschichtsphilosophie'. Friedrich Meinecke, 'Über Spenglers Geschichtsbetrachtung' in: idem, *Zur Theorie und Philosophie der Geschichte* (Stuttgart 1965) 181-195, there 194.

143   Friedrich Meinecke, 'Kausalitäten und Werte in der Geschichte' in: idem, *Zur Theorie und Philosophie der Geschichte* (Stuttgart 1965) 61-89, there 72; also see: Meinecke, 'Straßburg – Freiburg – Berlin. Erinnerungen', 193.

144   'wo das politische beginnt, hört das gelehrte auf' instead of: 'wo das gelehrte beginnt, hört das politische auf'.

145   As described in his memoirs: Meinecke, 'Erlebtes', 22-23.

and he grew more and more convinced that with Bismarck's unification in 1871 a synthesis of *Macht* and *Kultur* had indeed been realized. In particular, his research on *Weltbürgertum und Nationalstaat* confirmed this conviction. His enthusiasm with regard to this synthesis of *Macht* and *Kultur* seems to be a result of his interest in the ideas of the clergyman and politician Friedrich Naumann (1860-1919).

Around 1900 the left-liberal Naumann, backed by Max Weber, Hans Delbrück, Ernst Troeltsch and Meinecke, pled for a social monarchy in which liberals and socialists would join forces for the purpose of solving the problems between the workers and the rest of society.[146] This too was a proposal for a synthesis between power and culture, albeit on a smaller scale. This requires some explanation. The *Bildungsideal* – the intellectual perfection of the free individual – had always been a prerogative of the bourgeoisie and in particular the *Bildungsbürger* (intellectual bourgeoisie). Although to many (mainly the older generations) of the *Bildungsbürger* the *Macht* was always of secondary importance to *Kultur*, this does not apply to the generation of Meinecke and Naumann.[147] For the benefit of the workers they laboured for an increase in democratization, political integration and cultural education.[148] They thought the upper class lacked spiritual and social empathy with regard to the lower classes. It was precisely this contrast between the upper class and the workers they considered a threat to *Kultur* (and the state). On the one hand this *Bildungsideal* could degenerate into a *Kulturreligion* (religion of culture) of the upper class, on the other hand the lower classes would perish by a form of *Zivilisation* – a materialistic utilitarianism aimed at the masses – which would nullify the *Bildungsideal* and hence disrupt the entire society.[149] Bismarck's synthesis of *Staatsnation* and *Kulturnation* was thus only realised in certain layers of society.

Due to the politics of Wilhelm II – who gained complete control in 1890, when Bismarck was dismissed – Meinecke feared the end of *Kultur*, as well as the end of the monarchy, which is why he agreed with Naumann's suggestion: 'Naumann's intelligent prophecy (...) has only conjectural value, but it is a conjecture that deserves serious consideration'.[150] Naumann's ideas led to little or nothing; a collaboration

---

146 Under the flag of German colonial and maritime politics Naumann suggested this collaboration, which, according to him, could be used to distract from internal unrest. Boterman, *Moderne geschiedenis*, 176, 195. Also: William O. Shanahan, 'Friedrich Naumann: A Mirror of Wilhelmian Germany', *The Review of Politics* 13 (1951) 267-301; Kossmann, 'Friedrich Meinecke', 213; Iggers, *The German Conception*, 12, 15.

147 Boterman, *Moderne geschiedenis*, 187.

148 Boterman, *Moderne geschiedenis*, 176, 187; Iggers, *The German Conception*, 127-128; Peter Gay, *Weimar Culture. The Outsider as Insider* (New York and London 1974) 40.

149 Boterman, *Moderne geschiedenis*, 187-188.

150 Meinecke, *Cosmopolitanism*, 371. 'So hat auch die geistvolle Naumannsche Konstruktion (...) nur den Wert einer Möglichkeit, aber allerdings einer sehr zu erwägenden und ernst zu

between liberals and socialists never really got off the ground.[151] Nevertheless, in his autobiography Meinecke states that Naumann and his circle of intellectuals were not completely unsuccessful: 'We intellectually bridged the gulf between the bourgeoisie and the community of workers, unfortunately it was only a narrow bridge – nonetheless it reassured the workers that we placed our trust in them, even those of us who were rooted in the old culture of Germany participated in this'.[152] The appeal Naumann's proposal had on Meinecke and other (bourgeois) intellectuals is characteristic of the synthetical way of thinking of this time. Similar to the generation of 1871, this new generation considered itself on the threshold of fulfilling a great event, decision or task.[153]

Meinecke's emphasis on the synthesis of 1871, which he describes in *Weltbürgertum und Nationalstaat*, can be explained against the background of Naumann's failed synthesis. *Weltbürgertum und Nationalstaat* can be considered a confirmation and justification of Bismarck's politics. There is another issue related to Meinecke's justification of Bismarck's politics, namely, his indebtedness to the Prussian School and in particular to Johann Gustav Droysen's view of history.

Meinecke had an ambivalent attitude towards Droysen. He admired Droysen's *Historik*, but was rather reserved with regard to his politically oriented history. Meinecke's objections were aimed in particular at the speculative or teleological aspects of Droysen's writings, which obviously were handed down from Droysen's teacher Hegel.[154] Meinecke expressed this in *Weltbürgertum und Nationalstaat*: 'In wanting to derive Prussia's historical calling for Germany from its goals in actual politics, he [Droysen, RK] failed to recognize the true nature of this state'.[155] In 1915 when Meinecke held his inaugural address at the Prussian Academy of Sciences in Berlin he maintained Droysen's view on Prussian politics was too idealistic.[156] In his address Meinecke also mentioned or admitted that his own early works

nehmenden'. Meinecke, *Weltbürgertum*, 447; Kossmann, 'Friedrich Meinecke', 213; on the politics of Wilhelm II, see: Boterman, *Moderne geschiedenis*, 193ff.

151   Boterman, *Moderne geschiedenis*, 176.

152   Meinecke, 'Straßburg – Freiburg – Berlin. Erinnerungen', 317. 'Wir haben eine geistige Brücke geschlagen zwischen Bürgertum und Arbeiterschaft, eine leider nur schmale, – aber sie hat der Arbeiterschaft das Vertrauen gegeben, daß sie verstanden und geachtet würde auch von solchen, die in der alten Kultur Deutschlands wurzelten'.

153   Kossmann, 'Friedrich Meinecke', 213-214.

154   Friedrich Meinecke, 'Johann Gustav Droysen, sein Briefwechsel und seine Geschichtsschrei-bung' in: idem, *Zur Geschichte der Geschichtsschreibung* (Munich 1968) 125-167, there 146-147, 150-151.

155   Meinecke, *Cosmopolitanism*, 323. 'Indem Droysen die Neigung hatte, den geschichtlichen Beruf Preußens für Deutschland aus den Zielen seiner tatsächlichen Politik nachzuweisen, verkannte er das wahre Wesen dieses Staates'. Meinecke, *Weltbürgertum*, 391.

156   Friedrich Meinecke, 'Antrittsrede in der Preußischen Akademie der Wissenschaften' in: idem, *Zur Geschichte der Geschichtsschreibung* (Munich 1968) 1-4, there 1.

were also written in a similar idealistic vein, but he claims he liberated himself from this in his later (major) works.[157] Nevertheless, next to Ranke's influence, Meinecke's affinity with Droysen's idealism is also a hallmark of *Weltbürgertum und Nationalstaat*. Meinecke's confirmation of Bismarck's unification of Germany and especially the alleged synthesis of *Macht* and *Kultur* are in line with Droysen's idealism, for Meinecke also aimed at a synthesis, which was in fact nothing less of a proclamation of a past ideal, which to the present was intellectually satisfying, and moreover it served as a (political) substitution for failed syntheses, like the one Naumann had suggested.

Conclusion

Moral optimism is at the heart of *Weltbürgertum und Nationalstaat*. Meinecke considers every political act, whether or not inspired by power politics, as essentially morally just. As a result, *Macht* will generally triumph over *Kultur*. In *Weltbürgertum und Nationalstaat* he still assumes the 'absolute good' to act *in* and *through* the nature of the individual state. As a consequence, all political acts of the state will therefore be ethically just. This idea of a presence of a God-related source or nature (the *Geistnatur*) within the individual is something Meinecke partially borrowed from the German idealistic thinkers, which he analyses in *Weltbürgertum und Nationalstaat*. Walther Hofer characterized Meinecke's harmonious world view – moulded and developed by German Idealism – as 'objective idealism'. In my discussion of Hofer's views, I emphasized that this objective idealism was essentially pantheistic and therefore a 'development' is impossible, for everything is a priori determined.[158] Meinecke's interpretation of this idealism corresponds with his interest in philosophical-religious views, like Neoplatonism, pietism, Schelling's philosophy of nature, and the ideas on the 'individual' derived from Humboldt and Goethe. In these views there is more room for the unique development of the individual and individuality. This dynamic primarily stems from Herder's concept of individuality as well as his organicism in which the natural development of the individual is realised in relative freedom. Meinecke's harmonious philosophy of history and political views thus consists of a combination of the ideas of Herder, Goethe, Humboldt and the early Romantics.

Meinecke's interest in these thinkers could also be an explanation for the political ideal of Germany that he had in mind. By means of the ideas of these thinkers he creates a synthesis between *Macht* and *Kultur*, which in his view was realised in 1871. For example, Meinecke's interest in Schlegel's ideal state fits perfectly in this

---

157   Meinecke, 'Antrittsrede', 1.
158   Cf. Chapter 5.

regard. Contrary to Schlegel however, Meinecke's ideal was to be realised *within* history. So he does not advocate a timeless idea or ideal. Despite his idealism Meinecke indeed starts from the reality of the state, which he considers a unique individuality with its own principle of life. His desire for an ideal is hence grounded in reality. This is why he denounces a Fichtean ideal that is projected beyond history, in a so-called 'realm of reason'. In the end Meinecke always relates everything to reality: 'The essence of history as a whole is and remains – when one may interpret Ranke's phrase in terms of modern experiences – "living life of the individual"', Meinecke writes.[159] As mentioned before, Meinecke's interest in Naumann's desire for a concrete political synthesis is also a clear indication of Meinecke's sense of reality. When Naumann could not realise his ideas, Meinecke (re)turned to the past and the ideal synthesis of *Macht* and *Geist* of 1871.

Weltbürgertum und Nationalstaat was Meinecke's scientific breakthrough in the philosophy of history. He positioned himself in one line with Ranke, Droysen, and to a lesser degree with Sybel. In addition, he fell back on great Romantic thinkers as well as on German Idealism, which means he also created a tradition of the great German poets and thinkers; not to isolate them from the past, but to confirm Germany's identity and tradition(s). This creation of a German canon of thinkers is also characteristic of *Weltbürgertum und Nationalstaat*.

In *Weltbürgertum und Nationalstaat* Meinecke not only positioned himself in line with historians, poets, and thinkers, he also founded a new form of historicism, which I just outlined. In the next chapter I will elaborate in more detail on the development of Meinecke's historicism and the differences between historicism and the Enlightenment; a distinction which Meinecke draws in all of his studies.

Weltbürgertum und Nationalstaat is about the individual and the state, literature and politics, classic-romantic growth and the German unification.[160] In his earlier studies – the biography of Boyen and *Das Zeitalter der deutschen Erhebung* – Meinecke touched upon the relationship between the individual and the state, but *Weltbürgertum und Nationalstaat* is the first detailed exposition of this issue. All of the notions central to Meinecke's later studies are in fact already apparent in this first major study – notions like: personality, politics, (ethical) relativism, individuality and development. Considering Meinecke's trilogy, *Weltbürgertum und Nationalstaat* in this regard is a key study, for all the themes which would crystalise

---

159   'Innerster Kern alles geschichtlichen Lebens ist und bleibt, wenn man dies Wort Rankes im Sinne moderner Erfahrungen interpretiert, "lebend Leben des Individuums"'. Meinecke, 'Zur Beurteilung Rankes', 55; also: Jensen, "Unity in Antinomy", 178.

160   Schulin, 'Friedrich Meinecke', 40.

in his oeuvre are already manifest in this seminal work.[161] The following quote from *Weltbürgertum und Nationalstaat* is proof of this:

> Where did our historical and political mode of thought originate, our awareness that supra-individual human unions have an individuality of their own? It originated in an individualism that kept probing the nature of the individual over a long period of time until it had discovered its deepest roots and the connecting elements that bind the life of the individual to the life of human groups and organizations. Everywhere we look we see individuality, spontaneity, and the urge for autonomy and expanding power. We find the same things in the state and nation, too.[162]

In his second major work, *Die Idee der Staatsräson*, Meinecke further analysed the origin of the historical and political mode of thought. Also he deepened the relationship between the individual and 'higher things', which he expounds even further in his study on the origins of historicism. First however, the outbreak of the First World War forced Meinecke to reconsider his intellectual position. What had remained latent in *Weltbürgertum und Nationalstaat*, became manifest in his next major study. The result of Meinecke's intellectual struggle that followed from the war, will be the topic of the next chapter.

---

161   Maarten Brands claims these themes are already manifest in Meinecke's biography of general Hermann von Boyen. I agree however with Carl Hinrichs, who asserts that these themes are not altogether articulated clearly in *Boyen*, and should by no means be isolated from the study as a whole. Brands, *Historisme als ideologie*, 132, 154-168; Hinrichs, 'Der Historismus als ein Lebensproblem', 379.

162   Meinecke, *Cosmopolitanism*, 140. 'Wo stammt überhaupt unsere historisch-politische Denkweise, unser Sinn für die Individualität auch der überindividuellen menschlichen Verbände her? Doch wohl ganz wesentlich mit aus einem Individualismus, der seine ursprünglich flache Ansicht vom Wesen des Individuums im Laufe einer säkularen Arbeit immer mehr vertieft hat, bis er zu seinen Wurzelschichten gelangte und damit zu den Zusammenhängen, die das Eigenleben des Einzelnen mit dem Eigenleben der höheren menschlichen Verbände und Ordnungen verknüpfen. Individualität, Spontanität, Drang nach Selbstbestimmung und Machtausdehnung überall, und so auch im Staate und in der Nation'. Meinecke, *Weltbürgertum*,169.

2. Struggling with a world view

'Schöne Welt, wo bist du?
Kehre wieder'[1]

Introduction

When Meinecke held his inaugural address in 1915, the First World War was already well under way.[2] In his lecture he outlined his future scholarly plans, which developed from *Weltbürgertum und Nationalstaat*.[3] In line with this study Meinecke delved into the emergence and development of the historical and political mode of thought.[4] In his address he puts it as follows: 'These days two tasks occupy my mind, and under the influence of recent developments they only gain more significance: understanding the changes in the nature and spirit of power politics since the days of the Renaissance, and ascertaining the origins of our modern conception of history'.[5] At first glance, these topics are very different from each other, however in German history they are closely connected. That is why Meinecke

---

1   Extract from Schiller's poem 'Die Götter Griechenlandes' in: Friedrich Schiller, *Sämtliche Werke in fünf Bänden* (Munich 2004) Volume I, 169-173, there 172.

2   After his professorship in Strasbourg Meinecke transferred in 1906 to Freiburg where he held a position until 1914. That year he went back to Berlin, and he would stay there until his death in 1954, with only a short interruption at the end of the Second World War when he had to flee the city.

3   Friedrich Meinecke, 'Straßburg – Freiburg – Berlin. Erinnerungen' in: idem, *Autobiographische Schriften* (Stuttgart 1969) 135-320, there 257.

4   Later Meinecke expressed it in his memoirs as follows: 'Bei der Arbeit am ersten Buche des Weltbürgertums war mir ferner die Einsicht gekommen, daß unsere moderne historische Denkweise ganz wesentlich auf einem neuen Sinne für das Individuelle beruhe. Es bestanden, wie ich mir jetzt sagen mußte, innere Zusammenhänge zwischen dem neuen realpolitischen und dem neuen historischen Sinnen'. Meinecke, 'Straßburg – Freiburg – Berlin. Erinnerungen', 202.

5   Friedrich Meinecke, 'Antrittsrede in der Preußischen Akademie der Wissenschaften' in: idem, *Zur Geschichte der Geschichtsschreibung* (Munich 1968) 1-4, there 2. 'Zwei Aufgaben fesseln mich heute und sind mir durch das Erlebnis unserer Tage immer wichtiger geworden: die Wandlungen in Wesen und Geist der Machtpolitik seit den Tagen der Renaissance zu verstehen und der Entstehung unserer modernen Geschichtsauffassung nachzugehen'.

initially intended to write a book entitled 'Staatskunst und Geschichtsauffassung' (statecraft and conception of history), in which he would elaborate on the relationship between politics and historicism.[6] It is the concept of individuality that both have in common: that is, the recognition of the individual interest of a state is highly important for statecraft or *Staatsräson* (reason of State).[7] Political interests differ from one state to the next, and change in every situation. Now, whenever political actions of a state or a statesman are based on principles of *raison d'état*, it is a prerequisite to have insight and knowledge of the individual past of the state in question. By means of this idea the history of *raison d'état* laid a solid foundation for historicism. So, historicism is connected to *raison d'état* on both the level of practice and method.

Even before the end of the First World War Meinecke decided to examine *Staatsräson* and historicism separately, because in his view historicism was not entirely rooted in political theory, for it also emanated from more *weltanschaulicher* and intellectual changes. Besides, the current political circumstances asked for a revision of his former harmonious view of history and politics. In his second major study, *Die Idee der Staatsräson*, Meinecke decided to approach politics and history from the perspective of the tensions between power and ethics. It is this study of 1924 which is central to this and the next chapter. The origins of historical thinking, which Meinecke also mentions briefly in *Die Idee der Staatsräson*, is discussed later in great detail in *Die Entstehung des Historismus* (1936).[8]

The theme of *Die Idee der Statsräson* is in line with his first study, since in *Weltbürgertum und Nationalstaat* he also discussed power and ethics.[9] A crucial difference between both studies is the tone: *Die Idee der Staatsräson* is gloomy and alarming compared to *Weltbürgertum und Nationalstaat*. Not for nothing, *Die Idee der Staatsräson* is said to be his most tormented work, for, in coming to terms with the disillusionment of the First World War, he struggled with the apparent irreconcilability of power and ethics. Also, in this study Meinecke for the first time expressed his doubts concerning the tradition of Leopold von Ranke, by which he was highly influenced. More importantly, in contrast to his first major study, Meinecke now questioned the synthesis of power and culture. Moreover, the war had made very clear that his view of power did not tally with political reality any more. In the post-

---

6   Meinecke, 'Straßburg – Freiburg – Berlin. Erinnerungen', 257.

7   When it comes to Meinecke's ideas of statecraft I will use his notion of 'Staatsräson', for all other occasions I will use *raison d'état* or 'reason of State'.

8   I will discuss this in Chapters 4 and 5.

9   At the beginning of *Die Idee der Staatsräson* Meinecke states: 'Bekennen wir (...) auch die persönlichen Motive, die zur Auswahl der hier erörterten Probleme geführt haben. Daß sie herauswuchsen aus denen, die in "Weltbürgertum und Nationalstaat" behandelt wurden, wird dem Leser beider Bücher nicht entgehen'. Friedrich Meinecke, *Die Idee der Staatsräson in der neueren Geschichte* (Munich 1963, first published 1924) 25.

war climate there was no longer a rationale for moral optimism, which Meinecke had propagated in *Weltbürgertum und Nationalstaat* – for he had claimed that actions of a state were always ethically just and could never be *unsittlich* (unethical). Meinecke now saw himself confronted with a political situation which he had to adjust to and/or find a solution to on the basis of his own intellectual-historical background.

Die Idee der Staatsräson received both approval and criticism.[10] The criticism focussed mainly on the absence of particular thinkers. For instance, the German historian Gerhard Ritter – an expert on Thomas More – claims Meinecke overemphasized the role of Germany, for Meinecke gave little attention to England, and More was completely absent.[11] Ritter's criticism of Meinecke's image of Machiavelli is in this context more important, and I will come back to this in the next section. With regard to *Die Idee der Staatsräson* and its place in Meinecke's oeuvre Walther Hofer's view still plays a decisive role. In my introduction, I already discussed that Hofer was the first to consider Meinecke's post-war work as a clear shift in his thought. In addition to this, Hofer claimed Meinecke also abandoned his optimistic and monistic world view in favour of a strict dualism. The First World War indeed forced Meinecke to reconsider his position. Meinecke's Rankean resignation with regard to the harmony of *Macht* and *Kultur* as well as his optimistic view on power made room for doubt, disappointment, and disillusionment. In this I can agree with Hofer. I disagree however, with his statement that Meinecke's shift leads by definition to a strict dualistic view of history and politics. What is clear from Meinecke's argument in *Die Idee der Staatsräson* is his new emphasis on the tragic dimension of history. Nevertheless, Meinecke still believed in an underlying harmony in history. In this chapter Meinecke's struggle with harmony and tragedy are central. In the next chapter, I will elaborate in detail on Meinecke's assumed underlying harmony, which kept him from a strict dualism.

This chapter starts with a prelude to my discussion of *Die Idee der Staatsräson*. It is an account of Meinecke's observations regarding the collapse of the German

---

10 For an extensive overview see: Michael Stolleis, 'Friedrich Meineckes Die Idee der Staatsräson und die neuere Forschung in: Michael Erbe ed., *Friedrich Meinecke heute. Bericht über ein Gedenk-Colloquium zu seinem 25. Todestag am 5. und 6. April 1979* (Berlin 1981) 50-75, there 54-65. And also: Alfred Vierkandt, 'Friedrich Meinecke, Die Idee der Staatsräson in der neueren Geschichte', *Kant-Studien* 33 (1928) 299-300; Max Rumpf, 'Friedrich Meinecke, Die Idee der Staatsräson in der neueren Geschichte', *Archiv des öffentlichen Rechts* 48 (1925) 340-349. Carl Schmitt's review will be discussed in section 3.2.

11 Similar arguments are discussed by Michael Stolleis. Stolleis, 'Friedrich Meineckes Die Idee der Staatsräson und die neuere Forschung', 50-75. Gerhard Ritter, 'Die Idee der Staatsräson', *Neue Jahrbücher für Wissenschaft und Jugendbildung* 1 (1925) 101-114; cf. Gisela Bock, 'Meinecke, Machiavelli und der Nationalsozialismus' in: Gisela Bock and Daniel Schönpflug ed., *Friedrich Meinecke in seiner Zeit. Studien zu Leben und Werk* (Stuttgart 2006) 145-174, there 163-170.

Empire after the First World War and his ideas for a possible recovery. Next, I will discuss his conception of the history of the principles of *raison d'état*. In my analysis the focus is on Meinecke's discussion of Machiavelli, Frederick the Great, Hegel and Ranke. These thinkers, and in particular Meinecke's analysis, offer us insight into his philosophy of history, his view on politics, and his struggle with the doctrine of *raison d'état*. Finally, all this will elucidate why Meinecke considered the history of *Staatsräson* as tragic.

2.1   Prelude

Prior to the First World War many felt German culture had already fragmented.[12] That is why Meinecke, in *Weltbürgertum und Nationalstaat*, had claimed he realised the synthesis of *Macht* and *Kultur*; he looked for something to hold on to. This quest for new certainties, new values and 'purpose' was not only widely popular among intellectuals, but also among the rest of the German population.[13] This search for certainty can also explain (paradoxically) the initial enthusiasm by which the war was welcomed. In Germany the outbreak of the war caused a feeling of solidarity, unity, and 'meaning': the nation as a unifying medium.[14] Initially, Meinecke, like many other intellectuals, was also enthusiastic about the war.[15] He hoped for the synthesis of *Macht* and *Kultur*, or Naumann's 'national-social' synthesis to be realised.[16] When the war turned out to be a catastrophe, Meinecke changed his mind and suggested, together with Max Weber and Hans Delbrück among others, a so-called *Verständigungsfrieden* (rapprochement-peace): a peace which would confirm

---

12   Patrick Dassen, *De onttovering van de wereld. Max Weber en het probleem van de moderniteit in Duitsland, 1890-1920* (Amsterdam 1999) 232-233.

13   Dassen, *De onttovering*, 192-193.

14   Ibidem, 232, 244-247.

15   Kurt Flasch, *Die Geistige Mobilmachung. Die deutschen Intellektuellen und der Erste Weltkrieg. Ein Versuch* (Berlin 2000).

16   In 1914 Meinecke had published an article in which he connected past, present and ideal with each other: Friedrich Meinecke, 'Die deutschen Erhebungen von 1813, 1848, 1870 und 1914' in: idem, *Brandenburg, Preußen, Deutschland. Kleine Schriften zur Geschichte und Politik* (Stuttgart 1979) 509-531. The fact that Meinecke considered '1914' as an 'Erhebung' is a classification, but also a warning. See: Daniel Schönpflug, 'Revolution und "Erhebung". Friedrich Meinecke über 1789 und die deutsche Geschichte' in: Gisela Bock and Daniel Schönpflug ed., *Friedrich Meinecke in seiner Zeit. Studien zu Leben und Werk* (Stuttgart 2006) 21-49, there 32. Naumann's 'national-social' synthesis should not be confused with 'National Socialism', which was focussed on the organisation of the masses of workers and mostly against the intellectuals, the *Bildungsbürger*. Instead, Naumann hoped for a synthesis of the workers and the *Bildungsbürgertum*.

the status quo. This meant they were against one-sided German annexations.[17] They all proved to be voices in the wilderness.

The war had failed to realise a synthesis, and it was unlikely to still happen in the near future. Meinecke, therefore, began to reflect (again) on the German political and cultural past. For him this was not a glorification of the past or a form of (historical) escapism, rather a search for something that could heal the present German state and culture and make it sound again.

An article of 1920 proves Meinecke's concern regarding the German nation, which was in a deep crisis. In this article of 1920 entitled 'Wilhelm von Humboldt und der deutsche Staat' Meinecke comes straight to the point: 'Currently, we have alienated from ourselves'.[18] And: 'A coarse feeling is what we have; we feel mortal and are more realistic'.[19] But he also comes up with some solutions. According to him, the 'Germans' should not rob themselves of their present (1920) 'typical characteristics' (Eigenart). To perserve the Kultur, Meinecke deemed it necessary to restore or fall back on the spirit of 1800. Of course, Meinecke was well aware of the differences between the two eras, but Meinecke was convinced of the strong inner ties that could bridge the gap: 'we are aware of the rift which separates us from Goethe and Schiller's idealistic Germany. However, we also feel his [Humboldt's, RK] words resonate, and fill us with desire; an indissoluble tie with the intellectual world which he represents', Meinecke writes.[20]

According to Meinecke the 'modern German' should strive for the same spiritual heights as Humboldt, Schiller, and Goethe did. At the same time, he wonders if this is indeed realistic for the twentieth-century German.[21] He immediately adds that the loss of these ideals of Humboldt, Schiller, and Goethe could also be a blessing in disguise, for this loss of ideals is 'something we lost, and of which we do not immediately know if this loss really is a loss or maybe something we gained, a liberation

---

17   Ernst Schulin, 'Friedrich Meinecke', 44; Boterman, Moderne geschiedenis, 241-242.

18   Friedrich Meinecke, 'Wilhelm von Humboldt und der deutsche Staat' in: idem, Brandenburg, Preußen Deutschland. Kleine Schriften zur Geschichte und Politik (Stuttgart 1979) 279-296, there 279. 'Wir sind heute irre geworden an uns selbst'.

19   Ibidem, 280. 'Wir fühlen uns derber, realistischer, irdischer (...)'. The gist of this article anticipates what Meinecke would propose after the Second World War in Die deutsche Katastrophe, namely: he urged a return to the intellectual past, the Kultur of the Goethezeit.

20   Meinecke, 'Wilhelm von Humboldt', 280. 'Wir spüren die Kluft der Zeiten, die uns von dem idealistischen Deutschland Goethes und Schillers trennt. Aber wir spüren zugleich auch an der Resonanz, an der Sehnsucht, die sich bei seinen [Humboldt's, RK] Worten ergreift, einen unzerreißbaren inneren Zusammenhang mit der geistigen Welt, die er repräsentiert'.

21   Ibidem, 280. It is striking that Meinecke identifies himself with the German nation in this article. This identification with the German nation becomes a recurring theme later on. I will particularly elaborate on this in the last chapter.

from something which we should consider as impossible and unrealisable'.[22] At first glance, Meinecke seems to abandon his optimistic world view, as expressed in *Weltbürgertum und Nationalstaat*. Still, this remark on 'gaining something by the loss of ideals' reminds us of that other observation in *Weltbürgertum und Nationalstaat* that Meinecke put forward, namely: 'the liberation of universal ideals in favour of the formation of the national state'. In that sense, Meinecke's position has not changed that much, only the confidence in realising these ideals has diminished. And yet, he still points at the possibility of synthesising the '*Macht* of 1920' with the '*Kultur* of 1800'.

In searching for the causes of cultural loss, Meinecke suggests in his Humboldt article that Germany should try to find the answers in its inner self. It should not take this too far, for too much self-criticism leads to a loss of 'faith in the spirit of the nation at large'.[23] In other words, Meinecke, after the First World War, calls for a preservation or recovering of German culture, the cultural nation (*Kulturnation*).[24] The Germans should, in his view, reflect on the value of their own collective spiritual-intellectual past. This is a rather remarkable suggestion, for Meinecke reverts to an era – the *Goethezeit* – in which Germany was not yet a political unity.[25] So, when in 1920 – Germany had just become a Republic – he suggests (re)connecting with the nineteenth-century cultural nation (*Kulturnation*), this essentially refers to a desire for a synthesis of *Macht* and *Kultur*. In contrast with Meinecke's argument in *Weltbürgertum und Nationalstaat*, he now, in 1920, suggests connecting the *Kultur* with the Weimar Republic, instead of relating it to the hierarchic state of the German Empire. To realise this (ideal) synthesis, Meinecke needed to also 'convert' to the Weimar Republic. And that is what he did, albeit on merely rational grounds, by learning to live with it. He never became a passionate advocate of the Weimar Republic: 'I did not become a Republican out of love for the Republic, but on rational grounds, and in particular for love of my fatherland', Meinecke writes.[26] So, he became a *Vernunftrepublikaner*, but remained a *Herzensmonarchist* – that is,

---

22   Meinecke, 'Wilhelm von Humboldt', 280. 'was uns abhanden gekommen ist und von dem wir nicht gleich wissen, ob wir seinen Verlust als reinen Verlust oder nicht auch als Gewinn, als Befreiung von etwas vielleicht Unmöglichem und Undurchführbarem auffassen sollen'.

23   Ibidem, 279. 'Glauben an den Gesamtgeist der Nation'.

24   Meinecke was not alone in this; cf: Gay, *Weimar Culture*, 73-80, 94-96.

25   During the *Goethezeit* the 'nation' had a 'stateless' culture in 'Weimar' and a 'cultureless' state in Prussia. As a result, the individual was, with regard to politics, subordinated to the Prussian state, but with regard to (Weimar) culture, it was at the centre. Meinecke, 'Wilhelm von Humboldt', 282.

26   Friedrich Meinecke, 'Republik, Bürgertum und Jugend', in: idem, *Politische Schriften und Reden* (Darmstadt 1958) 369-383, there 377; Schulin, 'Meineckes Leben und Werk', 121. For a similar argument: Thomas Mann, *Von deutscher Republik* (Berlin 1923). He was accused of 'defecting' to the Republic. Mann claimed his argument was in a similar vein as the rest of his political writings; it was always aimed at the same thing: humanity. See: Kurt Sontheimer, 'Thomas

he was a republican on rational grounds, and not necessarily emotionally involved, for he always remained a convinced monarchist.

Meinecke's solution is striking. Even though he realised that a synthesis and the ideal were no longer viable, he nonetheless insists on striving for (universal) ideals.[27] With respect to the 'spiritualisation' and 'moralisation' of power and politics, Meinecke's idealism did, however, decrease around 1920. In *Weltbürgertum und Nationalstaat* Meinecke had spiritualised power, in the 1920's he warned about such a veiled power politics, for: 'time and again even the highest political idealism must in the end acknowledge that to some extent a Leviathan is part of every state, and that all spiritualisation and moralisation will not rescind its hard and rude essence'.[28] By spiritualising the essence of power we deny the dangerous aspects of power. In practice, this principle of *Staatsräson* turned out to be more complex than Meinecke's account of it in this article, for the 'hard essence' was not by definition equal to evil; it also could bring about something good. Good and evil are, in short, interwoven in the principle of *raison d'état*. In *Die Idee der Staatsräson* Meinecke analysed the history of the struggle with, and the insight into this principle, and it is also this study which reflects Meinecke's own struggle with his philosophy of history and political theory.

2.2   Staatsräson according to Meinecke

In my discussion of *Weltbürgertum und Nationalstaat* Meinecke considered the ideas of great personalities to be fundamental to history. In his inaugural address of 1915, he confirmed this once more: 'Personalities and ideas seemed to me ever more valuable bearers of historical life'.[29] In his search for answers to pre-war issues and his attempt to gain insight into the principle of *Staatsräson*, Meinecke also found inspiration in a series of great personalities. In this section I will discuss Meinecke's

---

Mann als politischer Schriftsteller' in: Helmut Koopman ed., *Thomas Mann* (Darmstadt 1975)165-226, there 187; also: Gay, *Weimar Culture*, 24.

27   The concrete political problems of the Weimar Republic and Meinecke's view on them, cf: Stefan Meineke, 'Parteien und Parlamentarismus im Urteil von Friedrich Meinecke' in: Gisela Bock and Daniel Schönpflug eds., *Friedrich Meinecke in seiner Zeit. Studien zu Leben und Werk* (Stuttgart 2006) 51-93; Nikolai Wehrs, 'Demokratie durch Diktatur? Friedrich Meinecke als Vernunftrepublikaner in der Weimarer Republik' in: Gisela Bock and Daniel Schönpflug eds., *Friedrich Meinecke in seiner Zeit. Studien zu Leben und Werk* (Stuttgart 2006) 95-118.

28   Meinecke, 'Wilhelm von Humboldt', 286. 'Und immer wieder wird auch der höchste politische Idealismus sich eingestehen müssen, daß in jedem Staate ein Stück von Leviathan steckt, und daß alle Vergeistigung und Versittlichung, die man mit ihm vornehmen mag, seinen harten und groben Kern nicht zu durchdringen vermag'.

29   Meinecke, 'Antrittsrede', 2. 'Persönlichkeiten und Ideen traten mir immer deutlicher als die wertvollsten Träger des geschichtlichen Lebens entgegen (...)'.

analysis of Machiavelli, who had uncovered the tragic character of power politics, and Frederick the Great, who struggled heavily with this issue. In Machiavelli's views Meinecke recognized the tragedy, which he saw himself confronted with in the German power politics after the First World War, and Frederick the Great's conflict resounds in Meinecke's own struggle with the principle of *Staatsräson*.

Meinecke's definition

In the introduction to *Die Idee der Staatsräson*, Meinecke describes *Staatsräson* as the principle of the state's actions: the state's 'law of motion'. To this he adds: 'So the 'intelligence' of the State consists in arriving at a proper understanding both of itself and its environment, and afterwards in using this understanding to decide the principles which are to guide its behaviour'.[30] The principle always concerns the individual state, for it depends on the situation and the environment of the individual state. Hence, a political decision is limited to the unique and individual state and its environment. As Meinecke puts it: 'For each State at each particular moment there exists one ideal course of action, one ideal *raison d'état*'.[31]

To the statesman as well as the historian it is important to recognize and acknowledge the *Staatsräson*. It is important for the statesman, because in this manner he keeps his state sound, and for the historian it is important to interpret the state's past. In Meinecke's words: 'Any historical evaluations of national conduct are simply attempts to discover [the secret to, RK] the true *raison d'état* of the States in question'.[32] That is exactly what Meinecke tried to do in *Die Idee der Staatsräson*: to find out what has been said in the past about the principle of *Staatsräson*, so as to uncover the core – the 'secret' – to this principle. Besides the *Staatsräson* of every individual state, there is also a general repetitive theoretical principle – the 'secret' – in all political actions. In this context, Meinecke claims the statesman's actions are directed by *Staatsräson*, which is both individual and general in character.[33]

---

30   Friedrich Meinecke, *Machiavellism. The Doctrine of Raison d'État and its Place in Modern History*, transl. Douglas Scott (New Haven 1962) 1. 'Die "Vernunft" des Staates besteht also darin, sich selbst und seine Umwelt zu erkennen und aus dieser Erkenntnis die Maximen des Handelns zu schöpfen'. Meinecke, *Die Idee der Staatsräson*, 1.

31   Meinecke, *Machiavellism*, 1. 'es gibt für jeden Staat in jedem Augenblicke eine ideale Linie des Handelns, eine ideale Staatsräson'. Meinecke, *Die Idee der Staatsräson*, 1.

32   Meinecke, *Machiavellism*, 1. 'Alle historischen Werturteile über staatliches Handeln sind nichts anderes als Versuche, das Geheimnis der wahren Staatsräson des betreffenden Staates zu entdecken'. Meinecke, *Die Idee der Staatsräson*, 1-2.

33   Meinecke considered *Staatsräson* to be imperative and at the same time a motive by which the state is able to autonomously and independently grow and 'live'. This reminds us of Kant's categorial imperative – the moral law – according to which man is able to act in freedom. Kant's general formulation of the categorial imperative runs as follows: act according to the principle that should serve as a general law. 'Maxim' is the subjective principle of willing. The

'Individual' because there is only one way for a state to act (this manner can also transgress general law and morality) and 'general' for, according to Meinecke, it concerns a fundamental drive (*Trieb*) which all states are forced to follow: 'So the individual element in actions prompted by *raison d'état* appears as the necessary outcome of a general principle'.[34]

Staatsräson's individual and general character has repercussions for the states-man's actions. In an arcticle on Meinecke the Dutch historian Kossmann explained this twofold character of *Staatsräson*. He claims that when a statesman acts along the lines of *raison d'état* and acts rationally in favour of the state's interest, he on the one hand 'tries to keep the basic, natural power instincts of the state and himself under control', but on the other hand, Kossmann claims that the statesman who is determined to hold on to the rational state's interest, 'can be forced to act against ethical norms by which the rational state power claims the right to go against prevailing morality and as a result become immoral'.[35] Hence, in favour of the state's interest, *Staatsräson* can force the statesman to act against prevailing ethics. In this context, Meinecke states: 'Raison d'état is a principle of conduct of the highest duplicity and duality (...)'.[36] This duality arose, according to him, with

---

Will is good if it is in fact the case. An act, which is not instigated by urge, but for the sake of duty is a good act; the duty is the necessary act for the sake of respect for the law. Immanuel Kant, *Fundering voor de metafysica van de zeden* transl. Thomas Mertens (Amsterdam 1997) 36, 41, 46-47 and 95. According to Meinecke Kant fails to see the individuality of morality. See: Meinecke, *Die Entstehung des Historismus*, 288; Meinecke, *Die Idee der Staatsräson*, 502.

34   Meinecke, *Machiavellism*, 2. 'Das Individuelle im Handeln nach Staatsräson erscheint so als notwendiger Ausfluß eines generellen Prinzips (...)'. Meinecke, *Die Idee der Staatsräson*, 2. Meinecke refers to an 'individual-universalism' which is effective in the actions according to *Staatsräson*. He derived these principles from Ranke, who expressed such an individual-universalism in his well-known expression: 'jede Epoche ist unmittelbar zu Gott'. Here Enlightenment and historicism go together or historicism even transcends the Enlightenment. Since it is *every* Epoche and every individual era that is unique, there is no progress in history. Besides, 'jede Epoche' implies that each and every epoch is equal to God. Again, this also refers to an individual-universalism. In this interpretation historicism thus transcends the Enlightenment. In that sense, historicism is a radicalization of the Enlightenment. I will come back to this issue in my discussion of Meinecke's *Die Entstehung des Historismus*. Also: Gadamer, *Wahrheit und Methode*; F. R. Ankersmit, *Macht door representatie. Exploraties III: politieke filosofie* (Kampen 1997) 208; F. R. Ankersmit, *De sublieme historische ervaring* (Groningen 2007) 209-210.

35   Kossmann, 'Friedrich Meinecke', 219. 'de elementaire, in de natuur gegeven machtsinstincten van hemzelf en van zijn staat in toom [tries, RK] te houden' (...) 'gedwongen worden dingen te doen die in strijd zijn met ethische normen zodat de rationeel beheerste staatsmacht voor zichzelf het recht opeist in te gaan tegen gangbare zedelijke opvattingen en op die manier zelf onzedelijk wordt'.

36   Meinecke, *Machiavellism*, 5. 'Die Staatsräson ist eine Maxime des Handelns von höchster Duplizität und Gespaltenheit (...)'. Meinecke, *Die Idee der Staatsräson*, 6.

the advent of Christianity, which had set up a universal moral command, which the state also had to obey. Furthermore, he points at the Germanic jurisprudence (*Rechtsgedanke*) of the Middle Ages, which had set the law above the state; it was a means to implement the law. According to Meinecke, as a result of Christianity and the Germanic jurisprudence the realization dawned in Europe that a ruthless *Staatsräson* was in fact a sin against God and divine standards, and also a sin against the sanctity and inviolability of ancient laws.[37] The history of the idea of *Staatsräson*, in Meinecke's view, had to start with a heathen who did not know the fear of hell, but who, because of his 'ancient naivety', thought through the essence of *Staatsräson*.[38]

Power and ethics

Niccolò Machiavelli (1469-1527) is the 'heathen' with whom Meinecke starts his history of *Staatsräson*. Despite Machiavelli's ironic and critical respect for the church and Christianity (and being influenced by both) Meinecke considers him a heathen who reproached Christianity of having made man humble, unmanly and feeble.[39] Machiavelli's recommendations in *Il Principe* – meant for the political ruler – can, against the background of Christian ethics, be considered as heathen. Yet, Machiavelli was not interested in whether certain recommendations were right in a Christian sense; he was concerned with the principle of power politics. He was not interested in questions concerning the Christian do's and don'ts of a ruler; he only advised how a ruler could retain absolute power.

Machiavelli distinguished between a Christian morality and state morality.[40] The Latvian-British historian of ideas Isaiah Berlin (1909-1997) describes in his essay 'The Originality of Machiavelli' that a state's morality is neither amoral nor immoral, for it is a morality which is completely different from a Christian morality. This state morality does in fact show the limits of Christian morality. Long before Berlin, it was the theologian Ernst Troeltsch who expressed the same thought in. one of his last lectures on politics, patriotism and religion.[41] According to Troeltsch, Machiavelli's thought was not immoral, but a 'political emancipation from religious morality and universalism'.[42] With regard to ethics, this meant for Machiavelli 'a

---

37   Meinecke, *Machiavellism*, 28-29; Meinecke, *Die Idee der Staatsräson*, 33.

38   Meinecke, *Machiavellism*, 29; Meinecke, *Die Idee der Staatsräson*, 34.

39   Meinecke, *Machiavellism*, 31; Meinecke, *Die Idee der Staatsräson*, 36.

40   Isaiah Berlin, 'The Originality of Machiavelli' in: idem, *Against the Current. Essays in the History of Ideas*, (Oxford, Toronto, Melbourne 1981) 25-79, there 45.

41   Berlin does not mention Troeltsch in this context.

42   Ernst Troeltsch, *Der Historismus und seine Überwindung. Fünf Vorträge von Ernst Troeltsch. Eingeleitet von Friedrich von Hügel-Kensington* (Berlin 1924) 90. 'Emanzipation der Politik von der religiösen Moral und dem religiösen Universalismus'.

conscious break with Christian morality and its political ideals, the return to the heathen virtues of the Roman virtù (...)', Troeltsch writes.[43]

The notion of *virtù* to which Troeltsch refers, is at the heart of Machiavelli's idea of the state. It points at power, skills, persuasiveness, charisma, and a feeling for timing. It is the ability of some individuals to control historical or political circumstances at the right time. The opposite of this notion of *virtù* is the notion of *fortuna*: luck or chance. According to Meinecke, in Machiavelli's view there were only two choices within this conflict between *virtù* and *fortuna*. On the one hand one could leave everything to chance or fate (*fortuna*), which would, on a political level result in a more or less 'natural' order. On the other hand, when people deliberately want a certain society, virtù is needed. Wherever people have little *virtù*, *fortuna* will be dominant.[44] Another fundamental notion in this context is *necessità*, necessity. *Virtù* is the power that pushes states forward, *necessità*, according to Meinecke, is the causal force. *Necessità*, in his view, is the means to adjust the state to the form desired by *virtù*.[45]

Machiavelli was convinced that man only makes 'good' decisions out of *necessity*. That is, man only acts out of necessity at a certain moment in a machiavellistic 'right' way. Machiavelli claims a man or in this case a ruler sometimes has to act in a 'wrong' manner in order to achieve something 'good'. In *The Prince* he recommends this in many passages:

> I believe this arises from the cruelties being used well or badly. Well used may be called those (if it is permissible to use the word well of evil) which are committed once for the need of securing oneself, and which afterwards are not persisted in, but are exchanged for measures as useful to the subject as possible. Cruelties ill-used are those which, although at first few, increase rather than diminish with time.[46]

Machiavelli adds to this another piece of advice to the ruler:

> A man who wishes to make a profession of goodness in everything must necessarily come to grief among so many who are not good. Therefore it is necessary for a prince, who wishes to maintain himself, to learn how not to be good, and to use it and not use it according to the necessity of the case.[47]

---

43   Troeltsch, *Der Historismus und seine Überwindung*, 91. 'den bewußten Bruch mit der christlichen Moral und ihren politischen Idealen, die Rückkehr zu den heidnischen Tugenden der römischen virtù (...)'.

44   Meinecke, *Die Idee der Staatsräson*, 37, 42, 47; Ankersmit. *Denken over geschiedenis*, 295-296; Ankersmit, *Macht door representatie*, 107-108.

45   Meinecke, *Die Idee der Staatsräson*, 44.

46   Niccolò Machiavelli, *The Prince* trans. Luigi Ricci (London 1921) 35-36.

47   Machiavelli, *The Prince*, 60.

According to Machiavelli, a ruler should act in favour of the individual state. If part of the population is harmed by the state's actions or if 'general' norms and values are violated, then there is nothing one can do about that. Machiavelli even adds that a ruler who does or did wrong, should not be concerned with his reputation, which his vices had caused: '(...) for if one considers well, it will be found that some things which seem virtues would, if followed, lead to one's ruin, and some others which appear vices result, if followed, in one's greater security and wellbeing'.[48] Hegel later developed this idea into the concept of the 'cunning of reason'; I will come back to this in the next section.

A ruler who acts upon the recommendations of Machiavelli is likely to 'sin' instead of obey Christian ethics. Meinecke's definition of *Staatsräson* already indicated that he was also convinced that a ruler could be forced by the necessity of the state (*Staatsnotwendigkeit*) to act against Christian ethics. In this kind of decision making, *Kratos* (to act in accordance with power) triumphs over *Ethos* (to act morally responsible).[49] The role of the individual, the statesman, is not completely undone by the necessity of state, for it is the *virtù* that creates room for individuals to take control of the state('s actions).[50] Meinecke's interest in Machiavelli's doctrine of the state was particularly aroused by the (double) dilemma between the ethics of *virtù* and the Christian morality, and between *virtù* an *fortuna*.[51]

According to Meinecke, Machiavelli's doctrine of the state was new or revolutionary in the sense that it permitted going against Christian moral laws, and moreover, these violations were justified by the necessity of the state.[52] Meinecke puts it as follows: 'Necessity was therefore the spear which at the same time both wounded and healed'.[53] In short, in the 'realm' of *raison d'état* all is fair for the purpose of the preservation of freedom and the life of the state. *Raison d'état* sets the course for what is right in a Machiavellian fashion, however from a Christian

---

48   Machiavelli, *The Prince*, 61.

49   Meinecke, *Die Idee der Staatsräson*, 14 and 34; Walther Hofer, 'Einleitung des Herausgebers' in: Friedrich Meinecke, *Die Idee der Staatsräson in der neueren Geschichte* (Munich 1976) ix-xxxii, there xix.

50   Meinecke, *Die Idee der Staatsräson*, 39-40; Humboldt identified this as 'Zeugung': a sudden explosion of power by which earlier developments ended and new one's began. The statesman (but also artists, philosophers, the Genius) who manage to control history are the product of this 'Zeugung'. Cf. Labrie, *'Bildung' en politiek*, 89.

51   Meinecke, *Die Idee der Staatsräson*, 59.

52   Felix Gilbert puts the 'intellectual' difference between Machiavelli and his humanistic predecessors in perspective: Felix Gilbert, *History. Choice and Commitment* (Cambridge Mass. and London 1977) 91-114, particularly 110; Contrary to this view: Quentin Skinner, *Machiavelli. A Very Short Introduction* (Oxford and New York 2000; first published 1981) 40, 44.

53   Meinecke, *Machiavellism*, 40. 'Die Notwendigkeit war also der Speer, der zugleich verwundete und heilte'. Meinecke, *Die Idee der Staatsräson*, 47.

perspective these actions are not right by definition, and *vice versa*.[54] According to Meinecke, evil obtained a place next to (or even within) the good: '(...) the devil forced his way into the kingdom of God'.[55] Hence, the 'devil' (partially) triumphs over the kingdom of God, the Christian ethics.

During the 1940's Gerhard Ritter in *Machtstaat und Utopie* – which was renamed *Die Dämonie der Macht* after the Second World War – severely criticized Meinecke's interpretation of Machiavelli.[56] According to Ritter, Meinecke ascribes too much 'ethical meaning' to Machiavelli. Ritter, who was influenced by Carl Schmitt, claims Machiavelli assumed a 'politics of struggle' to which every morality was subordinated.[57] In Meinecke's view, Machiavelli, 'merely' brought the tyranny to light and – contrary to Ritter's claim – he warned about this 'poison'. Ironically, Machiavelli is generally accused of injecting this 'poison', for someone who abuses his power is often called a machiavellist. Machiavelli chiefly tried to make the 'Renaissance-man' aware of the fact that there is more than one perspective from which 'true' ethics can be considered. Christian conscience rebelled against this idea and condemned Machiavelli; his work was put on the *Index Librorum Prohibitorum*.[58]

Meinecke's understanding of the tragedy of *Staatsräson*, which Machiavelli had revealed, helped him to better understand the political situation of Germany after the First World War. It became clear to him that *Macht* was not as ethical as he had thought it to be. Moreover, he acknowledged that 'good' and 'evil' in both *Ethos* and *Kratos* were not easy to distinguish. Meinecke struggled with this tragic character of *Staatsräson* (that was already clear from his 'conversion' to the Weimar Republic) and Germany's fate, which was inextricably bound up with it. Meinecke recognized his own struggle with the doctrine of *Staatsräson* in the ideas of the politician, ruler and thinker Fredrick the Great – the topic of the next section.

---

54   Cf.: Frank Ankersmit, 'Politieke stijl: Schumann en Schiller' in: Dick Pels and Henk te Velde ed., *Politieke stijl. Over presentatie en optreden in de politiek* (Amsterdam 2000) 15-42, there 27.

55   Meinecke, *Machiavellism*, 39. '(...) der Teufel drang in Gottes Reich ein'. Meinecke, *Die Idee der Staatsräson*, 46.

56   Gisela Bock describes this in detail: Gisela Bock, 'Meinecke, Machiavelli und der Nationalsozialismus', 145-175.

57   Meinecke was not really shaken by Ritter's criticism: 'Das gegen mich gesagte erkenne ich nur zum kleinsten Teile als berechtigt an'. Cited in: Bock, 'Meinecke, Machiavelli und der Nationalsozialismus', 166. As I already stated in my introduction, it seems Meinecke was not easily thrown off balance by his contemporaries. He was mainly in discussion with the (German) historicist tradition, and he tried to justify his views and convince his time of their truth. Cf. Meinecke's reaction to Carl Schmitt (section 3.2) and Benedetto Croce (section 4.6).

58   Meinecke, *Die Idee der Staatsräson*, 53-54.

State interest and an ideal of humanity

The seeds of the 'historical-political individualizing view of the state' were, according to Meinecke, already present in Machiavelli's thought. He had shown the statesman ought to know the historical and statistical facts of the state in question in order to act prudently in the present political reality. In *Die Idee der Staatsräson* Meinecke demonstrates how many different thinkers from the sixteenth century onwards had struggled with Machiavelli's ideas; thinkers such as: Bodin, Boccalini, Campanella, Richelieu, Rohan, Hobbes, Spinoza, and Pufendorf. All of them searched for a way to (intellectually) control or justify power (politics).

With regard to Germany, it was primarily Hermann Conring (1606-1681) who was highly important regarding the introduction of *raison d'état*. He integrated 'history' into natural law. Meinecke considers natural law in terms of universality of human nature, uniformity of reason, and the ideal of the 'best state'.[59] Conring attempted to reconcile Thomas Hobbes' (1588-1679) natural law theory with the demands of the doctrine of *raison d'état*. Conring made a deliberate choice to use Hobbes' system, because the basic assumption of this system (like Machiavellianism) is self-preservation. Conring claims that what is 'good' within natural law is not necessarily prudent within the doctrine of *raison d'état*. He made clear that acting predictably and ethically was the best way to serve the goal of self-preservation. It apparently will be in our own machiavellistic interest to not commit any *flagitia* (self-interested shameful acts), and not be tempted by the more radical recommendations of Machiavellism.[60] Since the personal interests of a ruler are not always in agreement with the interest of the state itself, only a certain amount of Machiavellism is granted in favour of the state's interest. Even though in this example natural law and *raison d'état* coexist, in the end they will both go into different directions.[61]

Conring, in following Machiavellism, considered history the best advisor for the statesman. From history the statesman can deduce the state's interests and find out how he can govern in accordance with the needs or demands of *raison d'état*. In short, Conring is considered the 'teacher' of *raison d'état* and the father of the German history of law and statistics. The notion of 'statistics' is derived from the word 'state' and is in fact the 'cognitive apparatus' which serves the politics of

---

59   Ibidem, 22; Friedrich Meinecke, 'Aphorismen' in: idem, *Zur Theorie und Philosophie der Geschichte* (Stuttgart 1965) 215-263, there 222, 225.

60   For the remainder of this Chapter I will use the word Machiavellism instead of Machiavellianism, because the former is mostly used with regard to Meinecke and his study, which is also translated as 'Machiavellism'.

61   F. R. Ankersmit, *Political Representation* (Stanford 2002) 28-29. For a revaluation of natural law theory: Leo Strauss, *Natural Right and History* (Chicago 1953).

raison d'état.[62] By means of statistics the state collects data to inform itself on its interests. One can think of information on constitutional and legal organisation, geography, population numbers, and industries. On the basis of such (historical) data the statesman can be a good Machiavellist in the tradition of the doctrine of *raison d'état.*[63]

To Meinecke the most interesting figure with regard to the history of *raison d'état* is not Conring, but the king of Prussia, Frederick the Great (1712-1786), to whom he devoted roughly eighty pages. Meinecke considers Frederick's work to be 'one of the most important for the history of European thought'[64], because Meinecke views Frederick as a pioneer in the transition from the Enlightenment to historicism. Moreover, as shown by Meinecke, Frederick tried to get even with Machiavelli.

Meinecke's argument clearly demonstrates Frederick's struggle with the ambiguous character of the principle of *Staatsräson*. It was not only this ambiguity which confronted him with dilemmas. He was also both a statesman and a philosopher, and that caused a lot of inner conflicts and dilemmas. That is why, according to Meinecke, he personally experienced the moral dilemmas of good and evil. As a philosopher Frederick had difficulties with the necessity of the state which can force a ruler to make ethically reprehensible decisions; but as a statesman who wants to protect his state he had to accept and act on the same necessity of the state.[65]

What is remarkable about Frederick is his desire to reconcile the demands of the state's interest with an enlightened ideal of humanity (*Humanitätsideal*). This notion of humanity had a more practical meaning in his time: the happiness of one's fellow men, and to be of value for the society. Frederick considered the ideal of furthering the common good part of the state's interest. In other words, he was convinced that he could reconcile the state's interest with this ideal of humanity. Frederick did not want to elevate (*veredeln*) power with this ideal of humanity, but he tried to improve both power and the happiness of his people.[66] According to Meinecke, together with the idea of a power state the notion of a cultural state (*Kulturstaat*) thus arose.[67]

Frederick's writings also give evidence of his pursuit of both a strong power state and a humanist state. As a prospective statesman contemplating the future, he wrote *Considérations sur l'état présent du corps politique de l'Europe* (1737-1738). In

---

62   A. Th. van Deursen, *Geschiedenis en toekomstverwachting. Het onderwijs in de statistiek aan de universiteiten van de achttiende eeuw* (Kampen 1971) 8-9.

63   Ankersmit, *Political Representation*, 30-31.

64   Meinecke, *Machiavellism*, 275. 'Einer der wichtigsten für die europäische Geistesgeschichte'. Meinecke, *Die Idee der Staatsräson*, 324.

65   Ibidem, 325-327.

66   Ibidem, 342.

67   Ibidem, 332-333

this work the emphasis is on machiavellist politics with a small doses of morality, whereas Frederick's *Réfutation du prince de Machiavel* (1739) opposes the recommendations of Machiavelli and is more focussed on morality leaving little room for the *Realpolitiker*.[68] It is Frederick the philosopher who wrote *Réfutation du prince de Machiavel*, which from 1740 onwards became better known as the 'Antimachiavell'. Frederick claims that the ruler – the 'first servant of the State'[69] – should consider his people not only as his equal but as his 'ruler'. 'The people', or the needs and will of the people, more or less referred to the demanding characteristics of the *Staatsräson*.[70] According to Meinecke, it was not a plea in favour of democracy, but a result of the idea that the ruler should follow the interests of the state. Put differently, Fredrick was not guided by the people, but by the interest of the state in order to serve his people.[71]

Frederick's dilemma regarding ethics and politics, and humanity and the state's interest remained. Fredrick had experienced 'that the man of action may be led beyond the boundaries which the man of thought has set up for himself', Meinecke writes.[72] It seems Meinecke paraphrases Goethe's famous maxim: 'The person engaged in action is always unconscionable; no one except the contemplative has a conscience'.[73] That is exactly the case with Frederick the statesman and the philosopher: in his actions he was an empiricist and realist, but intellectually he was a conscientious universalist.

The German historian Theodor Schieder (1908-1984) argues that Meinecke sketched a simplistic image of Frederick. Schieder modifies this image, for he claims Machiavellism and Anti-Machiavellism existed 'separately'. Frederick's Anti-Machiavellism was, according to Schieder, only an intellectual matter, while Frederick's Machiavellism stemmed from his 'instinctive ability'.[74] Nevertheless, Meinecke emphasizes the continuous struggle between both; though he does demonstrate that Frederick's later work was more centered on the state instead

---

68   Ibidem, 343.

69   Cited in Meinecke, *Machiavellism*, 280. 'erster Diener des Staates'. Meinecke, *Die Idee der Staatsräson*, 330.

70   Meinecke, *Die Idee der Staatsräson*, 330-331.

71   The Duke Henri de Rohan (1597-1638) already came up with this idea. Meinecke quotes his famous expression: 'Die Fürsten kommandieren den Völkern, und das Interesse kommandiert den Fürsten'. Meinecke, *Die Idee der Staaträson*, 198.

72   Meinecke, *Machiavellism*, 298. 'daß der handelnde Mensch über die Grenzen hinausgeführt wird, die der denkende Mensch sich errichtet'. Meinecke, *Die Idee der Staatsräson*, 352.

73   Johann Wolfgang von Goethe, *Maxims and Reflections* transl. Elisabeth Stopp (London etc 1998) 27; 'Der Handelnde ist immer gewissenlos; es hat niemand Gewissen als der Betrachtende'. Johann Wolfgang von Goethe, 'Maximen und Reflexionen' (posth. 1833) in: idem, *Goethes Werke. Hamburger Ausgabe in 14 Bänden* ed., Erich Trunz (Munich 1982) Volume XII, 399.

74   Theodor Schieder, *Frederick the Great* ed. and translated Sabina Berkeley and H.M. Scott (London and New York 2000) 76.

of the humanitarian ideal of the people. This transition in Frederick's thought is, in Meinecke's view, very important: 'The transition from 'people' to 'State' thus signified the transition from a humanitarian and moral ideology of power-policy to that other historical and political ideology of power-policy which afterwards came to be developed chiefly in nineteenth-century Germany'.[75] Meinecke adds to this that the final triumph of Machiavellism over Anti-Machiavellism in Frederick's political thought and action is only one side to the story, for Prussia did not become a pure power-state (*Machtstaat*), but also a civilized and constitutional state (*Recht-* and *Kulturstaat*). In that sense, Anti-Machiavellism triumphed over Machiavellism.

In the end, Frederick failed to overcome the dualism of power and humanity. Meinecke concludes that Frederick's enlightened thought had its shortcomings: 'The weapons of the philosophy of the Enlightenment revealed themselves as still incapable of solving the problem in such a manner that reality and the ideal could be harmonized together'.[76] Meinecke thought historicism could solve these problems, namely *his* interpretation of historicism; this will be discussed in the next chapter.

Meinecke's interest in Frederick the Great and especially his attempts to reconcile polarities or contrasting ideals within the state were very close to Meinecke's thought. It is the tragedy of the irreconcilability of these contrasting ideals which particularly appeals to Meinecke. This also fits his own situation, for his ideal expounded in *Weltbürgertum und Nationalstaat* – the synthesis of *Macht* and *Kultur* – seems no longer viable in the present, post-war political reality. That is why Meinecke concludes his discussion on Frederick with the following statement: 'What was ideal yielded to what was elemental in the king's actions, but still maintained itself in his thought'.[77] Initially, this also seems to be the case with Meinecke: the war interfered with his ideals, however he cherished a spark of hope regarding a synthesis. The tragedy which follows from this is the subject of the next section.

---

75   Meinecke, *Machiavellism*, 309. 'Der Übergang vom "Volke" zum "Staate" bedeutete also den Übergang von einer humanitären und moralischen Ideologie der Machtpolitik zu jener historisch-politischen Ideologie der Machtpolitik, die dann vor allem im Deutschland des 19. Jahrhunderts ausgebildet wurde'. Meinecke, *Die Idee der Staatsräson*, 366.

76   Meinecke, *Machiavellism*, 275. 'Die Waffen der Aufklärungsphilosophie erwiesen sich als noch nicht geeignet, das Problem so zu lösen, daß Ideal und Wirklichkeit miteinander harmonierten'. Meinecke, *Die Idee der Staatsräson*, 325; Cf.: Reinhart Koselleck, *Kritik und Krise. Eine Studie zur Pathogenese der bürgerlichen Welt* (Freiburg en Munich 1976, first published 1959) in particular 132-157.

77   Meinecke, *Machiavellism*, 339. 'Das Ideale wich dem Elementaren im Handeln des Königs, aber behauptete sich in seinem Denken'. Meinecke, *Die Idee der Staatsräson*, 399.

2.3   The tragic character of Meinecke's Staatsräson

The first two parts of *Die Idee der Staatsräson* are on thinkers and politicians who more or less belong to the tradition of a cyclical and Enlightened and natural law philosophy. In the third and last part of his work Meinecke discusses several philosophers and historians who break with this tradition; instead they took a view that is particularly aimed at the 'individual'. The multiplicity of 'life forces' and individual reason is at the heart of their thought instead of an enlightened universalism.

In the last part of *Die Idee der Staatsräson* Meinecke concentrates on a few thinkers who belong to the tradition of German Idealism. This movement, which came after the Enlightenment and existed parallel to Romanticism, longed for a harmony between idea and reality. According to Meinecke, certain thinkers veiled the brutal reality of the *Staatsräson* as a result of this harmonising philosophy. Meinecke claims it was Hegel in particular who should be held responsible for this veiling of harsh power politics. Hegel's position is not as unambiguous as Meinecke wants us to believe. Hegel was well aware of the evil dimensions of power; in fact, his ideas are quite close to Machiavelli. Meinecke, however, thought that veiling of power politics – due to Hegel – resulted in the fact that later generations, under the influence of different political situations, were again confronted with the tragedy already revealed by Machiavelli. Meinecke explains Ranke too was not able to prevent himself from this tragedy. Despite Meinecke's insight in the history of *Staatsräson*, which in his view 'presents one aspect to physical nature and another to reason'[78], he too could not avoid this tragedy of *Staatsräson*.

When a statesman wishes to maintain power no matter what, he will get entangled in ethical difficulties.[79] A power that violates laws and morality will eventually become a threat to itself. Yet, who decides when 'power' oversteps certain boundaries? Where are the boundaries between a *Staatsräson*, which a statesman follows in favour of the state's interest, and where does the individual and personal motives of a ruler take over? According to Meinecke, both are closely connected with each other; for 'the advantage of the State is always at the same time blended too with the advantage of the rulers'.[80] The consequence of this coalescence is that *Staatsräson* is constantly in danger of 'becoming a merely utilitarian instrument

---

78   Meinecke, *Machiavellism*, 5. 'eine der Natur und eine dem Geiste zugekehrte Seite' Meinecke, *Die Idee der Staatsräson*, 6.

79   Meinecke, *Die Idee der Staatsräson*, 3.

80   Meinecke, *Machiavellism*, 7. 'der Nutzen des Staates ist auch immer zugleich irgendwie mit dem Nutzen der Herrschenden verschmolzen'. Meinecke, *Die Idee der Staatsräson*, 8.

without ethical application'.[81] In Meinecke's view, the polarity (*Doppelpoligkeit*) of *Staatsräson*, the spirit or culture in man and thus the statesman is equally at risk of relapsing into 'the element of physical nature', the realm of sins that conflict with ethical norms.[82] In other words, there is a close connection between statesman and *Staaträson*. Besides the ambiguous character of the *Staatsräson* itself, there is also a conflict on the level of statesman and *Staatsräson*.[83] The tragedy of this ambiguity on both levels is described by Meinecke as follows:

> It is a tragic process, a continuously repeated combat against insuperable forces of destiny, which we have to present. In and out among all the other bright threads of the historical weft, there twines uninterruptedly (and everywhere immediately recognizable) the red, only too often blood-red, thread of raison d'état.[84]

It is in the nature of *Staatsräson* to continually besmirch law and ethics: 'It is apparently the case that the State must do evil'.[85] That is why Meinecke considered the history of *Staatsräson* a tragic history. This characterization points to a second aspect of this tragedy, for it seems Meinecke still holds on to an idealistic or even monistic conviction, since – wholly in line with *Weltbürgertum und Nationalstaat* – he again tries to achieve a harmony, in this case between *Ethos* and *Kratos*. When Meinecke was faced with the polarity of the *Staatsräson*, and he no longer thought a synthesis was possible, tragedy also crept into his own conviction. This is especially clear from his discussion of Hegel's veiling of power politics.

The veiling of evil in power politics

In *Weltbürgertum und Nationalstaat* Meinecke characterizes Hegel as one of the liberators of the state (*Staatsbefreier*), because he put an 'end' to the 'universal ideas' which obstructed the formation of the German national state. In *Die Idee der Staatsräson*, Meinecke condemns Hegel. Especially his view of *Staatsräson*, for it had led to a veiling of the 'true nature' of power politics.

---

81   Meinecke, *Machiavellism*, 7. 'ein bloß utilitarisches Instrument ohne ethische Anwendung zu werden'. Meinecke, *Die Idee der Staatsräson*, 8.

82   Meinecke, *Machiavellism*, 12. Meinecke, *Die Idee der Staatsräson*, 14.

83   Meinecke, *Die Idee der Staatsräson*, 11.

84   Meinecke, *Machiavellism*, 21. 'Es ist ein tragischer Hergang, ein immer wiederholter Kampf gegen unüberwindliche Schicksalsmächte, den wir darzustellen haben. Durch alle übrigen bunten Fäden des historischen Gewebes schlingt sich ununterbrochen und sofort überall zu erkennen der rote, nur zu oft blutigrote Faden der Staatsräson'. Meinecke, *Die Idee der Staatsräson*, 25.

85   Meinecke, *Machiavellism*, 12. 'Der Staat muß, so scheint es, sündigen'. Meinecke, *Die Idee der Staatsräson*, 14.

Hegel once summarized his view of the state aptly and to the point: 'the actual and existent State is also at the same time the rational State'.[86] Reason necessarily progresses in history, and everything in reality is an integral part of this. To put it differently, the history of the state's development is part of the self-realization of divine reason; hence the actual state is also the rational state. It follows that Hegel, in Meinecke's view, had a 'fluid' or dynamic conception of reason, which apparently was not 'available' in the age of the Enlightenment.[87] According to Meinecke, Hegel no longer considered reason as static, in the sense of universal and interchangeable, rather he historicized reason. He assumed an individual reason, which is not tied to a particular period.[88] In that sense, he agrees with Herder, Goethe, and the Romantics with regard to the discovery of the 'individual'. Meinecke claims these thinkers freed themselves from the 'absolute validity and uniformity of reason', and concentrated on 'the individual manifoldness of all the forces of life', in which the focus is on a better understanding of the 'special individual reason'.[89] According to Meinecke, this transition of the general to the individual is not completely applicable to Hegel's historicism, for in his conception of history the individualities are a means to an end: the world-spirit (*Weltgeist*) or world-reason (*Weltvernunft*).[90]

History, in Hegel's philosophy, is guided by a supra-individual Spirit or Reason. 'Everything individual', says Meinecke when speaking of Hegel's Spirit, 'serves to realize the single and unique reason, which has the particular skill of enticing into its service the evil elements as well as the good, the elemental as well as the intellectual and spiritual'.[91] So, Hegel does not consider history from the perspective of the acting individual but from the perspective of the interweaving of human actions. History is, according to Hegel, the *result* of a complex network of different individual human motives and actions, and not the *sum* of individual actions and motives. These human actions and motives are arranged in such a way that no one is able to fathom them – except for the philosopher who searches

---

86   Meinecke, *Machiavellismus*, 349. 'Der wirkliche Staat (...) ist auch der vernünftige Staat'. Meinecke, *Die Idee der Staatsräson*, 409.

87   Meinecke, *Die Idee der Staatsräson*, 410.

88   Hegel's historicism is in this sense rather close to Meinecke's, although Meinecke disagrees with Hegel's 'goal in and of history' and likewise Meinecke opposes the concept of the 'cunning of reason'. Cf.: Frederick C. Beiser, 'Hegel's Historicism' in: idem ed., *The Cambridge Companion to Hegel* (Cambridge 1993) 270-300, there 289, 296.

89   Meinecke, *Machiavellismus*, 361. 'Allgemeingültigkeit und Gleichheit der Vernunft', 'die individuelle Mannigfaltigkeit aller Lebensmächte', 'besondere individuelle Vernunft'. Meinecke, *Die Idee der Staatsräson*, 424.

90   Meinecke, *Die Idee der Staatsräson*, 427.

91   Meinecke, *Machiavellismus*, 364. 'Alles Individuelle dient der Realisierung der einen und einzigen Vernunft, deren List es eben ist, Böses wie Gutes, Elementares wie Geistiges für sich arbeiten zu lassen'. Meinecke, *Die Idee der Staatsräson*, 427

for reason in the historical process. The outcome of an individual action can be completely different from its initial intention and vice versa: an action which at first glance seems 'irrational' can have a rational outcome.[92] This is the principle of unintended consequences of intentional actions. Contrary to Meinecke's view, Hegel's historicism does in fact consider the individual, but in the end the supra-individual is indeed the dominant factor in his thought.[93]

In the process of the self-realization of reason, as presented by Hegel, reason makes use of 'irrational' elements or 'passions' with regard to the acts of the individual. 'Passions', here, should not be interpreted as a reckless urge in which reason is overruled; rather it is a desire for power and fame given by reason. Hegel focusses on the passions of great historical personalities who elevate history to a higher level. These personalities feel, as it were, what is required at a given moment. They cannot intervene at will in the process of history; they rather act on authority of their time: reason makes use of these individuals to realize its own goal.[94] In this regard, Meinecke states: 'Everything, absolutely everything serves to promote the progressive self-realization of divine reason; and what is peculiarly subtle and cunning about it is that it forces into its service even what is elemental, indeed even what is actually evil'.[95] When the goal has been achieved, these 'world-historical personalities' quickly make their exit.[96] This is what Hegel refers to as the 'cunning of reason'. Meinecke strongly disagrees with Hegel on this, because it assumes a goal in history and it forces us to consider all individualities as merely a means to this end. Meinecke therefore emphasizes the unique individuality of all periods in history, without a definite end in history.

---

92 Ankersmit, *Denken over geschiedenis*, 41. This idea of evil which leads to a good outcome reminds us of a scene of Goethe's *Faust* in which Mephistopheles makes himself known to Faust. In this scene he drops Faust a hint: 'Ein Teil von jener Kraft, die stets das Böse will und stets das Gute schafft'. Johann Wolfgang von Goethe, 'Faust' in: idem, *Goethes Werke. Hamburger Ausgabe in 14 Bänden*, ed. Erich Trunz (Munich 1981) Volume 3, 47.

93 Cf: section 4.6, in which I will discuss the origin of Meinecke's Hegel-image.

94 In this context, one should think of personalities like: Socrates, Caesar, Christ and Napoleon. The point is that the (political) creations of these *welthistorische Individuen* will continue to exist after their creators have perished. Ankersmit, *Denken over geschiedenis*, 42.

95 Meinecke, *Machiavellism*, 349. 'Alles, alles dient der fortschreitenden Selbstverwirklichung der göttlichen Vernunft, und ihre List ist es, auch das Elementare, ja selbst das Böse für sich arbeiten zu lassen'. Meinecke, *Die Idee der Staatsräson*, 410.

96 Ankersmit, *Denken over geschiedenis*, 42; Ankersmit, *Macht door representatie*, 107-108. The actions of the so-called *welthistorische Individuen* remind us of Machiavelli's notion of *virtù*: the statesman's ability to oversee complex situations and to quickly gain insight into these situations in order to act at the right moment. The main difference between the two is that the 'world-historical individuals' act unconsciously on authority of their time, whereas individuals who act on the basis of *virtù* are well aware of what they want to accomplish.

The essence of Hegel's philosophy consists of the idea of the self-realization of reason or spirit. The state's actions, regardless of being good or evil, are an integral part of this self-realization, but Hegel's philosophy is also aimed at *understanding* this self-realization or -revelation of reason in the world. Whoever succeeds in this understanding will really be free. In Meinecke's view this freedom is in a way misleading, for it is a 'freedom granted to the "appearance", and 'true freedom' in Hegel's thought lies 'only in the almost mystical union between the observing and thinking mind and the world-spirit'.[97] This is a remarkable claim, for Hegel always denounced all forms of 'mysticism'. Yet, Meinecke claims, good and evil are merely means to an end in Hegel's transcendent perspective. It is this almost indifferent attitude which shields Hegel from the tragedy of *Staatsräson*, for: 'Thereby he seemed to accomplish the remarkable achievement of managing both to grant all the assertions of a pessimistic view (which doubted the goodness in the world) and yet simultaneously to oppose it with a transcendental optimism, which looked down on this world with heroic superiority and calm', Meinecke writes.[98] Despite this seemingly optimistic conclusion, Meinecke still maintains that Hegel's philosophy retains its 'evil' character. Meinecke comes to the conclusion that Hegel's philosophy has had a fatal effect on German power politics, for 'in this connection [the relationship between good and evil, RK] it was possible to forget the sinister element (...), and that palliating light was capable of being shed also on the primitive, bestial and nocturnal aspect of *raison d'état*'.[99] And yet, Hegel did in fact refer to these 'dark' aspects of *Staatsräson*; however, he did not consider them to be 'dark'. For example in the *Philosophy of Right* he states: 'The view that politics in this assumed opposition [the tension between politics and ethics, RK] is presumptively in the wrong depends on a shallow notion both of morality and of the nature of the state in relation to morality'.[100] In short, with this statement Hegel is very close

---

97   Meinecke, *Machiavellism*, 368. 'Erscheinung gelassene', 'mystischen Vereinigung des betrachtenden und denkenden Geistes mit dem Weltgeiste'. Meinecke, *Die Idee der Staatsräson*, 432.

98   Meinecke, *Machiavellism*, 368. 'Dadurch schien ihm das Merkwürdige zu gelingen, alle Behauptungen eines an der Güte der Welt verzweifelnden Pessimismus gleichzeitig zuzugeben und zu widerlegen durch einen transzendentalen Optimismus, der heroisch gelassen und überlegen auf diese Welt herabsah'. Meinecke, *Die Idee der Staatsräson*, 432.

99   Meinecke, *Machiavellism*, 368. 'das Unheimliche dieses Zusammenhanges vergessen werden (...)', 'auf die Natur- und Nachtseite der Staatsräson ein beschönigendes Licht fallen konnte'. Meinecke, *Die Idee der Staatsräson*, 432.

100   G.W.F. Hegel, *Philosophy of Right*, third part, third section, B, paragraph 337. 'Die Ansicht von dem vermeintlichen Unrechte, das die Politik immer in diesem vermeintlichen Gegensatze haben soll, beruht noch viel mehr auf der Seichtigkeit der Vorstellungen von Moralität, von der Natur des Staates und dessen Verhältnisse zum moralischen Gesichtspunkte'. Georg Wilhelm Friedrich Hegel, 'Grundlinien der Philosophie des Rechts' in: idem, *Hauptwerke in sechs Bänden* (Hamburg 1999) Volume 5, 287. This and other remarks on the tension between politics and ethics are discussed by Hegel in the paragraphs 337 and 340.

to Machiavelli's distinction between the irreconcilability of Christian ethics and political morality.

What Meinecke considers the tragedy of *Staatsräson* is in fact inspired by Hegel's view, or better: Meinecke's interpretation of Hegel's view. According to Meinecke, Hegel reveals the tragedy of *Staatsräson* precisely at the moment when he covers it up it by means of the 'cunning of reason'. Meinecke strongly disagrees with this 'solution' to the problems of the interweaving of power and ethics. In that sense, Hegel's idealism has its shortcomings in Meinecke's view. Meinecke claims Hegel's conception of history fails to do justice to the individual or the manifold individualities in history. That is why Meinecke shifts his attention to the ideas of Leopold von Ranke, who in part inspired him to focus on the individual in history. Meinecke asserts Ranke was the 'greatest genius in realizing all the possibilities which were offered to consideration by historicism and the principle of individuality'.[101] Moreover, he hoped to find a better solution in Ranke's philosophy of history to counter the tragedy of *Staatsräson*.

The spiritualization of power

Leopold von Ranke's historicism and its principle of individuality has a strong religious character. 'God' to Ranke was the personal Christian God. He considered notions like God, God's ideas (*Gedanken Gottes*), and divine reason as manifestations in history.[102] According to Ranke, in Meinecke's interpretation, the revelation of God is partly realised through history. Since God as such is unfathomable, this revelation, in Ranke's view, is essentially an externalization. Ranke could therefore not agree with Hegelian pantheism in which 'God' and history merged. Ranke explicitly distinguished between God and history, which means pantheism is out of the question. Instead, as Meinecke states, we should identify Ranke's view as pan*en*theism.[103]

Panentheism has its roots in the philosophies of Plato, Aristotle, and Plotinus' Neoplatonism. Furthermore, it merged with Christianity and in particular with pietism. It is a view in which God and the world ontologically differ from each other. In panentheism the idea is that although God and the world are ontologically different and God even transcends the world, the world is essentially 'in' God ontologically.[104] In this respect God is at the same time transcendent and

101 Meinecke, *Machiavellism*, 378. 'der genialste Verwirklicher aller Möglichkeiten, die der Historismus und das Individualitätsprinzip dem Denken boten'. Meinecke, *Die Idee der Staatsräson*, 443.

102 Meinecke *Die Idee der Staatsräson*, 444.

103 Ibidem, 444.

104 John W. Cooper, *Panentheism. The Other God of the Philosophers. From Plato to the Present* (Michigan 2006) 18. According to Niels Gregersen, there are three closely related views

immanent, eternal and temporal. Contrary to this, pantheism (especially Spinoza's pantheism) presumes God and the world to be identical. In the following chapters I will frequently come back to Meinecke's panentheism. For now, it is important that Meinecke himself uses this notion with regard to Ranke.[105]

Ranke's panentheistic conviction is clearly expressed in apparently paradoxical notions like the 'real ideal' (Real-Geistige), and when he, for example, refers to states as God's ideas (Gedanken Gottes).[106] To Ranke ideas existed both 'in' and 'above' the world. He regards reality as a manifestation or reflection of the 'divine idea' in reality.[107] Wherever Ranke observed the hand of God in history, it should, according to Meinecke, be understood as a 'believing in' and a 'sensing' (ahnen), not a 'knowing' or a scientific explanation.[108] This sensing or experiencing refers to something which has no concrete individual form: sensing the divine in reality. In Über die Epochen der neueren Geschichte Ranke expressed this idea as follows: 'From the standpoint of the divine idea, I cannot think of the matter differently but that mankind harbours within itself an infinite multiplicity of developments which manifest themselves gradually according to laws which are unknown to us and more mysterious and greater than one thinks'.[109] Put differently, the divine idea manifests itself in reality

---

of panentheism: 'Three Varieties of Panentheism' in: Philip Clayton and Arthur Peacocke ed., In Whom We Live and Move and Have Our Being. Panentheistic Reflections on God's Presence in a Scientific World (Cambridge 2004) 19-35. I will elaborate in detail on panentheism in my discussion of Goethe and his importance in Meinecke's view of history (chapter 5) The German historian Carl Hinrichs demonstrated the Neoplatonic influences on Ranke's philosophy of history. Hinrichs claimed that Neoplatonism, apart from being related to Plato's philosophy of ideas and Aristotle's notion of entelechy, it also merged with the high points of Greek culture and Christianity. Carl Hinrichs, Ranke und die Geschichtstheologie der Goethezeit (Göttingen, Frankfurt and Berlin 1954) 249-250. Recent literature on Panentheism: R.T. Mullins, 'The Difficulty with Demarcating Panentheism', Sophia vol 55 issue 3 (September 2016) 325-346. DOI 10.1007/s11841-015-0497-6; Loriliai Biernacki and Philip Clayton, eds., Panentheism across the World's Traditions (Oxford 2014).

105   As far as I know Meinecke uses the notion of panentheism only a few times with regard to Ranke's views: Meinecke, Die Idee der Staatsräson, 444; also: Meinecke, Die Entstehung des Historismus, 597; Meinecke, 'Deutung eines Rankewortes', 133.

106   Hinrichs, Ranke und die Geschichtstheologie, 118-123.

107   Hinrichs offers a similar thought, however, he only relates it to Neoplatonism: 'Wie in der organischen Gestalt die Summe der Teile nicht das Ganze ist, sondern dieses erst durch ein Verbindend-Transzendentes, die innere Formkraft, die Idee, hergestellt wird, so wird die Summe der Weltphänomene zu einem harmonischen Ganzen erst durch Gott, die göttliche Idee, die zugleich in und über der Welt, eins mit ihr und doch nicht mit ihr identisch ist'. Hinrichs, Ranke und die Geschichtstheologie, 120-121.

108   Meinecke, Die Idee der Staatsräson, 444-445.

109   Leopold von Ranke, The Theory and Practice of History, Georg G. Iggers ed. New translations by Wilma A. Iggers (London and New York 2011) 22. 'Vom Standpunkte der göttlichen Idee kann ich mir die Sache nicht anders denken, als daß die Menschheit eine unendliche

in a different and individual manner, but at the same time remains connected to its origin. It is a dynamic of 'everything flows' from God, and everything aims at a reunion with God.

Meinecke claims that thanks to panentheism, the preliminary work of Romanticism and of German Idealism, Ranke managed to reconcile object (*Ding*) and idea (*Idee*). Unlike Hegel, Ranke does not search for a higher abstract synthesis of *Ding* and *Idee*, he only considers them as a unity of 'existence' (*Lebendigen*). Meinecke explains it as follows:

> Individual life in history, incapable of being derived from universal ideas, but imbued with special ideas by which it is shaped, so that in the process idea and body, soul and flesh, become essentially one, and the whole enwreathed in the breath of original divine creativity – this was the particular synthesis of the ideas of individuality and identity which Ranke was able to provide.[110]

This dualism of spirit and matter was overpowered by a monistic view in which all contrasts or polarities emanated from God. In practice this 'monism' leaves room for a pluralistic view of reality.[111] This pluralism or the emphasis on individualities in Ranke's outlook did not lead to (ethical) relativism or an 'anarchy of values', for an absolute moral authority (God) served as a harmonious basis.[112] That is why Ranke claims that 'every Epoch is immediate to God', and explains his disapproval of Hegel's self-realization of the spirit in which individualities are mere phases that serve the self-realization of the spirit.[113] Ranke expresses his criticism on Hegel as follows: '(...) that therefore in every epoch the life of mankind improves, that every generation completely surpasses the previous one, and as a consequence the most recent one is preferred over the previous ones, and that the previous ones

---

Mannigfaltigkeit von Entwicklungen in sich birgt, welche nach und nach zum Vorschein kommen, und zwar nach Gesetzen, die uns unbekannt sind, geheimnisvoller und größer, als man denkt'. Leopold von Ranke, *Über die Epochen der neueren Geschichte* (Darmstadt 1959) 9.

110 Meinecke, *Machiavellism*, 383-384. 'Das individuelle Leben in der Geschichte, unableitbar aus allgemeinen Ideen, aber erfüllt von besonderen Ideen, die es gestalten, Idee und Leib, Geist und Körper dabei wesenhaft eins, und das alles umwittert von dem Hauche ursprünglicher göttlicher Schöpferkraft, – das war diejenige Synthese des Individualitäts- und Identitätsgedankens, die Ranke geben konnte'. Meinecke, *Die Idee der Staatsräson*, 450.

111 Marcel F. Fresco, 'Platonisme. Naar hoger Honing?' in: Marcel F. Fresco en Rudi van der Paardt ed., *Naar hoger honing? Plato en platonisme in de Nederlandse literatuur* (Groningen 1998) 9-51, there 22-23.

112 Meinecke, *Die Idee der Staatsräson*, 445-446.

113 This is obviously an outdated interpretation of Hegel; I only use it here as Ranke's view. On the mutual misunderstanding of Hegel and Ranke, see section 4.6.

are considered to only support the most recent epoch'.[114] This, according to Ranke, reveals an 'injustice to the Godhead'.[115]

A similar aversion to Hegel's 'injustice to the Godhead' is expressed in Ranke's view of the state, which carries the same philosophical-religious character.[116] Ranke considers states as individualities and original creations of the human spirit, inspired by the divine. For that reason, he characterizes states as God's ideas, because they are guided by spiritual-intellectual tendencies (*Tendenzen*).[117] According to the historian Carl Hinrichs, Ranke considered states as 'real ideal forms with their own particular idea, which originated from a hidden law of inner intellectual development and stemmed from a divine idea'.[118] The eternal, the divine idea – 'originated from a hidden law' – will not completely reveal itself in reality, but will be merely visible as a reflection, a so-called 'Abglanz' of the true divine.[119]

Meinecke also points out that Ranke's view of the state fits a panentheistic view. In an article of 1924 on Ranke's 'political conversation' (*politisches Gespräch*) Meinecke explicitly elaborates on this philosophical-religious basis of Ranke's state theory. In this article, Meinecke claims it was Leibniz' monadology, Spinoza's pan(en)theism, and Schelling's philosophy of nature, which all had a strong influence on Ranke's philosophy of history.[120] Meinecke immediately adds that Ranke had always re-

---

114   Cited in Hinrichs, *Ranke und die Geschichtstheologie*, 165.

115   Idem.

116   On this issue: Reinbert A. Krol, 'Friedrich Meinecke: Panentheism and the Crisis of Historicism', *Journal of the Philosophy of History* 4 (2010) 195-209.

117   Meinecke *Die Idee der Staatsräson*, 443, 448.

118   Hinrichs, *Ranke und die Geschichtstheologie*, 126. 'realgeistige Gestalt mit einem eigenen, aus verborgenem Grunde entsprungen Gesetze innerer Bildung, einer besonderen Idee, die ihren Ursprung in der einen göttlichen Idee hat'.

119   'Abglanz' is a Platonic-Neoplatonic symbol and refers to the sun as a symbol of unity, the divine. Hinrichs, *Ranke und die Geschichtstheologie*, 143. Goethe also frequently uses this image in his work. At the beginning of *Faust II*, for example, at daybreak Faust is not able to look at the sun directly, instead he looks at the landscape on which the 'Abglanz' of the divine manifests itself. In other words, as Goethe asserts in 'Versuch einer Witterungslehre': 'Das Wahre ist dem Göttlichen identisch, läßt sich niemals von uns direkt erkennen. Wir schauen es nur im Abglanz, im Beispiel, Symbol, in einzelnen und verwandten Erscheinungen. Wir werden es gewahr als unbegreifliches Leben und können dem Wunsch nicht entsagen, es dennoch zu begreifen. Dieses gilt von allen Phänomenen der faßlichen Welt (...)'. Cited in Erich Trunz, 'Das Vergängliche als Gleichnis in Goethes Dichtung' in: idem, *Ein Tag aus Goethes Leben* (Munich 2006, first published 1999) 167-187, there 167. In my discussion of 'conscience' I will come back to this issue.

120   Friedrich Meinecke, 'Rankes "Politisches Gespräch"' in: idem, *Zur Geschichte der Geschichtsschreibung* (München 1968) 72-82, there 79; Cooper, *Panentheism*, 19, 71-72, 78. According to Cooper, we should count Spinoza as a panentheist. Even though Spinoza distiguishes between God and the world, ontologically God and his creation do not strictly differ from each other, Cooper writes. Leibniz considered Spinoza's 'God as nature'-identification as atheism

jected Hegel's pantheistic or 'panlogistic' philosophy. Through the philosophies of Leibniz and Schelling, Ranke apparently realized that God's creation of the world had also created polarities such as light and dark, good and evil, which in the end were all one in God.[121] Meinecke asserts that Ranke was 'however, too realistic to spiritualize and deify all of historical life'.[122] The spirit of the state, Meinecke cites Ranke: 'to him was "indeed the breath of the Lord, but simultaneously human impulse"'.[123] This of course corresponds with the panentheistic *Weltanschauung* in which the Idea is the manifestation of the divine in reality, and this Idea can also develop spontaneously within this reality.[124] And this obviously reminds us of Humboldt's notion of the theory of ideas (*Ideenlehre*) discussed in the previous chapter.

Notwithstanding Meinecke's claim that Ranke was too realistic to completely spiritualize everything in history, Meinecke does state in *Die Idee der Staatsräson* that Ranke's respect for the (panentheistic) unfathomable divine and moral law kept him from situating *raison d'état* above Christian morality.[125] To Ranke Christian morality outweighs a power political morality. Meinecke claims Ranke's strict distinction between good and evil is also evidence of this: 'Good must remain good and evil be evil', says Meinecke discussing Ranke's view.[126] In this regard, Ranke opposed Hegel's view of the cunning of reason. Ranke expressed his aversion to this concept clearly in *Über die Epochen der neueren Geschichte*.[127] It is also clear from Ranke's account that he was influenced especially by Spinoza's panentheism, which is why he could not agree with Hegel's pantheism:

---

and reintroduced finite individual substances (monads) to guarantee the ontological difference with God. Cf.: Chapter 4, section 4.2.

121   Meinecke, 'Rankes "Politisches Gespräch"', 79; Erich Trunz, 'Das Vergängliche als Gleichnis', 171

122   Meinecke, 'Rankes "Politisches Gespräch"', 79. 'zu realistisch, um alles geschichtliche Leben schlechthin zu spiritualisieren und zu vergöttlichen'.

123   Meinecke, 'Rankes "Politisches Gespräch"', 79.

124   The same thought Meinecke expressed already in an article of 1916 which was on Ranke's 'Große Mächte': 'Alle Kenntnis der Dinge aber steigert sich bei Ranke zu Anschauung und Mitgefühl, die das Besondere in seinen geheimsten Falten und das Allgemeine in seinen höchsten Beziehungen umfaßt. Weil beides bei ihm in jedem Augenblicke ineinanderlebt, ist auch das Besondere immer etwas von allgemeiner Bedeutung und das Allgemeine niemals eine bloße Abstraktion, sondern nur die höchste der verschiedenen ineinander verkapselten Individualitäten. Und über der höchsten Allgemeinheit der Geschichte, die sich schauen läßt, liegt immer noch ein geistiger Äther philosophisch-religiöser Ahnungen, der alles umhüllt. Keinem Historiker der Welt ist es gelungen, zugleich so realistisch und so transzendent die Dinge zu behandeln'. Meinecke, 'Rankes "Große Mächte"', 70.

125   Meinecke, *Die Idee der Staatsräson*, 456-459.

126   Meinecke, *Die Idee der Staatsräson*, 445. 'Gut mußte gut und böse böse bleiben' (sic).

127   Recently Beiser has shown the mutual misunderstanding between Ranke and Hegel: Beiser, *The German Historicist*, 261. Also: section 4.6 of this book.

The doctrine according to which the world spirit produces things, as it were, through deceit and uses human passions to achieve its goals is based on an utterly unworthy idea of God and mankind. Pursued to its logical conclusion consistently, this view can lead only to pantheism. Mankind is then God in the process of becoming, who gives birth to himself through a spiritual process that lies in his nature.[128]

In Meinecke's view, Ranke conceived the world as being ordered according to moral laws.[129] The aforementioned quote expresses Ranke's panentheistic view. His polarities of good and evil are not overcome by a Hegelian 'cunning of reason', but by a divine harmony. In that sense, Ranke's view of the state tends towards a veiled dualistic outlook.

Ranke's distinction between good and evil is not only conveyed by his rejection of Hegel's view, but also by a modest sympathy for Machiavelli. Early on Ranke delved into Machiavelli's view of the state. Meinecke refers to Ranke's 'Anhang über Machiavelli' of 1824 in which he expressed his admiration for the Florentine thinker. However, next to this admiration Ranke also expresses his disapproval of Machiavelli's 'terrifying' recommendations for the ruler. He could not reconcile himself with the idea that Machiavelli, who had fought for the freedom of Florence, could prescribe such recommendations.[130] On the other hand, Ranke was also well aware of the fact that his own view of history was in fact very close to Machiavelli's view, so if Ranke would really take notice of the problem of ethics he would most certainly, according to Meinecke, run into conflict with his own 'historical genius' and his conscience.[131] Ranke circumvents this problem in Meinecke's view by stating that Machiavelli's recommendations are very much a product of his time and strictly individual, so one should not derive any universality from them.[132] Moreover, Ranke consoled himself with the idea that Machiavelli was very close to the well-respected philosopher Aristotle.[133] When Ranke discusses the relationship between Aristotle and Machiavelli it is striking when he states: 'Strong minds, individuals, high

---

128   Ranke, *The Theory and Practice of History*, 22; Ranke, *Über die Epochen*, 9; Meinecke, *Die Idee der Staatsräson*, 444, 449. 'Der Lehre, wonach der Weltgeist die Dinge gleichsam durch Betrug hervorbringt und sich der menschlichen Leidenschaften bedient, um seine Zwecke zu erreichen, liegt eine höchst unwürdige Vorstellung von Gott und der Menschheit zugrunde; sie kann auch konsequent nur zum Pantheismus führen. Die Menschheit ist dann der werdende Gott, der sich durch einen geistigen Prozeß, der in seiner Natur liegt, selbst gebiert'.

129   Meinecke, *Die Idee der Staatsräson*, 449.

130   Ibidem, 446; Leopold von Ranke, 'Anhang über Machiavelli' in: idem, *Sämmtliche Werke* 54 vols. (Leipzig 1874) Volume 34, 151-174, there 169.

131   Meinecke, *Die Idee der Staatsräson*, 445-446.

132   Ibidem, 446.

133   Ranke, 'Anhang über Machiavelli', 169

above the crowd, they shake hands spanning several centuries'.[134] This might be interpreted as his own affinity with both these thinkers.

Viewed from the perspective of Ranke's relationship with Machiavelli it is striking that Meinecke in *Die Idee der Staatsräson* fails to discuss Ranke's famous inaugural Berlin address of 1836, entitled: 'Über die Verwandtschaft und den Unterschied der Historie und der Politik'.[135] Not only does this lecture fit Meinecke's initial plan to discuss the relationship between history and politics, but also reveals that Ranke was very close to Machiavelli. Like the 'Anhang über Machiavelli', in this lecture Ranke again claims that it is crucial to know and study the past of a state in order to know the essence and interest of the state in the present. He expressed it as follows in this famous lecture of 1836: 'It is the task of the historian to demonstrate what is the essence of the state on the basis of the events of the past and to further an adequate understanding of these events; it is, next, the task of the politician to develop and to complete the historian's insights after they have been understood and appropriated by the mind'.[136] Thus, Ranke thought the state's history could provide insight into the 'necessity' of the state's interest, to inform the politician who should base his decisions on these insights.[137] Ranke was convinced that a politician's prudent act, in line with the past, would result in favour of the state.[138] At the same time – and this is where Ranke is really close to Machiavelli – he claims that the state should not at all costs stick to certain laws if they did not correspond to the 'inner drive' (*inneren Trieb*) of life.[139] This obviously dovetails with Machiavelli's recommendations for the ruler. As well as in the 'Anhang über Machiavelli' Ranke also fails to address the moral dimension in his lecture of 1836. Meinecke states in *Die Idee der Staatsräson* that if Ranke would exclusively follow his historicizing insight and experience, he would probably give in to Hegel's reconciliation of evil and the ultimate goal in history.[140] That could be the ultimate conclusion to Ranke's address and also the explanation why Meinecke did not mention it, for he rejected

---

134   Ibidem, 167.

135   Leopold von Ranke, 'Über die Verwandtschaft und den Unterschied der Historie und der Politik' in: idem, *Sämtliche Werke* 54 vols. (Leipzig 1877) Volume 24, 280-293.

136   Cited in: F.R. Ankersmit, *Aesthetic Politics. Political Philosophy Beyond Fact and Value* (Stanford 1996) 212. Ranke, 'Über die Verwandtschaft und den Unterschied der Historie', 288-289. 'Demnach ist es die Aufgabe der Historie, das Wesen des Staates aus der Reihe der früheren Begebenheiten darzuthun und dasselbe zum Verständniß zu bringen, die der Politik aber, nach erfolgtem Verständniß und gewonnener Erkenntniß es weiter zu entwickeln und zu vollenden'.

137   Ibidem, 293.

138   Ibidem, 293.

139   Ibidem, 290; This 'inner drive' refers also to Aristotle's notion of entelechy, for it refers to an inner force (in this case the state) which will develop itself, no matter what.

140   Meinecke, *Die Idee der Staatsräson*, 449.

the idea of a goal in history. When Ranke realized he was in fact very close to Hegel's philosophy, he remained optimistic and found a new, less abstract (and less 'evil') way to preserve his philosophy of history.

Notwithstanding Ranke's characterization of the state as 'God's ideas', he, in Meinecke's view, neither spiritualized power politics (or veiled them like Hegel did), nor did he place them above Christian morality. In this context, Ranke emphasized that *Staatsräson* was not driven by reason, but by forces which he referred to as 'moral energy' (*moralische Energie*). Ranke considered this moral energy the source of power politics. By means of this moral-spiritual power, which manifests itself in state egoism, the *Staatsräson* also obtained in Ranke's view the dignity of a moral energy, Meinecke writes.[141] Again, Ranke tends towards a spiritualization (or 'moralization') of power politics; he even goes a step further in his spiritualization, for despite his emphasis on the real-spiritual character of states, he in the end considers the state's individual interest and their moral energy as part of a 'universal life-stream of history', which in essence was, as opposed to *Staatsräson*, all-embracing in its 'necessary force'. Meinecke therefore concludes: '(…) in the case of Ranke too, it is also the universal movement of historical life which evokes and justifies the developments of *raison d'état*'.[142] Thus, Ranke's 'universal life-stream' is more or less comparable with Hegel's 'cunning of reason', since both justify *Staatsräson*. Hegel's pantheistic conclusion was, however, unacceptable to Ranke.[143]

Meinecke disagrees with Ranke's 'solution' to the problem of the interweaving of *Kratos* and *Ethos*. He considers this solution a veiled dualism and he explicitly states that this is not a solution at all: 'It is clear that this nebulous and wavering dualism could not constitute the final possible solution of the problem'.[144] *Staatsräson* is 'no longer (as Ranke regarded it) the guiding principle, the leader and director of State existence, which, even when it fights and overthrows its adversaries, awakens new life there, or at least concedes it', Meinecke writes.[145] He even characterizes Ranke's 'spiritualisation' of power as dangerous, for Ranke's optimistic and idealistic disposition with respect to the world (*Weltstimmung*), and the philosophy of identity

---

141   Ibidem, 451-453.

142   Meinecke, *Machiavellism*, 391. '(…) die allgemeine Bewegung des geschichtlichen Lebens ist es auch bei Ranke, die die Entfaltungen der Staatsräson hervorruft und rechtfertigt'. Meinecke, *Die Idee der Staatsräson*, 459.

143   Meinecke, *Die Idee der Staatsräson*, 456-459.

144   Meinecke, *Machiavellism*, 391. 'Es ist klar, daß dieser verhüllte und schwebende Dualismus nicht die letzte mögliche Lösung des Problems bedeuten konnte'. Meinecke, *Die Idee der Staatsräson*, 459.

145   Meinecke, *Machiavellism*, 423. 'nicht mehr, wie Ranke es anschaute, das gestaltende Prinzip, die Leiterin und Ordnerin des Staatslebens, die auch da, wo sie kämpft und ihre Gegner niederringt, neues Leben weckt oder doch zuläßt'. Meinecke, *Die Idee der Staatsräson*, 498.

enabled the 'nocturnal aspects of life' to hide behind the *Gottnatur*.[146] This leads to Meinecke's far-reaching conclusion:

> that even though power politics of the state has a very brute and elemental side, in Ranke's thought the spiritual dimension is more prominent than the elemental one. "Power as such is not the cause" Ranke claimed in his 'Politischen Gespräch'. But God's ideas which reveal themselves in 'state-personalities' even ennoble their elemental struggle for power. And German Idealism as well as Ranke, who supported this far-reaching spiritualization and sanctioning of elemental violence and power, have set up conditions for the subsequent coarsening of the theory of power politics. (...) It is a tragic [and disconcerting] course.[147]

On the basis of a divine, moral-spiritual source Ranke sanctioned the force and violence of *Staatsräson*. In considering states as God's ideas Ranke veiled the brute essence of power politics and in that sense he is close to Hegel's spiritualization of evil in power.[148]

Disillusionment with, and a distrust of Ranke are at the heart of Meinecke's argument in *Die Idee der Staatsräson*. What Machiavelli revealed became a struggle for Frederick the Great and challenged Hegel and Ranke to solve or veil it. Meinecke thought the interweaving of *Kratos* and *Ethos*, of *Macht* and *Kultur* was no longer tenable. He considers the relationship between both as tragic: 'The historical world seems to us more obscure and, with respect to its further progress, more dangerous and uncertain than it did to him, [Ranke, RK] and to the generations that believed in the triumph of reason in history'.[149] The 'nocturnal and elemental' aspect of (political) history imposed itself with more force on Meinecke and his generation than

---

146   I will explain the content of this notion in my discussion of Goethe and Meinecke's notion of nature (*Natur*) (chapter 5).

147   Meinecke, 'Rankes politisches Gespräch', 81-82.

148   Meinecke noticed a similar idea in Fichte's thought. In *Weltbürgertum und Nationalstaat* Meinecke claimed that Fichte considered the power of the state to be a force with its own *Lebenstrieb*, which was in harmony with a moral *Weltanschauung*. However, in *Die Idee der Staatsräson* Meinecke takes the view that Fichte had 'moralized' the power state; Meinecke cites this notion on purpose and shows it still had a positive ring to it in *Weltbürgertum und Nationalstaat*. In *Die Idee der Staatsräson* his view had changed: he now thought such a view dangerous to politics and ethics: 'Aber diese Versittlichung konnte, wie wir jetzt hinzufügen müssen, dermaleinst zu neuer Unsittlichkeit führen, wenn die nationale Idee über ihre Ufer trat und zum modernen Nationalismus entartete'. Meinecke, *Die Idee der Staatsräson*, 441. Meinecke expressed a similar argument in 1914: Friedrich Meinecke, 'Nationalismus und nationale Idee' in: idem, *Politische Schriften und Reden* (Darmstadt 1958) 83-95.

149   Meinecke, *Machiavellism*, 424. 'Die geschichtliche Welt liegt dunkler und in dem Charakter ihres ferneren Verlaufs ungewisser und gefährlicher vor uns, als er [Ranke, RK] und die Generationen, die an den Sieg der Vernunft in der Geschichte glaubten, sie sahen'. Meinecke, *Die Idee der Staatsräson*, 499.

was the case with Hegel and Ranke.[150] Yet, it is striking that Meinecke reproached Ranke for a spiritualization of power, for – as will become clear in the next chapter – even though *Die Idee der Staatsräson* ends dejectedly, Meinecke does come up with a solution which is comparable to Ranke's spiritualization. Meinecke could no longer advocate a Rankean or Hegelian harmony, which caused for a shift in his thought, but his desire for such a harmony did not diminish. It will become clear that, in relation to a Rankean sensing (*ahnen*) of a harmony – and sensing the divine in reality – or a Hegelian *vision*, Meinecke will emphasize a *belief* in a harmony 'behind' history. This is a fundamental correction to the authoritative interpretation of Walther Hofer, who claimed that after the war Meinecke's harmonious philosophy of history was substituted by a pessimistic and strict dualism.

150   Meinecke, *Die Idee der Staatsräson*, 499.

3. A conscience akin to God

'Das Ew'ge regt sich fort in allen'[1]

Introduction

The collapse of the German Reich, the loss of the old ideal of a synthesis between power and culture, and the loss of that *Kultur* resulted in Meinecke's rejection of veiling constructions such as Hegel's and Ranke's. They were optimistic about the tensions between power and ethics, but for Meinecke this was no longer an option. Nonetheless, Meinecke's suggestion initially seems to be no less a veiled solution in a Rankean or Hegelian sense. Meinecke found his answer to the issue of the interweaving of power and ethics in aesthetics, and in particular in Schiller's aesthetics. To my knowledge this parallel between Schiller's aesthetics and Meinecke's 'Staatsräson as aesthetic category' has so far never been noticed, which is remarkable since both resemble each other very much. Analogous to Schiller's ideas I will demonstrate that Meinecke was able to reconcile contrasts in a 'dualistic harmony'. In this regard the question presents itself whether Meinecke is also guilty of a veiling of power politics.

In *Die Idee der Staatsräson* Meinecke explicitly states that he starts from a 'new dualism'. This was the reason for Walther Hofer to characterize Meinecke's post-war thought as dualistic. I will however demonstrate that 'behind' this 'new dualism' hides a panentheistic harmony. Meinecke gave a clear indication for this in the introduction of *Die Idee der Staatsräson*. He claims an historian – he probably refers to himself – should stand firm in order to survive the major issues of the interweaving of politics and ethics, of good and evil, and also of spirit (*Geist*) and nature (*Natur*): 'Here, if anywhere, he [the historian, RK] is in need of his own guiding world view', Meinecke warns.[2] It is the world view (*Weltanschauung*) of panentheism which Meinecke points at. In that sense, I correct Walther Hofer's influential view that Meinecke apparently replaced his harmonious monism after

---

1   Johann Wolfgang von Goethe, 'Vermächtnis' in: idem, *Goethes Werke. Hamburger Ausgabe in 14 Bänden*, ed. Erich Trunz (Munich 1981) Volume I, 369.

2   Meinecke, *Die Idee der Staatsräson*, 13. 'Hier wenn irgendwo braucht er eine eigene Weltanschauung'.

the First World War with a pessimistic dualism. I claim this apparent shift did not end in a strict dualism – that is the proposition put forward in this chapter.

In Meinecke's attempt to reconcile power and ethics two factors play an important role: the 'necessity of state' (*Staatsnotwendigkeit*) and the statesman. The necessity of state, as demonstrated in the previous chapter, informs the ruler or statesman, according to Meinecke: 'Thus must you act, if you wish to preserve the power of the State whose care is in your hands; and you may act thus, because no other means exist which would lead to that end'.[3] Analogous to Schiller's aesthetics Meinecke attempted to reconcile the clash with ethics that followed from this. This reconciliation does not exclude the possibility that the statesman, who may or may not wish to follow the necessity of the state, can still come into conflict with his personal hunger for power and individual interests. In this case another entity plays an important role in the solution to the tragedy of *Staatsräson*, namely: conscience. To decide and act on the basis of our conscience a certain belief in an underlying absolute morality is crucial, Meinecke states. I will demonstrate that conscience is enclosed in an ethical shell of a panentheistic philosophy.

In the first section I will discuss the similarities between Schiller's aesthetics and Meinecke's conception of *Staatsräson*. In the second section the notion of 'personality' and Meinecke's ideas of the statesman's conscience are central. I will focus on Meinecke's panentheistic conviction, which, as a political thinker and historian gave him the tools to cope with the tragedy of *Staatsräson* and the new political situation after the First World War.

3.1   Staatsräson as a 'solution'

In the introduction of *Die Idee der Staatsräson* Meinecke states that it lies in the nature of power to rule blindly. According to Meinecke, this is an exception in real life, for unregulated, blind rule of power will in the end destroy itself. After all, when power violates morality and or law, this same power could become a threat to its own existence.[4] With regard to Meinecke's view on power it is immediately clear that his opinion had changed from his pre-war conviction. In his previous studies his conception of power was still morally optimistic; now, after the war Meinecke takes the view that power needs rules and standards to preserve and develop itself. To keep a state healthy, it is important to find a balance between power and ethics. Despite its twofold character Meinecke is convinced that it is

---

3   Meinecke, *Machiavellism*, 10. 'Du mußt so handeln, um die in deinen Händen liegende Macht des Staates zu erhalten, und du darfst so handeln, weil es keinen anderen Weg zum Ziele gibt'. Meinecke, *Die Idee der Staatsräson*, 12.

4   Meinecke, *Die Idee der Staatsräson*, 12.

indeed *Staatsräson* which could establish this balance between power and ethics: '*Raison d'état* (...) presents one aspect to physical nature and another to reason. And it also has (...) a middle aspect, in which what pertains to nature mingles with what pertains to the mind'.[5] This 'middle aspect' (*Mittelstück*) is 'dominated equally by light and darkness'[6]; however, it will simultaneously reconcile or even transcend an 'unregulated power politics' with ethics. Meinecke's conception of *Staatsräson* thus differs considerably from the general interpretation of *Staatsräson*. That is, *Staatsräson* as power politics always opposed Christian ethics; as demonstrated with Machiavelli's example. Meinecke's notion of *Staatsräson* however encloses power and ethics, and he even leans toward the ethical dimension and not power politics. What this shift means in practice and how Meinecke thought this a (preliminary) solution to the problem of the twofold character of *Staatsräson* will be expounded in this section by means of Schiller's aesthetics.

In the previous chapter it became apparent that power and ethics, good and evil, but also *Kultur* and *Natur* are interwoven and these contrasts not only occur within the *Staatsräson*, but also within the statesman. Meinecke considers *Macht* related to *Natur*. 'Nature' is, according to Meinecke, understood to mean 'animality', uncontrolled (striving for) power. Meinecke disagrees with Jacob Burckhardt who claimed that power as such was 'evil in itself'.[7] To Meinecke it is 'naturally indifferent both towards evil and good'[8], which is why he considers power to be an 'amoral motive': 'The striving for power is an aboriginal human impulse, perhaps even an animal impulse (...)'.[9] According to Meinecke, it is in particular the ruler or state which is continuously subjected to a temptation to misuse power.[10] Meinecke distinguishes power from the rational and moral human being, because it is an amoral motivation; however, this characterization leads back to the 'animality' and 'nature' in man. The precise relation between animality and morality in man – and how 'nature' possibly becomes *Kultur* – will, against the background of Schiller's aesthetics, also become clear in this section. From this line of thought Meinecke's statement that *Macht* is ethically indifferent also raises the question if he is not

---

5   Meinecke, *Machiavellism*, 5. 'Die Staatsräson (...) hat eine der Natur und eine dem Geiste zugekehrte Seite und hat (...) ein Mittelstück, in dem Naturhaftes und Geistiges ineinander übergehen'. Meinecke, *Die Idee der Staatsräson*, 6.

6   Meinecke, *Machiavellism*, 6. 'von Licht und Finsternis zugleich beherrscht'. Meinecke, *Die Idee der Staatsräson*, 7.

7   Meinecke, *Machiavellism*, 13. 'an sich böse'. Meinecke, *Die Idee der Staatsräson*, 15.

8   Meinecke, *Machiavellism*, 13. 'naturhaft und indifferent gegen gut und böse'. Meinecke, *Die Idee der Staatsräson*, 15.

9   Meinecke, *Machiavellism*, 4. 'Das Streben nach Macht ist ein urmenschlicher, ja vielleicht animalischer Trieb (...)' Meinecke, *Die Idee der Staatsräson*, 4-5.

10 Meinecke, *Machiavellism*, 13; Meinecke, *Die Idee der Staatsräson*, 15; Trunz, 'Das Vergängliche als Gleichnis', 175.

also veiling the evil aspects of power. This question is important, for in *Die Idee der Staatsräson* Meinecke considers Ranke's spiritualization and veiling of power as problematic. So it remains to be seen whether and how Meinecke tried to solve this problem.

Schiller's Spieltrieb

Meinecke's conception of the relationship between the 'natural strive for power', the *Kultur*, and the 'middle section' of the *Staatsräson* corresponds with the ideas of the poet and philosopher Friedrich Schiller (1759-1805) as expounded in *Über die ästhetische Erziehung des Menschen in einer Reihe von Briefen* (1795). In this study, written against the background and aftermath of the French Revolution, Schiller suggested, with regard to the relation between art and politics, to attribute a key role to art in the transformation of the 'natural' political state to the modern rational state. According to him, the aesthetic education of man would enable to overcome its own 'animality' in favour of a moral and political freedom.[11]

For Schiller every human being carries a timeless, immutable ideal (*Idee*) of humanity in itself. According to Schiller, it is the task of man to reconcile his temporal existence with his immutability.[12] This unchanging status of the ideal of humanity is, in Schiller's view, represented by the state: the community in which the multiplicity of individualities attempt to form a unity.[13] He further claims that the reconciliation between 'the man of time' ('der Mensch in der Zeit') and 'the man of idea' ('dem Menschen in der Idee') can be realized (within the state) in two ways. On the one hand, the state can cancel out the individual as a result of the 'ideal man' transcending the 'temporal man'. On the other hand, the individual 'man of time' can elevate himself – Schiller speaks of *veredeln* – to the 'ideal man' and in that sense coincide with the idea(l) of humanity.[14] In short, the tension between universality and individuality, which has its effect on humans as well as on the state, is at the heart of Schiller's aesthetics.

---

11   Because of this, *Über die ästhetische Erziehung* is, in Janz' view, considered a political allegory, in which reason and nature are equal to the ruler and the subject. Rolf-Peter Janz, 'Über die ästhetische Erziehung des Menschen in einer Reihe von Briefen' in: Helmut Koopman ed., *Schiller Handbuch* (Stuttgart 2011) 649-666, there 651-652, 664; Stephen Boos, 'Rethinking the Aesthetic: Kant, Schiller, and Hegel' in: Dorota Glowacka and Stephen Boos ed., *Between Ethics and Aesthetics. Crossing the Boundaries* (New York 2002) 15-27, there 21.

12   Friedrich Schiller, 'Über die ästhetische Erziehung des Menschen in einer Reihe von Briefen' in: idem, *Sämtliche Werke in fünf Bänden* (Munich 2004) Volume 5, 570-669, there 577.

13   Schiller, 'Über die ästhetische Erziehung', 577.

14   Friedrich Schiller, 'Letters Upon the Aesthetic Education of Man' in: Charles W. Eliot ed., *Literary and Philosophical Essays. French, German, And Italian* Volume 32 (New York 1910) 221-313, there 216; Schiller, 'Über die ästhetische Erziehung', 577.

It is 'reason' that demands unity, whereas 'nature' creates diversity, but both are, according to Schiller, closely related and interwoven in man: 'The law of the former is stamped upon him by an incorruptible consciousness, that of the latter by an ineradicable feeling'.[15] Consequently education of man, Schiller claims, will always appear deficient 'when the moral feeling can only be maintained with the sacrifice of what is natural'; and likewise, 'a political administration will always be imperfect when it is only able to bring about unity by suppressing variety'.[16] Schiller wants the state not only to 'respect the objective and generic' but also 'the subjective and specific in individuals'.[17] In other words, the state, in its development and growth of the moral and rational, should dispose itself of the 'natural'. The solution to this problem of contrasts, in Schiller's view, lies in the aesthetic education of man.[18]

According to Schiller, the aesthetic education reconciles the universal with the individual, reason with feeling, morality with nature, the outer with the inner, and *form* with *material*. In Schiller's view, this is a subtle interplay or even a twofold task, because, to not simply remain 'material' man needs to give *form* to matter; and likewise, to not merely remain form he should attempt to actualize or materialize everything in himself.[19] In this actualization man is, according to Schiller, driven by two opposing forces, impulses or instincts: the sensuous drive (*sinnliche Trieb*) and the formal drive (*Formtrieb*).[20] The sphere of the sensuous drive includes the

---

15   Schiller, 'Letters Upon the Aesthetic Education of Man', 216. 'Das Gesetz der erstern ist ihm durch ein unbestechliches Bewußtsein, das Gesetz der andern durch ein unvertilgbares Gefühl eingeprägt'. Schiller, 'Über die ästhetische Erziehung', 577.
16   Schiller, 'Letters Upon the Aesthetic Education of Man', 216.
17   Schiller, 'Letters Upon the Aesthetic Education of Man', 216; 'wenn der sittliche Charakter nur mit Aufopferung des natürlichen sich behaupten kann', 'Staatsverfassung wird noch sehr unvollendet sein, die nur durch Aufhebung der Mannigfaltigkeit Einheit zu bewirken im Stand ist'. 'bloß den objektiven und generischen, er soll auch den subjektiven und specifischen Charakter in den Individuen ehren'. Schiller, 'Über die ästhetische Erziehung', 577.
18   Boos, 'Rethinking the Aesthetic', 21.
19   Schiller, 'Über die ästhetische Erziehung', 604.
20   Contrary to Kant, Schiller does not start from a radical opposition between both instincts. In Kant's ethics, which presumes universality and unity, sensuous instincts are subordinate to Reason (formal instinct). According to Schiller, this is the reason why Kant's view does not leave any room for harmony; instead only uniformity is possible. Schiller states that both instincts are not actually in conflict with each other. Only when one of them intrudes on the other a conflict emerges. It is, in his view, the aesthetic education which can tell them apart and ensure that both can develop in a mutual subjection and structuring. This way Schiller's view transcends Kant's uniformity and he also manages to find a harmonious solution to the tensions between both instincts. Schiller, 'Über die ästhetische Erziehung', 606-611; Boos, 'Rethinking the Aesthetic', 22; Peter Hanns Reill, 'Schiller, Herder, and History' in: Michael Hofman, Jörn Rüsen and Mirjam Springer ed., *Schiller und die Geschichte* (Munich 2006) 68-78, there 73-74.

physical, the finite in man and is concerned with the temporal: existence *in* time.[21] The formal drive issues from the rational nature of man and is concerned with the absolute, universal and it demands the real, existence, to be necessary and eternal. So, notwithstanding all the diversity and changes of its manifestations, the formal drive attempts to maintain the (inner) absolute. In other words, the formal drive strives for an actualization of the eternal, the absolute, and in that sense, it cancels out the temporal, the sensuous drive. Schiller states it is all about 'permanence in change'.[22] According to Schiller, the formal drive tends towards 'truth' and 'justice'.[23] The tension and interaction between the sensuous drive, which requires change and gives time its contents, and the formal drive, which supresses time and cancels out change, results in what Schiller calls the play drive (*Spieltrieb*) or the aesthetic state.[24] So from the tensions between 'supressing time' and 'giving contents to time' flows the play drive, and as a result this drive is, in Schiller's view, aimed at suppressing 'time within *time* to conciliate the state of *becoming* with the absolute being, change with identity'.[25]

To clarify his argument Schiller adds that the sensuous drive can also be expressed in terms of the general notion of 'life' (*Leben*), and the formal drive in terms of 'shape' or 'form' (*Gestalt*). Next, the so-called 'living form' is the object of the play drive. This object should in the broadest sense of the word be characterized as 'beauty'.[26] The best possible balance between the sensuous drive and the formal drive is found in beauty. However, and this is important with regard to Schiller and also to Meinecke, this equilibrium always remains an *idea*, which will never be fully achieved in reality. Schiller distinguishes between the idea(l) and the experience of beauty: 'Ideal beauty is therefore eternally one and indivisible, because there can only be one single equilibrium; on the contrary, experimental beauty will be eternally double, because in the oscillation the equilibrium may be destroyed in two ways – this side and that'.[27] In reality, one of these drives will always dominate

---

21   Schiller, 'Über die ästhetische Erziehung', 604; Boos, 'Rethinking the Aesthetic', 22.

22   Schiller, 'Letters Upon the Aesthetic Education of Man', 255. 'Beharrlichkeit im Wechsel'. Schiller, 'Über die ästhetische Erziehung', 603.

23   Schiller, 'Über die ästhetische Erziehung', 605.

24   Ibidem, 612.

25   Schiller, 'Letters Upon the Aesthetic Education of Man', 262. '(...) die Zeit *in der Zeit* aufzuheben, Werden mit absolutem Sein, Veränderung mit Identität zu vereinbaren'. Schiller, 'Über die ästhetische Erziehung', 613.

26   Schiller, 'Über die ästhetische Erziehung', 614.

27   Schiller, 'Letters Upon the Aesthetic Education of Man', 253. 'Die Schönheit in der Idee ist also ewig nur eine unteilbare einzige, weil es nur einziges Gleichgewicht geben kann; die Schönheit in der Erfahrung hingegen wird ewig doppelt sein, weil bei einer Schwankung das Gleichgewicht auf eine doppelte Art, nämlich diesseits und jenseits, kann übertreten werden'. Schiller, 'Über die ästhetische Erziehung', 619.

over the other, but this can continuously change and vary. In experience there is, to put it differently, an oscillation between the two drives; the one time reality, and the other form or shape will have the advantage over the other.[28]

Life and *form* (the sensuous drive and the formal drive) are in reality continuously in contrast with each other, but can be in harmony as *Spieltrieb*, without cancelling out one or the other. The play drive gives room for acting ethically in which the finite (actions) coincide with the universal (ethics). So, it is the *idea* of beauty which Schiller denotes as a condition in which both drives are in harmony and 'humanity' is perfected, but beauty, in his view, also enables the transition of 'nature' to 'reason' (morality).[29] Schiller's idea of the aesthetic education of man thus has two (opposing) goals, for on the one hand the aesthetic state clears the way for acting ethically, and on the other hand the aesthetic state is the highest completion of subjectivity because of the transition of nature to pure morality. In that sense, Schiller both holds on to a Kantian ethics and surpasses it; he holds on to it when ethical actions are based on reason, he transcends Kant when he considers the highest morality as beauty.[30] In short, ideally the play drive is the state in which the sensuous drive and the formal drive are in harmony; however in reality this harmony remains twofold. It is this idea which corresponds with Meinecke's conception of *Staatsräson*.

Staatsräson as an aesthetic category

Of course, Schiller's 'play drive', which forms the reconciliation of the contrasting drives, reminds us of Meinecke's conception of the 'Mittelstück' of the *Staatsräson* in which *Ethos* and *Kratos* are reconciled or in harmony. Did Meinecke have Schiller's aesthetics in mind? Even though Meinecke does not discuss Schiller's ideas in *Die Idee der Staatsräson*, there are enough clues which support the thesis that Meinecke did in fact have Schiller's aesthetics in mind when discussing his own idea of *Staatsräson*.[31]

---

28   Schiller, 'Über die ästhetische Erziehung', 619.

29   Boos, 'Rethinking the Aesthetic', 22.

30   Janz, 'Über die ästhetische Erziehung', 663.

31   In the 1930s, after the publication of *Die Entstehung des Historismus*, Meinecke wrote a few articles on Schiller for the first time. In these articles Meinecke appears to be ambivalent regarding Schiller's thought, in particular Schiller's contribution to historicism. Meinecke does not consider him a pioneer, rather a so-called *Zwischenstufe*. Schiller should be positioned, according to him, between normative natural law and the principle of individuality (*Individualitätsgedanke*). In the end, Meinecke considers him too much of a Kantian, for Schiller does not start from 'real' individualism, but from a 'generalizing individuality'. However, during the period in which Schiller wrote his aesthetic works, the principle of individuality is most apparent. This, says Meinecke, was mainly the result of Schiller's contact with Goethe and Humboldt. In short, 'the Schiller' of the aesthetic

Die Idee der Staatsräson carries the following stanza of Schiller's poem 'The Dance' as a motto:

Say, how's it done, that restless renews the supple formations
And that calmness endures even in moveable form?[32]

Schiller considered art as stemming from the play drive. A 'dance' is of course a form of art and therefore the highest form of 'play', since here the ideal reconciliation takes place between form and matter. This is also clear in Schillers stanza in which 'formations' continuously vary. In addition, Schiller refers to 'calmness' which is part of the movement of the form. This clearly evokes Schiller's idea of the reconciliation of 'change' (the sensuous drive) and invariability (the formal drive) within the aesthetic state, the play drive. Like the play drive, the principle of *Staatsräson* can also be viewed as a 'dance', for power as an expression of the sensuous drive involves change, whereas ethics, in 'forming' power, is always constant and absolute. Both power and ethics are thus involved in a 'dance'. Next, it is the 'middle section' (*Mittelstück*) of the *Staatsräson* – the 'overarching drive' – that reconciles and harmonizes both.

The fact that Meinecke's motto did indeed refer to this theme is clear from the introduction of *Die Idee der Staatsräson* in which he quotes another part of Schiller's poem:

Ever destroyed, creation rotating begets itself ever,
And an unspoken law guides the transformative play.[33]

The similarities with the principle of *Staatsräson* are clear: the 'eternal conflict' between *Kratos* and *Ethos*, which in all its changes and variations is guided by the 'silent law' of *Staatsräson*. The dance or play unites the rational with the 'natural', and according to Meinecke the same applies to the *Staatsräson*, since it has an ethical and a fundamental 'natural' dimension. The political state is in that sense

---

studies is indeed highly important to Meinecke. Friedrich Meinecke, 'Schiller und der Individualitätsgedanke. Eine Studie zur Entstehungsgeschichte des Historismus' in: idem, *Zur Theorie und Philosophie der Geschichte* (Stuttgart 1965) 285-322, there 292, 300, 312.

32   For this translation I used and combined two versions of Marianna Wertz of the Schiller Institute. Marianna Wertz and William F. Wertz Jr. ed., *Friedrich Schiller. Poet of Freedom Volume IV* (Washington 2003) 68-69; Marianna Wertz (13 July 2016). http://www.schillerinstitute.or g/transl/trans_schil_2poems.html Sprich, wie geschieht's, daß rastlos/Erneut die Bildungen schwanken/Und die Ruhe besteht in der bewegten/Gestalt? Cited in Meinecke, *Die Idee der Staatsräson*, title page.

33   Wertz, *Friedrich Schiller*, 69. Ewig zerstört, es erzeugt sich ewig die drehende Schöpfung/Und ein stilles Gesetz lenkt der Verwandlungen Spiel. Cited in Meinecke, *Die Idee der Staatsräson*, 21.

an amphibian, which both 'lives' in nature and in an ethical world. In Meinecke's view, this eternal contrast brings forth the state: 'Kratos and Ethos together build the State and fashion history'.[34]

Yet Meinecke goes even a step further, for like Schiller he searches for a cancellation (*Aufhebung*) of the contrasts. In this regard the notion of 'entelechy' is important. The entelechy – the ability to develop the essence of things which are already apparent within these things – manifests itself, according to Meinecke, in the *Staatsräson*. In this case entelechy refers to a higher ideal or goal. In Meinecke's words:

> (...) where the crystallization into nobler forms begins, where what was formerly no more than necessary and useful now begins to be felt also as beautiful and good. Until finally the State stands out as a moral institution for the provision of the highest qualities of life – until finally the impulsive will-to-power and to-life on the part of a nation is transformed into morally conscious national mode of thought, which sees in the nation a symbol of an eternal value. Thus, by imperceptible changes the *raison d'état* of the rulers becomes ennobled and forms a connecting-link between Kratos and Ethos.[35]

Analogous to Schiller's 'dynamics of contrasts', which the play drive reconciles, ethics and power are likewise harmonized in Meinecke's view of *Staatsräson*. Similar to Schiller's play drive, Meinecke's conception of *Staatsräson* also consists of two ideas. Schiller asserts that within the play drive the perfection of mankind took place; simultaneously it caused a transition of 'nature' to reason and morality. To Meinecke, as is clear from the quote above, *Staatsräson* is the 'reconciler' of *Kratos* and *Ethos*, and in that sense it is the highest completion of the state, because the state's actions are experienced as 'good' and as beauty. Also, *Staatsräson* fulfills the transition of 'nature' or power to ethics. That is what Meinecke means when he states that *Staatsräson* 'ennobles'.[36] Put differently, the ennobled *Staatsräson* not only gives room to act ethically, but the result of these actions is, in Meinecke's view, ethically just.

---

34 Meinecke, *Machiavellism*, 4; 'Kratos und Ethos zusammen bauen den Staat und machen Geschichte'. Meinecke, *Die Idee der Staatsräson*, 5.

35 Meinecke, *Machiavellism*, 11. '(...) die Kristallisierung zu edleren Formen beginnt, wo das, was zuerst nur als notwendig und nützlich galt, auch als schön und gut empfunden zu werden beginnt, bis schließlich der Staat als sittliche Anstalt zur Förderung der höchsten Lebensgüter erscheint, bis schließlich der triebhafte Lebens- und Machtwille einer Nation übergeht in den sittlich verstandenen Nationalgedanken, der in der Nation das Symbol eines ewigen Wertes sieht. In unmerklichen Übergängen veredelt sich so die Staatsräson der Herrschenden und wird zum Bindeglied zwischen Kratos und Ethos'. Meinecke, *Die Idee der Staatsräson*, 13.

36 Schiller applied the same term with regard to the *Spieltrieb*, which forms the transition and reconciliation of the sensuous drive and the formal drive.

In the concluding chapter of *Die Idee der Staatsräson* Meinecke claims: 'The State shall become moral and strive to achieve harmony with the universal moral law, even when one knows that it can never quite reach its goal, that it is always bound to sin, because hard and natural necessity forces it to do so'.[37] It is here that the tragedy of *Staatsräson* is clearly expressed. Meinecke attempts to bridge *Ethos* and *Kratos*, but in reality, *Ethos*, as opposed to *Kratos*, will always come off worst.[38] Yet Meinecke, like Schiller, tends towards aesthetics and ethics. *Staatsräson* embodies, in Meinecke's view, the highest ideal – the reconciliation of power and ethics – and fulfills the role of *Mittelstück* in the transition of *Kratos* to *Ethos*. Meinecke's conception of *Staatsräson* is therefore different form the other, older conception which was dominant in politics ever since Machiavelli. This older conception of *Staatsräson* is, after all, focussed on the contrasts between *Staatsräson* as brutal force (as power politics) and Christian ethics, whereas Meinecke elevates *Staatsräson* to a meta-level in which *Kratos* and *Ethos* are reconciled, and in which the *Staatsräson* simultaneously can become ethical itself. It is this transition that Meinecke, by means of Schiller's aesthetics, accomplishes and which suggests a new view on the notion of *Staatsräson*.

Considering the above it seems Meinecke, in *Die Idee der Staatsräson*, constructs a synthesis which is in line with *Weltbürgertum und Nationalstaat*. Like this first major study, in *Die Idee der Staatsräson*, Meinecke (partly) assumes a moral optimism with regard to the state's actions, for he claims in *Die Idee der Staatsräson* a synthesis has been realised between power and ethics in the form of a 'non-immoral' *Staatsräson*. The question remains if Meinecke considered this synthesis or 'solution' to the problem of the contrasts and intertwinement of power and ethics as an answer to the tragedy of *Staatsräson*. After all, he accused Ranke and Hegel of veiling the brutish reality of power politics, but does his alternative not exactly tend towards the same kind of veiling of the hard political reality?

In 1925, a year prior to the publication of *Die Idee der Staatsräson*, Meinecke had written an article (which was published only in 1928) in which he claims that a Rankean 'moralisation' of the state, unattainable as it might be, is an essential ideal for the perfection of the state:

> (...) the elevation of one's own personality spiritually and morally, the attempt to spiritualize and moralize the state in which one lives, even when one knows that this cannot be attained completely, is the highest demand that can be laid upon ethical activity. For the state will always constitute the most influential and

---

37   Meinecke, *Machiavellism*, 429. 'Der Staat soll sittlich werden und nach der Harmonie mit dem allgemeinen Sittengesetz streben, auch wenn man weiß, daß er sie nie ganz erreichen kann, dass er immer wieder sündigen muß, weil die harte naturhafte Notwendigkeit ihn dazu zwingt'. Meinecke, *Die Idee der Staatsräson*, 506.

38   Ankersmit, *Macht door representatie*, 88-117.

extensive community of life, and the man who aspires to perfection can breathe freely only in a state which aspires to perfection.[39]

It is man that should be educated to become an ethically and rationally acting individuality. This education, as shown by Meinecke, should also be an aesthetic education: 'If one asks for the realms in which man can raise himself highest over nature, these are undoubtedly the realms of religion, of art, of philosophy and science'.[40] This proves Meinecke did not tend towards a veiling power of the 'elemental'. Meinecke's interpretation of 'moralization' thus differs from Ranke and Hegel, since Meinecke considers it to be a 'surpassing' or a development from nature to *Kultur* and ethics, whereas Hegel and Ranke considered 'moralization' as a veiling of evil in power and the spiritualization of power. Meinecke indeed acknowledges the 'animality' – the brute power and tries to reconcile it with ethics. Thus, in *Die Idee der Staatsräson* he both departs from and holds on to a synthesis-thinking.[41] This becomes clear at the end of *Die Idee der Staatsräson* when Meinecke first states: '(...) there are only too many things in which God and the devil are intertwined. One of the most important of these (...) is *raison d'état*'.[42] Immediately, however, he urges the statesman to 'always carry State and God together in his heart (...)'.[43] Thus, the conflict between good and evil remains and Meinecke, in following Schiller's aesthetics, tries to reconcile and transcend them by means of *Staatsräson*. But in the end, the statesman's decision seems to prevail over the imperative 'State necessity'. And that seems to point at a new entity, namely 'conscience', which, in Meinecke's thought, rests upon a panentheistic conviction.

---

39   Friedrich Meinecke, 'Values and Causalities in History' in: Fritz Stern ed., *The Varieties of History. From Voltaire to the present* (New York 1956) 268-288, there 286. 'Den Staat, in dem man lebt, zu vergeistigen und zu versittlichen, auch wenn man weiß, daß es nie ganz gelingen kann, das ist, nächst der Forderung, die eigene Persönlichkeit geistig und sittlich zu erhöhen, die höchste der Forderungen, die an ethisches Handeln gestellt werden kann – weil der Staat die wirksamste und umfassendste aller Lebensgemeinschaften nun einmal bildet und weil der nach Vollendung strebende Mensch nur in einem nach Vollendung strebenden Staate frei atmen kann'. Meinecke, 'Kausalitäten und Werte', 87.

40   Meinecke, 'Values and Causalities in History', 285. 'Fragt man, in welchen Sphären sich der Mensch am höchsten über die Natur erheben kann, so sind es zweifellos die Sphären der Religion, der Kunst und Wissenschaft'. Meinecke, 'Kausalitäten und Werte', 86.

41   What we have here is a 'unity in differences', similar to what Jensen described. However, Jensen does not discuss the factor of transcendence (*Aufhebung*). See my Introduction.

42   Meinecke, *Machiavellism*, 433. 'Nur zu viele Dinge aber gibt es, in denen Gott und Teufel zusammengewachsen sind. Zu ihnen gehört voran (...) die Staatsräson'. Meinecke, *Die Idee der Staatsräson*, 510.

43   Meinecke, *Machiavellism*, 433. 'Staat und Gott zugleich im Herzen tragen müsse'. Meinecke, *Die Idee der Staatsräson*, 510.

3.2   The statesman's conscience

In his argument Meinecke tends to avoid the possibility of a statesman acting against the 'recommendations' of *Staatsräson*. [44] The statesman who acts responsibly, conscientiously, and in accordance with the state's necessity, and yet violates 'law and ethics', could, says Meinecke, 'still feel himself morally justified at the bar of his own conscience, if in doing so he has, according to his own personal conviction, thought first of the good of the State entrusted to his care'.[45] The statesman who goes against (Christian) ethics for the sake of the state's interest, can thus square this with his own conscience. Meinecke acknowledges that these actions of the statesman are 'problematical and dualistic, because the conscious infringement of morality and law must in any circumstances (whatever motives may have prompted it) be a moral stain, a defeat of Ethos in its partnership with Kratos'.[46]

---

44   Cf. Sterling, *Ethics in a World of Power*, 268-299; Richard Sterling, 'Political necessity and moral principle in the thought of Friedrich Meinecke', *The Canadian Journal of Economics and Political Science* 26 (1960) 205-214, there 214; Carl Schmitt, 'Zu Friedrich Meineckes "Die Idee der Staatsräson"' in: idem, *Positionen und Begriffe im Kampf mit Weimar – Genf – Versailles 1923-1939* (Hamburg 1940) 45-52, there 50. As far as I know only Sterling and Schmitt point this out. Carl Schmitt argues in a review on *Die Idee der Staatsräson* that this study lacks an 'Entscheidung'. Schmitt states: 'Das Problem liegt nämlich gar nicht in der inhaltlichen Normativität eines Moral- oder Rechtsgebotes, sondern in der Frage: Wer entscheidet?'. When something conflicts with general morals, Meinecke calls this 'tragische Schuld'. According to Schmitt, 'tragisch' is no category that can serve as an answer to a conflict. Thus, Schmitt claims Meinecke's work lacks a 'last word'. Carl Schmitt, 'Zu Friedrich Meineckes "Die Idee der Staatsräson"' in: idem, *Positionen und Begriffe im Kampf mit Weimar – Genf – Versailles 1923-1939* (Hamburg 1940) 45-52, there 50-51. Contrary to Schmitt, I will argue *Die Idee der Staatsräson* indeed has, albeit implicitly, a final word. Schmitt's claim that Meinecke submitted to a general morality is not very convincing, for Meinecke could not find himself in such a strict Kantian ethics; cf footnote 31. In reaction to Schmitt's criticism Meinecke states in a letter to him: 'Wir haben nun freilich beide "einen anderen Geist" – das ist nicht zu leugnen. Aber warum soll man sich nicht gegenseitig in seiner Eigenart gelten lassen?'. Later on, in a reaction to a lecture of Schmitt, Meinecke regrets the fact that *Staatsrechtler* and historians have 'so wenig innere Fühlung' with each other. Meinecke refers to a notion (*Staatspersönlichkeit*) which Schmitt attributes to a scholar of constitutional law who described this notion in 1837. Meinecke however argues that Ranke had already discussed this notion at great length in 1836. In addition, Meinecke casually mentions *Die Idee der Staatsräson*, in which Schmitt might find an argument against his idea of a 'passive Neutralität des Staates'; Meinecke reminds him also that his study on *Staatsräson* could not meet with Schmitt's approval. Meinecke, *Neue Briefe*, 287, 319-320.

45   Meinecke, *Machiavellism*, 6. 'vor dem Forum des eigenen Gewissens (...) sittlich gerechtfertigt fühlen, wenn er nach seiner subjektiven Überzeugung dabei in erster Linie an das Wohl des ihm anvertrauten Staates gedacht hat'. Meinecke, *Die Idee der Staatsräson*, 7.

46   Meinecke, *Machiavellism*, 6. 'weil die bewußte Verletzung von Sitte und Recht unter allen Umständen, aus welchen Motiven sie auch erfolgen mag, ein sittlicher Schmutzfleck bleibt,

Despite this downside, he places trust in *Ethos* and avoids the option to discuss the justification of a statesman who disregards the state's interest and only follows his own intuition, his conscience.[47]

The statesman is faced with a conflict between an individual and a general (political and Christian) morality: does he accept the state's interest or his 'own' ethics? Clearly, the difficulty lies in the fact that both are not strictly separated but intertwined, as is the case with *Staatsräson*. Yet, maybe the contrasts in the individual statesman are more deeply felt than in *Staatsräson*: '(...) how obscure and problematic', claims Meinecke, 'the relation between them [Kratos and Ethos, RK] is at each stage of development, and especially in the conduct of the statesman'.[48] When the statesman adjusts himself to *Staatsräson*, his inner conflict will, according to Meinecke, ease off. It is all the more striking that Meinecke states in his last chapter of *Die Idee der Staatsräson*, in which he points at the individuality and personality of the statesman: 'It depends on the personal manner in which the statesman resolves the conflict in himself between moral command and State interest, whether his decision in favour of State interest will be held to be a moral act or not (...) is justified to himself'.[49] Though Meinecke presumes that the statesman will act in line with *Staatsräson*, it is clear that to him the statesman's personality and his conscience do in fact play an important role in the state's political and ethical decisions.

In the next subsection I will discuss Meinecke's view on 'personality' and 'conscience'. Both are essentially based on a panentheistic conviction, on the grounds of which Meinecke finds in the statesman's conscience a solution to the contrast between good and evil, ethics and power, *Natur* and *Kultur*. It is this panentheistic conviction which forms the fundamental principle underpinning Meinecke's 'tragic *Staatsräson*'.

Personality

One can discern roughly between two views regarding the notion of 'personality'. Both originated at the end of the eighteenth and beginning of nineteenth-century

---

eine Niederlage von Ethos in seinem Zusammenhang mit Kratos'. Meinecke, *Die Idee der Staatsräson*, 7.

47   Meinecke, *Die Idee der Staatsräson*, 504.

48   Meinecke, *Machiavellism*, 4-5. '(...) wie dunkel und problematisch ist nun das Verhältnis der beiden [Kratos and Ethos, RK] zueinander auf jeder Stufe der Entwicklung und insbesondere im Handeln des Staatsmanns'. Meinecke, *Die Idee der Staatsräson*, 5.

49   Meinecke, *Machiavellism*, 428. 'Von der persönlichen Art, wie der Staatsman dabei den Konflikt zwischen Sittengebot und Staatsinteresse in sich austrägt, hängt es ab, ob man seine Entscheidung für das Staatsinteresse eine sittliche Tat nennen darf oder nicht (...) vor sich selber rechtfertigt ist'. Meinecke, *Die Idee der Staatsräson*, 504.

Germany. I focus here mainly on securalised views of personality.[50] Kant's ideas on reason as creator of a moral personality is one of them, the other is Goethe's and Humboldt's conception of the development of the 'whole self'. The difference between them lies in the fact that Humboldt and Goethe focus on a harmonious development of the self, whereas Kant emphasizes the dominance of reason over the self.[51]

Humboldt and Goethe's conception of personality is closely related to the notion of *Bildung*. In the first chapter I have already discussed that *Bildung* refers to the aspiration of a moral and spiritual completion. This ideal has left a considerable mark on nineteenth-century thought and in particular on German intellectuals and the *Bildungsbürgertum*.[52] Goethe and Schiller's conception of *Bildung* developed in the direction of an ethical category different from 'personality'. 'Duty' is central to their conception but not in a Kantian sense, for Goethe could not agree with Kant's strict abstract maxims. Goethe was convinced personality revealed itself from *within*, that is from the inner core, our nature, instead of considering reason to be the dominant factor in the development of personality, as Kant did. In reality personality is expressed in work and activity, which correspond with the duties set by these activities, according to Goethe.[53] Moreover, Goethe thought 'personality' could only be realized by both self-restraint and all 'forces' which the individual has at its disposal.[54]

Bildung is about inner harmony, which subsequently has an effect on the 'environment', our reality.[55] This attention to *Innerlichkeit* originates from religious traditions, in particular Lutheranism and pietism (which derived from Lutheranism).[56] Through pietism other traditions were passed down, especially Neoplatonism and Panentheism.[57] Meinecke too was brought up in a pietist environment, however in his twenties he broke with pietism, for he felt no affinity with the idea of a personal God. Panentheistic conceptions had nevertheless

---

50   Harvey Goldman, *Max Weber and Thomas Mann. Calling and the Shaping of the Self* (Berkeley, Los Angeles and London 1988) 116.

51   Goldman, *Max Weber and Thomas Mann*, 117 and 120; also: John H. Zammito, *Kant, Herder, and The Birth of Anthropology* (Chicago and London 2002).

52   Dassen, *De onttovering*, 310.

53   Goldman, *Max Weber and Thomas Mann*, 128-129. Later on, under the influence of Schiller, Goethe shifted his conception of duty towards Kant.

54   Goldman, *Max Weber and Thomas Mann*, 129.

55   Ibidem, 125-127.

56   Dassen, *De onttovering*, 312.

57   Dietrich Blaufuß, 'Pietism' in: W.J. Hanegraaff ed., *Dictionary of Gnosis and Western Esotericism* (Leiden 2005) 955-960; Christine Maillard, 'Johann Wolfgang von Goethe' in: W.J. Hanegraaff ed., *Dictionary of Gnosis and Western Esotericism* (Leiden 2005) 432-434, there 433.

apparently moulded his mind.[58] He retained his (panentheistic) belief in a divine *Welthintergrund* (substratum of the world) as manifesting itself in the world of ideals (*Welt der Ideale*).[59] This belief in a divine substratum of the world and the sense for inner feelings influenced Meinecke's ethical views and his thoughts on individual conscience. In short, there is a relation between pietism, panentheism, personality and individual conscience. Moreover, in Meinecke's time there was a revival of interest in the notion of personality. In the works of Friedrich Nietzsche, Max Weber, Ernst Troeltsch and Thomas Mann this notion is frequently mentioned whether or not in relation with notions such as 'life' and 'conscience'. According to Meinecke, this revival was inspired by the First World War, for that had caused a rethinking of the 'old problem' of the relation between state morality and a private morality[60], and thus a reorientation towards personality and conscience.

The question is: what is Meinecke's understanding of the concept of 'personality'? In 1918 he published an essay entitled 'Persönlichkeit und geschichtliche Welt' in which he discusses the meaning of the historical world for the *Bildung* of the individual personality. In his view, personality is a:

> world unto itself and yet organically related to the world as a whole; characteristic and replaceable, and yet only a specific expression of general powers of life; free in itself but dependent on the whole – and above all at the same time encompassing vitality and the most real substance we have, and what no critique of knowledge can rob us of – our own self-aware.[61]

The panentheistic dimension – as already elaborated on in the discussion of Ranke in the previous chapter – in Meinecke's view of personality is clearly expressed in the above quote. For example, the phrase 'free in itself, but dependent on the whole' cannot fail to remind us of the panentheistic idea of the emanation of the individual from God. Meinecke agrees with Goethe and Humboldt in that a personality is free to shape and to educate itself.[62] Man or the individual, according to Meinecke, can by means of his personality create a 'free world' in itself, of which the highest good

---

58   Meinecke, 'Straßburg – Freiburg – Berlin. Erinnerungen', 240; Meinecke, 'Erlebtes', 44–49.

59   Hinrichs, 'Der Historismus als ein Lebensproblem', 363.

60   Meinecke, *Die Idee der Staatsräson*, 499.

61   Friedrich Meinecke, 'Persönlichkeit und geschichtliche Welt' in: idem, *Zur Theorie und Philosophie der Geschichte* (Stuttgart 1965) 30- 60, there 31. '(...) Welt für sich und doch organhaft verbunden mit der großen Welt, eigenartig und ersetzlich und doch nur eine besondere Ausprägung allgemeiner Lebensmächte; frei in sich und doch abhängig vom Ganzen – und über das alles hinaus zugleich das Allerwirklichste und Lebendigste umfassend, was wir haben und was uns keine Kritik der Erkenntnis rauben kann – das seiner selbst gewisse Ich'.

62   Meinecke, 'Persönlichkeit', 31.

is 'character' (*Eigenart*), but this characteristic individuality remains also related to the whole.[63]

To Meinecke 'character' and 'freedom' do not refer to a personality who suppresses his egoistic urges in favour of a Kantian maxim of general laws, for this is, according to Meinecke, merely a formal freedom that has no real content. After all, this 'freedom' is imposed on 'from the outside'. What matters to Meinecke is a personally 'felt' freedom. That is, Meinecke suggests supplementing the categorical imperative with an individuality of personality (*Eigenart*).[64] In the words of Meinecke:

> (...) acknowledge the organic law [and simultaneously, RK] search for a leading principle, an idea of your own life within yourself, which only counts for you and none other than you, for to know one's duty in every critical step in life one can only confer with one's own conscience.[65]

At first glance, Meinecke acknowledges that the shaping of an individual 'idea of life' seems to also point at a conflict between egoistic drives; however it is not about suppressing these urges but ordering them. In fact it is important to acknowledge these 'lower, worldly drives' in our personality: 'the innate dowry of the worldly-spiritual collective nature [*Gesamtnatur*, RK] is and will be the fruitful basis of personality, and only because of its harmony, its mutual insight of senses and soul, a characteristic, beautiful and powerful individuality will develop'.[66] Meinecke refers to the accomplishments of Rousseau, Herder, Goethe, Wilhelm von Humboldt and the romantics, who 'discovered' the importance of the individual personality.[67] Against the background of the previous sections we should also remember the influence or at least the parallel with Schiller's aesthetics, which is also clear in the aforementioned notion of the 'sinnlich-geistigen Gesamtnatur'.

The inner self of the personality (of the statesman) is thus governed by a struggle between rational and 'natural' drives. Combined with the quandaries of the *Staatsräson*, the question rises whether a statesman is really able to act in line with

---

63   Ibidem, 31.

64   Ibidem, 38.

65   Idem. '(...) erkenne das organische Gesetz, (...), suche ein leitendes Prinzip, eine Idee deines Lebens in dir selbst, die nur für dich und keinen anderen in gleicher Weise gelten kann, weil du bei jedem entscheidenden Schritte im Leben dich und dein Gewissen allein befragen mußt nach deiner Pflicht'.

66   Ibidem, 39. 'Die naturhafte Mitgift der sinnlich-geistigen Gesamtnatur ist und bleibt der nährende Boden der Persönlichkeit, und nur aus der Harmonie, aus der gegenseitigen Durchdringung und Durchleuchtung von Sinnen und Seele erwächst ihre Eigenart, ihre Schönheit und Kraft'.

67   Ibidem, 39.

his conscience. How, apart from the state's interest and Christian morality, does he know what is right? Can conscience, in a panentheistic sense, bring any relief?

Meinecke's 'X'

In *Die Idee der Staatsräson* Meinecke turns against Ranke's spiritualisation of power, and especially against Hegel's 'veiled dualism'. Meinecke warns against a one-sided dualism and its dangers of deteriorating power politics. After all, in veiling power politics, Meinecke claimed, 'the way was cleared for the establishing of a crudely naturalistic and biological ethics of force'.[68] Meinecke therefore suggests to stop idealising or romanticising power politics, and to consider it in its duality. Only in this way 'will it be possible to reach a doctrine that is not only truer, but also better and more moral in its effects', says Meinecke.[69] His concrete proposal, next, is what he calls a 'new dualism', which is 'more complete and organic' than the earlier version(s). By this 'new dualism' Meinecke means the following:

> It takes over from monistic thought the part that is undeniably correct, the inseparable causal unity between mind and nature; but it holds fast to the equally undeniable and essential difference existing between mind and nature. The unknown quantity X, which serves to explain simultaneously both this unity and this opposition, we shall leave unsolved, because it is insoluble.[70]

It remains to be seen whether this is a dualism at all, and if it is really as 'new', as Meinecke claims it to be. Moreover, there is still the question what Meinecke actually means by this 'unknown X'. The fact that it remains outside of the contradiction between Spirit and Nature reminds us of the panentheistic principle as discussed in the previous chapter.[71] Within panentheism, and especially Plotinus' dialectic, which plays a major part in this, everything is an emanation of the One,

---

68 Meinecke, *Machiavellism*, 426. '(...) Raum gegeben für die Entstehung einer grob naturalistischen und biologischen Gewaltethik'. Meinecke, *Die Idee der Staatsräson*, 502.

69 Meinecke, *Machiavellism*, 426. 'wird man zu einer nicht nur wahreren, sondern auch besseren, sittlicher wirkenden Lehre gelangen'. Meinecke, *Die Idee der Staatsräson*, 502.

70 Meinecke, *Machiavellism*, 428. 'Vom monistischen Denken her übernimmt er [the new dualism, RK] das, was unabweisbar richtig an ihm ist, die untrennbare kausale Einheit von Geist und Natur, aber hält fest an der ebenso unabweisbaren wesenhaften Verschiedenheit von Geist und Natur. Das unbekannte X, das diese Einheit und diesen Gegensatz zugleich erklärt, lassen wir ungelöst, weil es unlösbar ist'. Meinecke, *Die Idee der Staatsräson*, 504.

71 Meineckes derived the notion of the 'unknown X' from Droysen. As a student Meinecke was deeply impressed by Droysen's lectures on the 'inexplicable', the spontaneous and individual x in history. That is, an historian can explain, but in the end an inexplicable x remains which is 'freedom' in history. Meinecke, 'Willensfreiheit und Geschichtswissenschaft', 8-10; Meinecke, 'Erlebtes', 72, 80; Reinbert A. Krol, 'Friedrich Meinecke: Panentheism and the Crisis of Historicism', *Journal of the Philosophy of History* 4 (2010) 195-209, there 205.

of God.[72] These emanations are free to develop themselves and always keep a spark, or *Abglanz* of the divine source – the unknown X – within themselves. In that sense they remain part of the divine, while God or the One transcends everything.[73] In short, God creates the world thus with polarities of good and evil, light and dark, spirit and matter. But these polarities all strive again to a unification and unity in God or the divine.[74] This is how we should understand Meinecke's proposal of a 'new dualism', which, according to him, 'must be a unified mode of thought, which is actually dualistic in principle'[75] – nevertheless, behind this dualism hides a panentheistic harmony.

In Meinecke's view the same panentheistic principles are active within ethics: the statesman's conscience.[76] Everything is in harmony in God and all moral individualities, which emanated from God, carry a spark of the absolute divine morality. That is why they can indeed strive to return to God; they have an 'imprint' of the eternal absolute Good, True and Beautiful. Analogous to this the statesman can intuit (*ahnen*) absolute morality by means of his conscience, which is an *Abglanz* of the absolute.[77] In other words, the statesman's conscience is the link between 'God'

---

72   In *Erlebtes* Meinecke describes his intellectual and spiritual-metaphysical struggle around 1878. During that time he preferred a pan(en)theistic view: '(...) ich glaube mit Freuden an einen tragenden göttlichen Urgrund alles Lebens und an eine Welt der Ideale, in denen das Göttliche für uns sich offenbart'. Meinecke, 'Erlebtes', 45. Around 1914-1915 Meinecke became familiar with 'Plotinus'; in his research for *Die Entstehung des Historismus* he relates Goethe and Herder to Plotinus. I will come back to this in my discussion of this major study on historicism.

73   Cooper, *Panentheism*, 90.

74   Trunz, 'Das Vergängliche als Gleichnis', 171.

75   Meinecke, *Machiavellism*, 426. 'einheitliche, prinzipiell eben dualistische Denkweise sein muß'. Meinecke, *Die Idee der Staatsräson*, 501.

76   Hegel too discussed the role of conscience in political decisions. Meinecke leaves this undiscussed, probably because he could not agree with Hegel's speculative view of history. In *Grundlinien der Philosophie des Rechts* Hegel devotes a whole chapter on conscience, in which he, for example, states that: 'Das wahrhafte Gewissen ist die Gesinnung, das, was *an und für sich* gut ist, zu wollen; es hat daher feste Grundsätze; und zwar sind ihm diese die für sich objektiven Bestimmungen und Pflichten'. Hegel, 'Grundlinien der Philosophie des Rechts', 121 (paragraph 137). For an overview on conscience within German Idealism and Romanticism: Willy Bremi, *Was ist das Gewissen? Seine Beschreibung, seine metaphysische und religiöse Deutung, seine Geschichte* (Zürich 1934) 118-128; Stefan Hübsch, *Philosophie und Gewissen. Beiträge zur Rehabilitierung des philosophischen Gewissensbegriffs* (Göttingen 1995) 92-220; a more philosophical and contemporary interpretation: Paul Dauner, 'Das Gewissen' (Stuttgart 2008), unpublished dissertation. Also: Mooij, *Het morele domein*. An overview of panentheism during the 18th and 19th century: Cooper, *Panentheism*, 64-120.

77   In panentheism and Neoplatonism God is often represented or compared to the sun. The sun in this regard is the center of the universe and also the core, the highest 'Gleichnis' of the idea of the Good. Also within man a center or sun is present, namely: conscience. The notion of the *Abglanz* is derived from this idea: the individual conscience is an *Abglanz* of the eternal divine moral law. Hinrichs, *Ranke und die Geschichtstheologie*, 143; Trunz, 'Das Vergängliche als

and the statesman. In Meinecke's statement at the end of the previous section, in which he refers to the statesman who should carry both the state and God in his heart, the 'God' he refers to is the panentheistic spark of absolute morality. Of course, the ethical decisions and actions of the statesman should be based on a panentheistic conviction, a belief 'in not only a mere subjective faith in righteousness of the individual, but a belief in a common divine ground from which all individuals rise up', Meinecke writes.[78] This is also the crucial difference with Ranke and Hegel (as mentioned in the previous chapter), for to them harmony was a matter of observing (*Anschauung*), whereas to Meinecke it was a question of belief.

This all seems very abstract, even though *Die Idee der Staatsräson* is particularly focussed on the *practice* of the statesman's actions – prudence is the main issue at hand. Nevertheless, it is Meinecke's panentheistic conviction that enables him to combine both practice and abstraction. With regard to prudence it is the practical philosophy of Aristotle the statesman should turn to, and for the moral dimension of his actions and decisions he should turn to Plato's philosophy of ideas. Aristotle's practical wisdom refers to the most suitable means to reach the goals set by 'philosophical wisdom'.[79] This way Meinecke manages to unify the application of ethics (Aristotle) with the formulation of ethics (Plato). Put differently, the statesman can act prudently on the basis of his individual conscience, which is closely related to a higher 'authority': the divine.[80] In short, Meinecke's panentheism is the shell which encloses the 'new dualism' and his interpretation of conscience.

As already stated in the previous chapter, *Die Idee der Staatsräson* covers both political decision making and the practice and method of historicism. The link

---

Gleichnis', 176; Trunz refers to Goethe's poem 'Vermächtnis' – which Meinecke partly quotes when discussing conscience – in particular the phrase: 'Sofort nun wende dich nach innen / Das Zentrum findest du da drinnen / Woran kein Edler zweifeln mag / Wirst keine Regel da vermissen / Denn das selbständige Gewissen / ist Sonne deinem Sittentag'.

78   Meinecke, 'Ernst Troeltsch und das Problem', 376. My italics. Posthumously a collection of Troeltsch's lectures were published in which conscience is also discussed and suggested as a solution to ethical problems: Ernst Troeltsch, *Der Historismus und seine Überwindung*, 21, 40; Also: Eduard Spranger, 'Das Historismusproblem an der Universität Berlin seit 1900' in: idem, *Gesammelte Schriften V. Kulturphilosophie und Kulturkritik* (Tübingen 1969) 430-446, there 438-439.

79   Philosophical wisdom (scientific knowledge combined with intuitive reason) shows us which goals to aim for to reach perfection and happiness. Practical wisdom or prudence gives us insight into the manner in which these goals can be realised. Ankersmit, *Macht door representatie* 47.

80   The reason why Meinecke refers to Schleiermacher as the one who pointed to the 'individual ethics', is that Schleiermacher is close to Plotinus' idea of an eternal dialectic of polarities that are one within God. And Schleiermacher considered God as unchanging, eternal – this also goes for Meinecke. This contrasts with Hegel, who considered God as developing and changing in time. Cooper, *Panentheism*, 110-112.

between the two is the concern with individuality. The foregoing does not only deal with the individual but also with the interest of a reconciliation of the individual and the universal. Meinecke's universal-individualism is the hallmark of his conception of both *Staatsräson* and historicism. By means of his panentheism he unifies the universal or absolute with the individual. He combines, so to speak, the divine and reality. Conscience here is the intersection between the absolute and the individual, and, as will become clear in the next chapters, it realizes the deepest relation between *Staatsräson* and historicism, between statesman and historian, between present and past.

Conclusion

Man struggles with polarities of light and dark, of ethics and power, *Kultur* and *Natur*, *Form* and *Stoff*, the absolute and the individual. In the present and previous chapter it became clear that this conflict is deeply felt within the principle of *Staatsräson*, and in particular within the statesman. To reconcile the contrasts, Meinecke considered *Staatsräson* as an aesthetic category in which polarities like *Kratos* and *Ethos* were reconciled or even transcended. In his proposal to elevate the *Staatsräson* Meinecke unified ethics with power and considered *Staatsräson* to be ethical. Both in his view of the 'Staatsräson-as-synthesis' and his idea of *Staatsräson* as a transition between power and ethics, he gave priority to *Ethos*.

The synthesis of *Ethos* and *Kratos* is of course an ideal, for in reality they will always be in conflict. Nonetheless, Meinecke is convinced of the statesman's last word in this matter. His panentheistic philosophy enables him, by means of his conscience (which is a reflection (*Abglanz*) of the moral absolute), to intuit the moral right and to act on these grounds. Since conscience is an *Abglanz* of the absolute, the stateman as well as the historian can never fully gain insight into God or the unknown X. But by striving for the highest (ideals) they, according to Meinecke, can catch a glimpse of the divine. Contrary to Ranke – who sensed the hand of God *in* history – or Hegel – who considered history to be the self-realization of the Spirit – Meinecke, after the First World War, could no longer afford such a perspective.[81] He substituted this vision (*Anschauung*) for a panentheistic harmony, which is located within the unknown X, behind history. In a profound panentheistic consideration Meinecke, at the close of *Die Idee der Staatsräson*, puts it as follows: 'In history we do not see God, but only sense His presence in the clouds that surround Him'.[82]

---

81   See Meinecke's attitude towards the crisis of historicism, in which 'conscience', for Meinecke, plays a major part. Section 5.5.

82   Meinecke, *Machiavellism*, 443.

In his first major study Meinecke considered the state an individuality with morally right intentions. That changed in *Die Idee der Staatsräson*. In that study he concluded that good and evil were intertwined within the state. Meinecke's harmonious and morally optimistic worldview was tested severely. Nonetheless, Meinecke does not abandon his philosophy, as Walther Hofer has claimed. Like in his pre-war *Weltbürgertum und Nationalstaat*, Meinecke also strives for harmony between *Kultur* and *Macht* after the First World War. When he realized that the reality of the state was far more complex, he claimed the history of the *Staatsräson* to be tragic; by means of his panentheistic conviction he considered the 'elevation' of *Staatsräson* to be a solution.

During the 1930s Meinecke's harmonious *Weltanschauung* is put to the test even more. In line with his panentheistic conviction he tried to resist the so-called 'crisis of historicism', which had reached its high point at that time. Again, Meinecke falls back on several thinkers from the past. The outcome of this research resulted in his magnum opus, *Die Entstehung des Historismus*. Meinecke initially intended to discuss in this study both historicism and the principle of *Staatsräson*, since both are closely related. Given that the history of *Staatsräson* was now written, Meinecke could now focus on the emergence of historicism. He shifts his attention from statesmen and historians to poets and thinkers, for in their ideas the transition had taken place which made historicism 'one of the greatest intellectual revolutions that has ever taken place in Western thought'.[83] And, according to Meinecke, it was Goethe who was central in this revolution.

---

83   Friedrich Meinecke, *Historism. The Rise of a New Historical Outlook*, transl. J.E. Anderson (London 1972) liv.

4. A world view in the making

'Man kann die Frucht nicht anders als durch
den Baum kennen lernen, auf dem sie ent-
sproß'[1]

Introduction

In 1928 the 65-year-old Meinecke retired.[2] Approximately two years later he began
his research on the emergence of historicism. During this research he got more
room to bury himself in this subject, for in 1934 due to the Nazi takeover he lost
his position as chair of the *Historische Reichskommission* and in 1935 he resigned as
editor-in-chief of the *Historische Zeitschrift*, of which he had been editor since 1896.[3]
It is striking that the Nazis left Meinecke alone, for he publicly distanced himself
from National Socialism shortly before Hitler's takeover.[4]

Die Entstehung des Historismus was published in 1936 in two comprehensive volumes. This study has been criticized mainly because it fails to discuss politics

1   Johann Gottfried Herder, 'Vom Geist der Ebräischen Poesie: Eine Anleitung für die Liebhaber
    derselben und der ältesten Geschichte des menschlichen Geistes' (vol. 2) in: idem, *Sämtliche
    Werke* 33 Volumes (Berlin 1877-1913) Volume 12, 107.
2   Schulin, 'Friedrich Meinecke', 50; Ritter states that Meinecke officially retired in the winter of
    1927-1928. Meinecke however lectured until 1931, which is why sometimes it is thought that
    Meinecke did not retire until 1932. See: Meinecke, *Akademischer Lehrer und emigrierte Schüler*,
    30.
3   Generally, it is thought that Meinecke had to give up his position as chair of the *Historische
    Zeitschrift*: Schulin, 'Friedrich Meinecke', 50. Recently, however, it has become clear that he
    resigned, for he could no longer agree with the course of the journal (which the newly
    appointed editors proposed). Meinecke, *Neue Briefe und Dokumente*, 604-605.
4   See among others: Friedrich Meinecke, 'Keine Fahnenflucht vor der Schlacht!' in: idem,
    Politische Schriften und Reden (Darmstadt 1958) 477-478; Friedrich Meinecke, 'Von Schleicher zu
    Hitler' in: idem, *Politische Schriften und Reden* (Darmstadt 1958) 479-482; Schulin, 'Meineckes
    Leben und Werk', 127. Also the volume of Bock and Schöpflug, Ritter's study, and the tenth
    volume of Meinecke's collected work show Meinecke's resistence against Nazism. Iggers is
    critical of this rehabilitation of Meinecke; see my Introduction. In the last chapter I will
    come back to Meinecke's attitude towards Nazism. Meinecke, *Neue Briefe und Dokumente*, 335;
    Meinecke, *Ausgewählter Briefwechsel*, 137, 343.

or that it even abandons politics altogether.[5] After his political studies Meinecke, according to Kossmann, withdrew into 'an aesthetic delight of the diversity of the world and his admiration of the great, mainly German, writers (...)'.[6] Iggers and Brands went a step further and claimed Meinecke's work had been written in reaction to the rise of Nazism.[7] Yet, Meinecke's interest in the emergence of historicism had been aroused long before the rise of the Nazis, and he also did not abandon politics as such.[8] As far back as 1905, during a Schiller commemoration, Meinecke had felt drawn to what he later described as 'the major problem (...) that I tried to master in the autumn of my life in *Die Entstehung des Historismus*'.[9] If we look at his previously mentioned 1915 *Antrittsrede* we can see that he had already conceived his plan to research 'our modern conception of history' during the First World War.[10] Besides, as mentioned, Meinecke initially intended to discuss both politics and historicism in a work entitled 'Staatskunst und Geschichtsauffassung'. Because of the time between the publication of *Die Idee der Staatsräson* (1924) and *Die Entstehung des Historismus* (1936) it might seem at first that both studies stand alone, but to Meinecke both arose from the same thought: *raison d'état* was the forerunner to historicism. This also explains why Meinecke left out certain thinkers

---

5   Kossmann, 'Friedrich Meinecke', 222; Vasco Groeneveld, 'Friedrich Meinecke en de Goethetijd', *Theoretische Geschiedenis* 22 (1995) 254-278, there 273; Schulin, 'Friedrich Meinecke', 51; Hofer, 'Einleitung des Herausgebers', x; Jaeger and Rüsen, *Geschichte des Historismus*, 104; Schulin, 'Friedrich Meineckes Stellung', 28; Iggers, *The German Conception*, 218, 222; Felix Gilbert, 'Review Essays', *History and Theory. Studies in the Philosophy of History* 13 (1974) 59-64; Brands, *Historisme als ideologie*, 221-222; Pois, *Friedrich Meinecke and German Politics*, 132; Pachter, 'Masters of Cultural History II: Friedrich Meinecke and the Tragedy of German Liberalism', 37, 40. cf. Borowksi, 'Friedrich Meinecke's *Historism*', 166, 168, 177.

6   Kossmann, 'Friedrich', 222.

7   Iggers, *The German Conception*, 218; Brands, *Historisme als ideologie*, 221-222.

8   Blanke states that a direct influence of Nazism on the structure and 'Zielrichtung' of *Die Entstehung des Historismus* either positive or negative cannot be determined. Horst Walter Blanke, *Historiographiegeschichte als Historik* (Stuttgart-Bad Cannstatt 1991) 589. Also: Eberhard Kessel, 'Friedrich Meinecke in eigener Sicht' in: Michael Erbe ed., *Friedrich Meinecke heute. Bericht über ein Gedenk-Colloquium zu seinem 25. Todestag am 5. und 6. April 1979* (Berlin 1981) 186-195, there 190-191: 'Man hat wohl gemeint, daß sein letztes Hauptwerk über die Entstehung des Historismus im Unterschied von seinen früheren Arbeiten zur politischen Geschichte und Ideengeschichte so etwas wie eine „Flucht" in die reine Geistesgeschichte unter dem Totalitarismus des Dritten Reiches gewesen wäre. Das hat Meinecke *nicht* so gesehen, und wenn wir ihn recht verstehen, ist es alles andere als „Flucht" gewesen, wenn er zuletzt als Krönung seines Lebenswerkes diese entscheidende geistige Revolution des abendländischen Denkens untersucht und dargestellt hat. (...) Er hätte das Werk geschrieben, ja schreiben *müssen*, auch wenn 1933 das Dritte Reich nicht ausgebrochen wäre (...)'.

9   Meinecke, 'Straßburg – Freiburg – Berlin', 163-164. In this chapter and the next it will become clear to what this 'problem' refers.

10   Meinecke, 'Antrittsrede', 2.

in *Die Entstehung des Historismus*; he had already discussed them in *Die Idee der Staats-räson*. Moreover, the consequences of the First World War, the turbulent Weimar Republic[11] and the questioning of historicism – as described by Ernst Troeltsch in his comprehensive work *Der Historismus und seine Probleme* (1922) – were 'merely' additional reasons for Meinecke to focus on historicism.[12] Contrary to Kossmann, I therefore think Meinecke did not retreat into an aesthetic delight, for, as expressed in a letter written during his research on the emergence of historicism: 'At the moment I in fact lead two different lives, one of great satisfaction, because I have had the fortune of studying Goethe for over a year now (...) and the other life, the political one, attacks us all and difficulties accumulate around us'.[13] In *Die Entstehung des Historismus* Meinecke did not intend to discuss politics; this study was rather intended to be about historicism as such, with Goethe as its high point.[14]

In the introduction to *Die Entstehung des Historismus* Meinecke states that this account adopts 'an affirmative attitude'. He immediately adds that this might seem strange, 'seeing that for years the cry has been sounding that histori[ci]sm must be transcended'.[15] But according to Meinecke the rise of historicism was one of the greatest intellectual revolutions in western thought, and because such revolutions cannot be undone, an affirmative history of historicism is not problematic in Meinecke's view.

First of all, Meinecke is interested in the rise of historicism, or as the German title more clearly suggests: the origin (*Entstehung*) of historicism. So, Meinecke is not interested in all the different debates on historicism, which were prominent at the time when he published his study. Yet, *Die Entstehung des Historismus* is predomi-nantly criticised for failing to discuss the so-called 'crisis of historicism'. This 'crisis' is not completely unambiguous; the same holds for the notion of historicism itself:

---

11   Gay, *Weimar Culture*, 13-14.

12   Michael Erbe, 'Das Problem des Historismus bei Ernst Troeltsch, Otto Hintze und Friedrich Meinecke' in: Horst Renz und Friedrich Wilhelm Graf ed., *Umstrittene Moderne. Die Zukunft der Neuzeit im Urteil der Epoche Ernst Troeltsch (Troeltsch Studien)* (Gütersloh 1987) 73-91, there 88.

13   '...Ich führe jetzt eigentlich zwei Leben nebeneinander in mir, eines der stärksten Befriedigung, denn ich habe ein Jahr hindurch das Glück genossen, mich am Arbeitstische nur mit Goethe zu beschäftigen (...) Und das andere Leben, das in der Politik, stürmt gegen uns alle an und häuft dunkle Wolken um uns'. Meinecke, *Ausgewählter Briefwechsel*, 131; also: Gisela Bock, 'Friedrich Meinecke und seine Briefe', 13; Daniel Fulda, *Wissenschaft aus Kunst. Die Entstehung der modernen deutschen Geschichtsschreibung 1760-1860* (Berlin and New York 1996) 462.

14   At the same time, Meinecke also claims: 'Eben die intensivste *vita contemplativa* ist es, die oft die *vita activa* am stärksten befruchtet'. Meinecke, *Die Entstehung des Historismus*, 369. And also in his Aphorisms: 'Wer Geschichte schreibt, muß auch neue Geschichte schaffen'. Meinecke, 'Aphorismen', 239. Cf. Chapter 6.

15   Friedrich Meinecke, *Historism. The Rise of a New Historical Outlook*, transl. J.E. Anderson (London 1972) liv.

both are described, defined, defended and attacked for many different reasons and in many different ways.[16] In the closing chapters of *Die Entstehung des Historismus* Meinecke indeed discusses the crisis of historicism – and, as mentioned, the prologue is an account of why he wrote this affirmative study of historicism in the first place. Also, several of his articles written at the time of his research on historicism dig deep into the aporias of historicism. The question remains whether Meinecke's conception of (the crisis of) historicism is comparable with that of his contemporaries. I will come back to this issue in the next chapter. In the present chapter Meinecke's conception, or better his construction or even invention of the tradition of historicism is central; a tradition which incidentally is still influential to this day.[17] The general debate(s) on historicism will be discussed only if they touch upon Meinecke's views.

Another issue concerning *Die Entstehung des Historismus* is aimed at Meinecke's apparent strict distinction between historicism and the Enlightenment. Yet, Meinecke searches for the roots of historicism, which means he tries to find them *within* the Enlightenment; he connects the rise of historicism with the Enlightenment.[18] I will show that the discussion about the boundaries between the Enlightenment and historicism are often a reflection of different interpretations of historicism. What is indeed remarkable about Meinecke's argument is his discussion of great personalities – poets and thinkers – for he hardly ever deals with (academic) historians. This again shows that Meinecke first and foremost is constructing a grand narrative of the tradition of historicism: he establishes a tradition. In addition to his choice of certain thinkers it is clear that historicism for him is more

---

16   Ernst Troeltsch, *Der Historismus und seine Probleme*; Karl Heussi, *Die Krisis des Historismus* (Tübingen 1932); Herbert Schnädelbach, *Geschichtsphilosophie nach Hegel. Die Probleme des Historismus* (Freiburg and Munich 1974) 19-30; Thomas Nipperdey, 'Historismus und Historismuskritik heute' in: idem, *Gesellschaft, Kultur, Theorie. Gesammelte Aufsätze zur neueren Geschichte* (Göttingen 1976) 59-73; Otto Gerhard Oexle, '"Historismus". Überlegungen zur Geschichte des Phänomens und des Begriffs' (1986) in: idem, *Geschichtswissenschaft im Zeichen des Historismus. Studien zu Problemgeschichte der Moderne* (Göttingen 1996) 41-72; F. R. Ankersmit, *Denken over geschiedenis. Een overzicht van moderne geschiedfilosofische opvattingen* (Groningen 1986) 169-180; Blanke, *Historiographiegeschichte*, 55-63; Jaeger and Rüsen, *Geschichte des Historismus*, 4-10; Georg G. Iggers, 'Historicism: The History and Meaning of the Term', *Journal of the History of Ideas* 56 (1995) 129-152; Fulda, *Wissenschaft aus Kunst*, 267-272; Schulin, 'Neue Diskussionen über Historismus', 109-118; Dunk, *De glimlachende sfinx*, 31-55; Beiser, *The German Historicist Tradition*, 1-26; Paul, *Het moeras van de geschiedenis*, 11-33. Some of these historians I will discuss at the end of this chapter.

17   Frederick Beiser wrote his study on historicism as an addition to and correction of *Die Entstehung des Historismus*. Beiser, *The German Historicist Tradition*.

18   Beiser also claims that this is in fact the central thesis of *Die Entstehung des Historismus*. Meinecke searched for the origin of historicism in the seventeenth and eighteenth century. Beiser, *The German Historicist Tradition*, 12 note 22.

than 'merely' a research method. To him it is a world view, a *Weltanschauung*.[19] For that reason he considers historicism not only a matter of a scientific principle and its application (*Fachwissenschaft*) but also a 'guiding principle of life in the highest sense of that expression'.[20] What exactly Meinecke means by this will become clear in the present chapter.

Meinecke's distinction between Enlightenment and historicism, the function of the idea of individuality in this, and why this according to Meinecke caused a revolutionary break in western thought is discussed in the first section. With regard to Meinecke's conception of historicism and the structure of *Die Entstehung des Historismus* it is important to briefly discuss in this first section that Meinecke was indebted to Herder. Next, by means of several themes connected to the great thinkers Meinecke considers important to (the rise of) historicism, I will discuss what historicism and Enlightenment meant to Meinecke. In other words, this chapter revolves around the question of what Meinecke exactly had in mind with his study on the origin of historicism.

4.1 The individualizing view

More often than not the antecedents of revolutions confirm or even strengthen the existing state of affairs, Meinecke states at the beginning of the first part of *Die Entstehung des Historismus*.[21] He refers to historical features, which are according to him already present in the Enlightenment thinkers. Meinecke describes the Enlightenment by means of natural law and typological models.[22] He identifies the Enlightenment with notions like 'static' and 'generalising'. In addition he states that the Enlightenment always assumes an 'eternal', 'fixed', and 'general' Reason, norms and values. These views stand in contrast to what Meinecke describes as historicism, which he says is 'dynamic', aimed at the individuality and uniqueness of periods, cultures, states, institutions, personalities, acts, morality and Reason.

---

19   Schulin, 'Friedrich Meinecke', 51.

20   Friedrich Meinecke, 'Zur Entstehungsgeschichte des Historismus und des Schleiermacher-schen Individualitätsgedankens' in: idem, *Zur Theorie und Philosophie der Geschichte* (Stuttgart 1965) 341-357, there 341; 'Lebensproblem im höchsten Sinne'. Meinecke, *Historism*, lv.

21   Meinecke, *Historism*, 3. This thought is already visible in Meinecke's study on Boyen: 'Man kann (...) ganz moderne Forderungen stellen und sie doch mit Gründen verteidigen, die den alten, absterbenden Zuständen entnommen sind'. Friedrich Meinecke, *Das Leben des Generalfeldmarschalls Hermann von Boyen. Band I, 1771-1814* (Stuttgart 1896) 122. Meinecke considered Boyen as an 'in-between figure': between Enlightenment and historicism. Also: Hinrichs, 'Der Historismus als ein Lebensproblem', 373.

22   Blanke, *Historiographiegeschichte*, 582.

Meinecke considers the opposition between Enlightenment and historicism first and foremost as an opposition between an individualising and a generalising view.

Meinecke's distinction between Enlightenment and historicism often has been rejected as being too simplistic, ahistorical, and even anachronistic, because this distinction implies that the Enlightenment is merely a predecessor of historicism.[23] Contrary to Meinecke's distinction, later historians emphasized a certain continuity between the Enlightenment and historicism.[24] For example, some historians claim that the historicist's focus on the individual is already present in the Enlightenment.[25] Meinecke, who does not renounce this view, states in his prologue that indeed the individual has already been observed in earlier times, but, according to him, it was always under the spell of a generalizing judgement.[26] That is why Meinecke considers the historicist view a revolutionary change in western thought; it was a break with a generalising pattern of thought. Nonetheless, Meinecke does acknowledge the persistence of the Enlightened world view and natural law (whether or not in combination with Christian salvation), and that both also exist alongside historicism (up to his day).[27] Although Meinecke recognized the

23   Iggers, *The German Conception*, 218-219; Jaeger and Rüsen, *Geschichte des Historismus*, 11. Beiser, *The German Historicist*, 11-12 Brands, *Historisme als ideologie*, 223; Blanke, *Historiographiegeschichte*, 582. Also: Blanke, 'Aufklärungshistorie und "Historismus" im Denken Friedrich Meineckes', 142-160.

24   Peter Hanns Reill, *The German Enlightenment and the Rise of Historicism* (Berkeley 1975). Gadamer claimed historicism to be the crowning glory of the Enlightenment: Gadamer, *Wahrheit und Methode*; Hans Erich Bödeker, Georg Iggers, Jonathan Knudsen and Peter Hans Reill ed., *Aufklärung und Geschichte. Studien zur deutschen Geschichtswissenschaft im 18. Jahrhundert* (Göttingen 1986); Horst Walter Blanke and Jörn Rüsen ed., *Von der Aufklärung zum Historismus. Zum Strukturwandel des historischen Denkens* (Paderborn etc. 1984); Hayden White, 'Historism: The Rise of a New Historical Outlook by Friedrich Meinecke (review)', *Pacific Historical Review* vol43, No. 4 (November 1974) 597-598. Georg G. Iggers, 'Review: Von der Aufklärung zum Historismus. Zum Strukturwandel des historischen Denkens', *History and Theory* 26 (1987) 114-121; Ernst Schulin, 'Aufklärung und Geschichte', *Storia della storiografia* 12 (1987) 108-113; Jaeger and Rüsen, *Geschichte des Historismus*, 21-24. Rüsen recently focussed more on a rehabilitation of the Enlightenment and humanism; also see my Introduction. Fulda, *Wissenschaft aus Kunst*, 70-76; Reill considers 'vitalism' to bridge the Enlightenment and historicism: Peter Hanns Reill, 'Aufklärung und Historismus: Bruch oder Kontinuität?' in: Otto Gerhard Oexle and Jörn Rüsen eds., *Historismus in den Kulturwissenschaften. Geschichtskonzepte, historische Einschätzungen, Grundlagenprobleme* (Cologne, Weimar and Vienna 1996) 45-68; Horst Walter Blanke, 'Aufklärungshistorie und Historismus: Bruch *und* Kontinuität' in: Otto Gerhard Oexle and Jörn Rüsen eds., *Historismus in den Kulturwissenschaften. Geschichtskonzepte, historische Einschätzungen, Grundlagenprobleme* (Cologne, Weimar and Vienna 1996) 69-97; Beiser, *The German Historicist*, 10-13.

25   Beiser, *The German Historicist*, 12. n 21; and see the previous footnote.

26   Meinecke, *Die Entstehung des Historismus*, 2-3; also: Beiser, *The German Historicist*, 11-12.

27   Meinecke refers to Troeltsch, who had shown that Christianity had adopted the philosophy of natural law. Friedrich Meinecke, 'Klassizismus, Romantizismus und historisches Denken im

value and importance of Enlightenment thought, he mainly wants to emphasize that the historicist world view had caused a revolution in western thought that broke with natural law and the Enlightened generalizing view. In Meinecke's view the nature of historicism becomes clear in contrast with the Enlightenment.[28]

Meinecke's distinction between the Enlightenment and historicism also determines his selection of personalities which he discusses in *Die Entstehung des Historismus*. Those thinkers that are still bogged down in natural law and Enlightenment thought or another view that differs from historicism are left untouched. Poets and thinkers like Schleiermacher, Hölderlin, Novalis, Fichte, Hegel, who all in one way or another focussed their attention on the individual, are not discussed in *Die Entstehung des Historismus*. Of course, some of these thinkers Meinecke had already addressed in his *Weltbürgertum und Nationalstaat* and *Die Idee der Staatsräson*.[29] So, Meinecke is primarily interested in those personalities who break with Enlightenment thought or at the very least attempt the first steps towards a more

---

18. Jahrhundert' in: idem, *Zur Theorie und Philosophie der Geschichte* (Stuttgart 1965) 264-278, there, 266.

28 Meinecke's contrast between Enlightenment and historicism also explains his use of the notion of 'Pre-Romantics' in his discussion of the English 'Pre-Romantics'. Meinecke, *Die Entstehung des Historismus*, 243-260. This notion is used to indicate a transition, but such a notion can only have meaning if and when the periods in which this transition take place are well defined. In reality it seems, however, that the use of such a term invites unnuanced contrasts that attribute a static character to these periods in which the transition takes place. The notion 'Pre-Romanticism' does not function as a neutral indication for a transitional phase; it does not mean *preceeding*, but in fact it indicates an *anticipation* of Romanticism. See: W. van den Berg, 'De preromantiekconceptie in de Nederlandse literatuurgeschiedenis' in: idem, *Een bedachtzame beeldenstorm. Beschouwingen over letterkunde van de achttiende en negentiende eeuw* (Amsterdam 1999) 13-39, there 14-15. Meinecke's use of the notion of pre-Romanticism fits this description perfectly, since he also strictly distinguishes – an unnuanced contrast – between Enlightenment and Romanticism, resulting in historicism. 'Enlightenment', 'Romanticism' and the rise of historicism: they all are simultaneously a transition and all are at the same time parallel and complementary to each other. That is why Meinecke only discusses those thinkers who were interested in polarities of, for example, rationality and irrationality, which they tried to transcend. Put differently, these thinkers carry both Enlightened and historicist ideas.

29 On Novalis, Fichte and Hegel see Chapters 1 and 2. Novalis' absence is indeed striking, for his *Blüthenstaub* is considered very historicist in thought. ('Nach Innen geht der geheimnißvolle Weg'. Novalis, *Werke, Tagebücher und Briefe* 3 volumes. (Munich 1978) Volume 2, 233). On the other hand, Meinecke did discuss Novalis in *Weltbürgertum und Nationalstaat*. As mentioned at the beginning of this chapter, we should keep in mind that both studies, *Die Idee der Staatsräson* and *Die Entstehung des Historismus*, are based on the same idea, and since Meinecke had already addressed Fichte and Hegel, he did not need to discuss them again. Hölderlin's absence throughout Meinecke's oeuvre remains a mystery; see my comment on this in the Introduction. Three years after *Die Entstehung des Historismus* Meinecke devoted an essay to Schleiermacher: Friedrich Meinecke, 'Zur Entstehungsgeschichte des Historismus

dynamic, historicist (world) view. Because, Meinecke claims, 'Historicism was the first to truly apply a temporal dimension to thinking, due to its inherent ideas of individuality and notions of development; thus managing to place every single phenomenon, every personality, states, and peoples at a definitive and never recurring point in the flow of time'.[30]

In the above quote Meinecke refers to the significance of the historicist principle of individuality and development. He considers the principle of individuality the most important. In an article that is regarded as Meinecke's *Historik* (historiology) he phrased it as follows: 'This understanding [of individuality, RK] brought with it a correct estimate of evolutionary thinking, which is often incorrectly taken as the key criterion of modern histori[ci]sm, but which is much too ambiguous and versatile for this'.[31] Meinecke's assertion that the principle of development is too versatile means that it is attributed to many different views. That is, he distinguishes between a biological development and an historical development: the growth of a foetus, for example, is purely biological and not historical, according to Meinecke. An historical development takes place – and this is where Meinecke agrees with the philosopher Heinrich Rickert (1863-1936) – when a spontaneously acting human being, in accordance with (good, true, and beautiful) values, realises an individual creation: 'With this an historical individuality "evolves" and all things that evolve historically are always individualities and through evolution alone they are revealed'.[32]

During his reseach for *Die Entstehung des Historismus* Meinecke wrote several 'aphorisms' on historicism, in which he clearly states that individuality is the heart of historicism. In one of these aphorisms he claims that the proper task of historicism is to understand 'historical phenomena as arising from within developing individualities, embedded in an overall current of development'.[33] In other words, 'development' is a means to an end: insight into individuality. Thanks to historicism, according to Meinecke, such an insight is possible. He expressed this in his famous definition of historicism in *Die Entstehung des Historismus*: 'The essence of histori[ci]sm is the substitution of a process of *individualising* observations for a

---

und des Schleiermacherschen Individualitätsgedankens' in: idem, *Zur Theorie und Philosophie der Geschichte* (Stuttgart 1965) 341-357.

30   Meinecke, 'Aphorismen', 222-223. 'Erst der Historismus mit seinem Individualitäts- und Entwicklungsgedanken hat wahrhaft zeitlich denken gelehrt und innerhalb dieser Zeit jede einzelne Erscheinung, Persönlichkeiten, Staatsgebilde, Völker, an einen bestimmten, nie wiederkehrenden, eigenartigen Punkt im Strome der Zeit gestellt'.

31   Meinecke, 'Values and causalities in history', 280.

32   Idem; Charles Bambach, *Heidegger, Dilthey and the Crisis of Historicism* (Ithaca and London 1995) 103-117.

33   Meinecke, 'Aphorismen', 240.

generalising view of human forces in history'.[34] It follows that Meinecke's origin of the history of historicism starts with a search for evidence of the idea of individuality within the Enlightenment. With his selection of Enlightened thinkers, poets, politicians and historians – the so-called Forerunners – Meinecke outlines a development of a weakening of natural law and Enlightenment thought. This makes room, according to Meinecke, for the 'great German minds' Möser and Herder to clear the path for the historicist world view, which would reach its climax with Goethe.[35]

Herder's contribution

Johann Gottfried Herder (1744-1803) is considered the 'father of historicism'.[36] At least the characteristics Meinecke associates with the historicist view are originally expounded by Herder in what Meinecke calls a 'cosmos of ideas' in *Auch eine Philosophie der Geschichte zur Bildung der Menschheit* (1774). In Meinecke's words: 'Here [Herder's study, RK] all at once explodes a universe of ideas'.[37] In his study Herder labels the Enlightenment as ahistorical.[38] He points at the lack of a sympathetic ability, a disregard for the individual, the unique and singularity of human nature, morality, and Reason.[39] An understanding of individuality – Herder introduced the notion of *Einfühlen* – is central to Herder's view.[40] In addition, according to him, the identity of individualities lies in their origin and their development.[41] Later on, in his unfinished study *Idee zur Philosophie der Geschichte der Menschheit* (1784-1791),

---

34   Meinecke, *Historism*, lv.
35   Goethe will be the key figure of the next chapter.
36   Beiser, *The German Historicist*, 98; Iggers, *The German Conception*, 34-35; Rüsen, *Geschichte des Historismus*, 25; Fritz Wagner, *Geschichtswissenschaft* (Freiburg im Breisgau 1951) 128. Dobbek claims that it is too simplistic an image of Herder, besides he is considered to be the father of many disciplines: Wilhelm Dobbek, *J. G. Herders Weltbild. Versuch einer Deutung* (Cologne 1969) 8; Also: John Zammito, '(Re)Discovering Johann Gottfried Herder. A Personal Memoir', *Groniek* 171 (2006) 191-214, there 200-201.
37   Meinecke, 'Aphorismen', 246-247.
38   Herder's historicism was inspired by or even a response to the scholastic philosophy of Christian Wolff, Kants early writings, and the French Enlightenment: in particular Rousseau. See: Beiser, *The German Historicist*, 101-103; on the influence of Kant: Zammito, *Kant, Herder, and The Birth of Anthropology*. Also, Hamann is important to Herder; on this cf. section 4.5.
39   Meinecke, 'Aphorismen', 226; Meinecke, *Die Entstehung des Historismus*, 385-410; Beiser, *The German Historicist*, 1, 132; Beiser, *The Fate of Reason*, 143; Also: Hans Dietrich Irmscher, 'Aspekte der Geschichtsphilosophie Johann Gottfried Herders' in: Marion Heinz ed., *Herder und die Philosophie des deutschen Idealismus* (Amsterdam and Atlanta 1997) 5-47, there 13-19.
40   Meinecke, *Historism*, 297.
41   Beiser, *The German Historicist*, 100-102, 105, 135; Irmscher, 'Aspekte der Geschichtsphilosophie, 36-44; Meinecke, *Die Entstehung des Historismus*, 373-377.

he distanced himself from these ideas. So, it is first and foremost the early Herder that is of importance to Meinecke, for in these early writings he traced ideas that were essential to the rise of historicism.

It seems Meinecke's distinction between 'generalising and individualising' can be traced back to Herder's ideas in *Auch eine Philosophie*[42] However, Herder's impact reaches further than these ideas. The *structure* of Meinecke's research is also clearly influenced by Herder's so-called 'genetic method'. The underlying idea of this method runs as follows: Herder starts from the relativity of all periods, cultures, values and norms. That means we should understand all different periods and cultures on the basis of the standards of the respective periods, culture, et cetera, and not apply the perspectives of later generations or an absolute standpoint. Put differently, if we want to attain insight into the essence of a certain period, we should, according to Herder, research this period in its individual development.[43] That is exactly what Meinecke did in *Die Entstehung des Historismus*; he concentrates on the process of the origin of historicism to finally reach the essence of historicism: he historicizes historicism.

There is a danger contained in this idea that the essence of a period, culture, et cetera can be traced back to its history. This has to do with a so-called internal dialectic of historicism. Initially, historicists like Meinecke emphasize a continuous process of historical development: everything has its own history. This emphasis on the individuality and the uniqueness of different stages and phases is also stressed by historicists like Meinecke. However, the more we emphasize the differences between the individual and unique phases of its history, the less it will be possible to follow this same entity through the different phases of its history, for it loses its unique character.[44] Meinecke was well aware of this 'ultimate conclusion to which a combination of the principles of individuality and development could lead'.[45] I will come back to the dialectic of historicism in my discussion of Meinecke's attitude towards the 'crisis of historicism'. For the moment it is important to recognize that by means of the *structure* of *Die Entstehung des Historismus* Meinecke circumvents the problems of overlap, for he combines the focus on the individual – the origin of the idea of individuality – while also respecting the historical development of the idea

---

42   Beiser agrees that Herder's ideas on the ahistorical Enlightenment were first expressed in this study of 1774. Beiser, *The German Historicist*, 132.

43   Beiser, *The German Historicist*, 105; Beiser, *The Fate of Reason*, 143; Irmscher distiguishes three views on 'development' in Herder's thought: Irmscher, 'Aspekte der Geschichtsphilosophie', 37-40.

44   F. R. Ankersmit, *Historical Representation* (Stanford 2001) 132-133; Ankersmit, *De sublieme historische ervaring*, 143.

45   Meinecke, *Historism*, 314.

of individuality, that reveals itself through different thinkers in different epochs.[46] This influential grand narrative of the historicist tradition will be at the heart of the next sections.

4.2   Pietism and Neoplatonism

Meinecke claims that the German pietistic tradition was of great importance for the rise of historicism.[47] It is difficult to pinpoint the scope of pietism, but it inspired many different movements, such as romanticism and nationalism among others. Moreover, thanks to pietism a revitalisation of Neoplatonism took place in Germany, which was also important for the rise of historicism.[48] In this section I will clarify why pietism and Neoplatonism, in Meinecke's view, played an important part in the genesis of historicism and which personalities he argues played an essential part in this rise.

The emergence of pietism is connected with two names: Johann Arndt (1555-1621), who wrote *Vier Bücher vom wahren Christentum* (1605/1610), and Philipp Jacob Spener (1635-1705), the leader of Lutheran pietism in Germany.[49] In his works Arndt emphasized a subjective profession of faith. He proposed replacing the outward, dogmatic and strict Christianity with a profession of faith aimed at the inner self. It is difficult to give a comprehensive and precise definition of pietism, but we

---

46   In this context Beiser criticized Meinecke's historicism. Beiser claims that certain thinkers in Meinecke's study are compared to Goethe. For example, when Meinecke states that Herder's philosophy of history still bears too much 'transcendence', he [Meinecke] trivializes Herder's theological views. What is striking regarding Beisers study – he considers his research a critical addition to *Die Entstehung des Historismus* – is that he reproaches Meinecke on many occasions of ignoring certain thinkers or cut them short, thinkers that according to Beiser belong within the tradition of historicism; however, Meinecke's study is, as mentioned in my introduction, not 'the' history of the rise of historicism, but an account of Meinecke's *own* view of historicism. Beiser, *The German Historicist Tradition*, 99. Also: Herman Paul, 'Historisme op een procrustusbed', *Tijdschrift voor Geschiedenis* 126.1 (2012) 134-136. Meinecke, for that matter, acknowledges the importance of Herder's theological views, but against the background of Meinecke's conception of historicism, it is not surprising that he cannot agree with this teleological view. Meinecke, *Die Entstehung des Historismus*, 408. Also: Meinecke, 'Klassizismus, Romantizismus und historisches Denken', 266.

47   Meinecke, 'Aphorismen', 220; Meinecke, *Die Entstehung des Historismus*, 50-51, 72, 359; Also: Lewis White Beck, *Early German Philosophy. Kant and his Predecessors* (Cambridge Mass. 1969) 159; Beiser points out the role of holism and nominalism: Beiser, *The German Historicist*, 4-6.

48   Meinecke, *Die Entstehung des Historismus*, 45-46, 359; Koppel S. Pinson, *Pietism as a Factor in the Rise of German Nationalism* (New York 1934); Friedrich Meinecke, 'Pietism as a Factor in the Rise of German Nationalism', *Historische Zeitschrift* 151 (1935) 116-117; Also: Nicholas Boyle, *Goethe. The Poet and the Age. Volume I. The Poetry of Desire (1749-1790)* (Oxford 1991) 12-13.

49   Blaufuß, "Pietism", 955–960; Johannes Wallmann, *Der Pietismus* (Göttingen 2005) 21-22.

can name some key features. Important are a personal experience of faith, individualisation, withdrawal from the world and from class distinctions; further a development of an emotional life, a deepening of the awareness of our personality and also a refinement of the self by means of 'self-control' and contemplation of the self are of great importance to pietism.[50] Especially the respect for the inner self and the individual are considered a breeding ground for the emergence of the historicist attitude.[51]

In *Die Entstehung des Historismus* Meinecke discusses the pietist and, according to him, the forerunner of historicism: Gottfried Arnold (1666-1714). Arnold had connections with Spener and was familiar with his work.[52] Arnold is considered a radical pietist. The difference to 'ordinary' pietism is that the followers of radical pietism were quite indifferent to the church as an institution. These institutions, in their view, stand in the way of a direct contact with God.[53] Arnold is of interest to Meinecke, for Arnold's pietism is aimed at the individual, or in this case the individual profession of faith. Meinecke also refers to the element of asceticism as part of the individual profession of faith. What is important in this case is the individual's *isolation* from society, by which it simultaneously acquires more room and worth, for it can directly get in contact with God. It is on this point that Meinecke identifies pietism as one of the main routes which run via *Sturm und Drang* to historicism. During the *Sturm und Drang* movement the emphasis was on feelings, the heart, the irrational, the *Ahnung* (sense), and the soul instead of the 'universal reason' (*Vernunft*) of the individual.[54] Meinecke does not regard this 'irrationality' of the soul as less important than human rationality; indeed, he claims that sometimes this irrationality transcends the rational capabilities. According to Meinecke, pietism 'helped finally mobilise all the forces of the spirit'[55], which are fundamental for *Sturm und Drang* and the rise of historicism.

---

50   For many pietists it was also important to do good *within* the world, and likewise to study life and nature to find the hand of God in these affairs. August Hermann Francke is a good example of this kind of pietism: Helmut Obst, *August Hermann Francke und sein Werk* (Halle 2013).

51   Blaufuß, "Pietism", 955–960; Wallmann, *Der Pietismus*, 21-22; Boterman, *Moderne geschiedenis van Duitsland*, 30. Wilhelm Dilthey indicated in his 'Leibniz und sein Zeitalter' the significance of pietism for the humanities. Wilhelm Dilthey, 'Leibniz und sein Zeitalter' in: idem, *Gesammelte Schriften* III (Stuttgart and Göttingen 1969) 3-80, there 74-74.

52   Peter Schicketanz, *Der Pietismus von 1675 bis 1800* (Leipzig 2001) 75.

53   Wallmann, *Der Pietismus*, 136; also: Hans Schneider, '"Mit Kirchengeschichte, was hab' ich zu schaffen?" Goethes Begegnung mit Gottfried Arnolds *Kirchen- und Ketzerhistorie*' in: Hans-Georg Kemper and Hans Schneider ed., *Goethe und der Pietismus* (Tübingen 2001) 79-110, there 85.

54   Meinecke, *Die Entstehung des Historismus*, 461; Boyle, *Goethe*, 152-157.

55   Meinecke, *Historism*, 35.

Pietism caused a revival of Neoplatonism in Germany. In this context two thinkers are important in Meinecke's view: Shaftesbury (1671-1713) and Leibniz (1646-1716). Shaftesbury's thought was a mix of Neoplatonism and natural law, according to Meinecke.[56] That is, the universal equality of human nature, morality, and Reason – the main features of natural law – are compatible with Plotinus' Neoplatonism in which all individualities are 'emanations from the original divine source, reflections and broken gleams from the original divine light'.[57] Shaftesbury takes these principles as a starting point. His Neoplatonism is, according to Meinecke, encapsulated in his concept of natural law.[58] Everything in life according to this Neoplatonism relates to a 'living and moving whole, though as much as ever in a supertemporal and really unhistorical sense'.[59] This is why Meinecke characterizes Shaftesbury's combination as an idea of natural law with a Neoplatonic ring to it.[60]

The connection between the individual, the divine, and the inner self, which are features of pietism and Neoplatonism, are also present in Leibniz' philosophy. This mathematician, theologian, philosopher, jurist and historian from Hannover, who was no pietist himself, but who had correspondence with Spener, had developed a philosophical system in which he tried to combine or even harmonize the individual and the absolute or God.[61] This system that he called 'monadology', revolved around the idea of an 'inwardness'. Meinecke states that Leibniz' philosophy is also a combination of Neoplatonism and natural law.[62] The philosopher, historian, and Leibniz expert Dietrich Mahnke had already shown that Leibniz' philosophy manoeuvres between Aristotelianism and Neoplatonism and less between natural law and Neoplatonism.[63] Meinecke does not elaborate on Leibniz' Aristotelianism; he merely points at it in the context of Leibniz' monadology.

---

56   Ibidem, 7.

57   Ibidem, 7.

58   Ibidem, 15.

59   Ibidem 7. So this is stricter than panentheism. Neoplatonism cannot be equated with panentheism, because there are several Neoplatonic elements that are not panentheistic. The former mainly refers to ideas that are not by definition one with, or emanated from 'God'. On the other hand, there are many philosophical elements present in panentheism that are not considered Neoplatonic (Spinoza, Gnosticism). And, finally, panentheism is also present in religions which are far beyond Neoplatonism (Hinduism, Buddhism). See: Cooper, *Panentheism*, 19.

60   Meinecke, *Die Entstehung des Historismus*, 18, 26.

61   F. P. M. Jespers, 'Inleiding' in: Gottfried Wilhelm Leibniz, *Monadologie of De beginselen van de wijsbegeerte* translation F. P. M. Jespers (Kampen 1991) 9-53, there 26; Boyle, *Goethe*, 13, 17.

62   Jespers, 'Inleiding', 27.

63   Dietrich Mahnke, *Leibnizens Synthese von Universalmathematik und Individualmetaphysik* (Stuttgart and Bad Cannstatt 1964, first published 1925) 319-320. Meinecke does not mention Mahnke.

According to Leibniz all concrete, material, and temporal things in the world are constructed from endless indivisible eternal quantities: monads. These monads are outside of time and space; they are spiritual entities for only then can they be one and eternal.[64] With regard to humans we can think of the soul as a monad. After its divine creation the monad, similar to the Neoplatonic principle, develops in complete freedom.[65] In the case of monads it is better to speak of 'evolving' instead of 'development', for the development takes place entirely within the monad; it is an internal, closed process, an entelechy – from this we can immediately infer Leibniz' Aristotelianism. Every monad evolves in a unique way. Leibniz', however, departs from Aristotelianism on the issue of divinity; he claims that the evolution of a monad is predestined by God. With regard to the inner process he also deviates from Aristotle by claiming that this evolution is also a closed process.[66] That means there is no contact between the myriad of monads, because the monads are 'windowless'.[67] All contact with other monads is in fact part of the inner windowless evolution of the monad itself. All changes are internal and the result of a predestined divine harmony, which ensures that all different monads are arranged in such a way that it seems as though they influence each other.[68] In the words of Leibniz: 'as the same town, looked at from various sides, appears quite different and becomes as it were numerous in aspects [*perspectivement*]; even so, as a result the infinite number of simple substances, it is as if there were so many different universes, which, nevertheless are nothing but aspects [*perspectives*] of a single universe, according to the special point of view of each monad'.[69] Put differently, every single monad reflects on every moment from its own perspective all other monads and with that it mirrors in itself the universe as a whole.[70]

This, in brief, is Leibniz' monadology. The emphasis on the unique, the individual and the dynamic, the relation with God, and also the idea of the inner evolution of the monad influenced to a great extent both Herder and Goethe. The idea of multitude and unity is, for example, clearly expressed in Goethe's famous phrase in his debut *Die Leiden des jungen Werther* (1774), in which Werther cries out: 'I return to myself, and find a world!'.[71] Leibniz' monad reflecting the universe resounds in this quote. After all, it is the (concrete) individual historical appearance, the monad –

---

64   Jespers, 'Inleiding', 27, 37.

65   Ibidem, 28, 32, 47.

66   Mahnke, *Leibnizens Synthese*, 390.

67   Jespers, 'Inleiding', 28; Ankersmit, *Denken over geschiedenis*, 171.

68   Jespers, 'Inleiding', 39-40; Boyle, *Goethe*, 14; Ankersmit, *Denken over geschiedenis*, 171.

69   Gottfried Wilhelm Leibniz, *The Monadology and other philosophical writings*, transl. Robert Latta (London 1948) 248.

70   Jespers, 'Inleiding', 28.

71   Meinecke does not mention this phrase. Johann Wolfgang von Goethe, 'Die Leiden des jungen Werther' (1774) in: idem, *Goethes Werke. Hamburger Ausgabe in 14 Bänden*, ed., Erich Trunz

in this case Werther – in which the individual and the whole coalesce. Leibniz' idea of the eternal (and also changing) monad likewise influenced Herder and Goethe. I will come back to this. First, I will discuss Meinecke's thoughts on Leibniz.

According to Meinecke Leibniz' monadology holds a revolutionary germ which transcends natural law and later on fully takes root in historicism, that is: 'This was the idea of specific individuality, spontaneously operating and developing according to its own particular laws, which is yet the offshoot of a single law-abiding universe'.[72] It is this 'loosening' of natural law which makes Leibniz' monadology interesting to Meinecke. It is telling that Meinecke uses the word 'germ' in this context. He seems to refer to a Leibnizian-Aristotelian development – an entelechy – of the idea of individuality that precedes historicism. To put it in historicist terms: the 'seed' of the idea of individuality, which had already been germinated during the Enlightenment, had grown into (entelechy) the tree that fully blossomed with historicism. At first glance, this does not sound like historicism, because it now seems that the history of historicism was a necessary or predetermined development. Meinecke, however, does not refer to a Leibnizian entelechy, but an 'open' (Aristotelian) entelechy as proposed by Herder. Consistent with the metaphor of the tree, Herder expressed it as follows: 'What I am is what I have become. Like a tree, I have grown into what I am: the seed was there, but air, soil, and all other elements around me had to contribute in order to form the seed, the fruit and the tree'.[73] So the tree is already present within the seed, but its environment is subject

---

(Munich 1982) Volume VI, 7-124, there 13. 'Ich kehre in mich selbst zurück, und finde eine Welt!'.

72   Meinecke, *Historism*, 18. Ernst Cassirer – Meinecke does not mention him in this context – expressed this thought already in 1932 in *Die Philosophie der Aufklärung*. Ernst Cassirer, 'Die Philosophie der Aufklärung' in: idem, *Gesammelte Werke. Hamburger Ausgabe*, ed. Birgit Recki (Hamburg 2003) volume 15, in particular 239-244. Benedetto Croce claims Meinecke shifts his idea of Leibniz' thought too much in the direction of Meinecke's own view of historicism: '(...) denn die Leibnizsche Monade ist genau das Gegenteil der historischen Individualität, die die Individualität der Taten ist und nicht die der substanzhaften Seelen, weshalb man die Idee der Monade unaufhörlich beiseiteschieben und sogar vernichten muß, um historisch den Vorgang der Individualisierung und Entindividualisierung zu denken, des Lebens, des Todes und des neuen Lebens der der Lauf der Geschichte ist'. Benedetto Croce, *Die Geschichte als Gedanke und als Tat* (Bern 1944) 121-122. On the differences between Croce and Meinecke see section 4.6. Troeltsch described the reconciliation of temporality and infinity in Leibniz' philosophy in his study on historicism: 'Die Monade (...) bedeutet die Identität des endlichen und unendlichen Geistes bei Aufrechterhaltung der Endlichkeit und Individualität des letzteren'. Troeltsch, *Der Historismus und seine Probleme*, 675; cf. Mahnke, *Leibnizens Synthese von Universalmathematik und Individualmetaphysik*.

73   Cited in Frank Ankersmit, *Meaning, truth, and reference in historical representation* (Ithaca 2012) 1-2. The original: Johann Gottfried Herder, 'Vom Erkennen und Empfinden der menschlichen Seele' in: idem, *Sämtliche Werke* 33 volumes (Berlin 1877-1913) volume 8, 198. 'Was ich bin, bin ich geworden. Wie ein Baum bin ich gewachsen: der Keim war da; aber Luft, Erde und alle

to continuous change which affects the outcome of the actual tree; thus uncertainty remains as to what the tree eventually will look like.

Herder's open entelechy thus leaves room for contact and influence from outside. Leibniz' monadology is the complete opposite of this which leads Meinecke to dismiss Leibniz' theory as too limiting. The different eternal, indestructible, windowless monads, which develop according to God's laws, remind Meinecke too much of the strictness of natural law; Meinecke concludes that Leibniz starts from eternal, necessary, innate truths.[74] But Meinecke in particular finds it difficult that the goal of Leibniz' system in the end is to reach perfection, since Leibniz interprets the monad's 'development' as a so-called *Vervollkommnungsprozeß* (a process of perfection): the human soul (the monad) will eventually grasp absolute values, which are already present within the monad, and fathom these values to reach a state of perfection.[75] Meinecke claims that Leibniz prefers the idea of development at the expense of the idea of individuality, because within a historical development individual forces affect each other and take on new forms; moreover, every individual state is in itself of value and therefore not merely a stage in the development to higher, better or perfected states of being.[76]

Meinecke concludes that Leibniz simultaneously holds on to and breaks with natural law. Even though with this middle position Leibniz influenced the Enlightenment, German Idealism, and historicism, Meinecke still concludes that Leibniz was not able to fully think through the idea of individuality. Natural law continues to be dominant in Leibniz' philosophy, and therefore Meinecke claims the principle of individuality remains restricted.[77]

Meinecke fails to do justice to Leibniz in this regard. The importance of the individual, the idea of a unique development, the relation between the divine and the problem of unity and multitude are all themes which are central to Meinecke's conception of historicism. Moreover, Leibniz' view of the monad as 'spiritual entity' reminds us of Humboldt's and Ranke's notion of the theory of ideas (*Ideenlehre*), in which different periods in history or a variety of nations are conceived of as the product of an idea – Leibniz considered the world as a by-product of the spiritual

---

Elemente, die ich nicht um mich setzte, mußten beitragen, den Keim, die Frucht, den Baum zu bilden'. Also see: Meinecke, *Die Entstehung des Historismus*, 377.

74   Meinecke, *Die Entstehung des Historismus*, 34.

75   For example, Leibniz claims in 'Von der Glückseligkeit': *'Vollkommenheit* nenne ich alle Erhöhung des Wesens; denn wie die Krankheit gleichsam eine Erniedrigung ist und ein Abfall von der Gesundheit, also ist die Vollkommenheit etwas, so über die Gesundheit steiget (...)' (sic). Gottfried Wilhelm Leibniz, 'Von der Glückseligkeit' in: idem, *Kleine Schriften zur Metaphysik*, ed. transl. Hans Heinz Holz (Frankfurt am Main 1965) 390-401, there 393.

76   Meinecke, *Die Entstehung des Historismus*, 34-35.

77   Ibidem, 44-45.

monad.[78] Besides, Leibniz' monadology clearly reminds us of pietism and Neo-platonism, and in that sense his thought is very close to Meinecke's interest in panentheism. Even though Meinecke admits all these factors, he still maintains that Leibniz' Neoplatonism and his regard for the individual is eventually over-shadowed by what he calls Leibniz' 'intellectualism'.[79] Evidently Meinecke searches for a dimension of the idea of individuality which is more focussed on feeling and the 'irrational'. This desire initially leads Meinecke to several French thinkers.

4.3   Reason and Unreason

After his discussion of the forerunners of historicism with their pietistic-Neoplatonic ideas, Meinecke shifts his attention to the 'French'. The first major Enlightened French thinker he discusses is Voltaire (1696-1778). His conception of history was, according to Meinecke, shaped by the influence of the optimism of his age, a 'feeling of satisfaction with this present life'.[80] Voltaire considered his own time better and more perfect than the dark past. This is in fact already a version of historicist thinking, for Voltaire points at being able to understand the difference between the present and the past: the 'otherness' of the past. However, he passes judgements – the present is better than the past – and that is anathema in historicism.

Voltaire suggested historical writing ought to consist only of facts. Everything that was based on myths and legends – despite the fact that in earlier centuries people wrote down their truths in these sources – should be ignored. Voltaire sets the Enlightenment against superstition and ignorance; Reason against Unreason: 'Thus the world of history now gave the appearance of a dualistic juxtaposition and opposition of reason and unreason', Meinecke writes.[81] Voltaire's world view and conception of history is aimed at the universal and eternal reason and morality, which means, according to Meinecke, that 'the irrational depths of the soul re-mained a sealed book to him'.[82] The singular, the inexpressibility of and within the individual remained hidden, but also Voltaire's view on 'development' was, according to Meinecke, still underestimated by Voltaire, because he understood development as the perfection which would reach its goal in his, Voltaire's, era.[83] Contrary to Voltaire, Meinecke claimed: 'Genuine historical development can never

---

78   Jespers, 'Inleiding', 34. On Ranke's and Humboldt's notion of the historical idea cf. Chapter 1.

79   Meinecke, *Die Entstehung des Historismus*, 45.

80   Meinecke, *Historism*, 55.

81   Ibidem, 61; cf. Hayden White, *Metahistory. The Historical Imagination in Nineteenth-Century Europe* (Baltimore and London 1973) 50-52.

82   Meinecke, *Historism*, 71.

83   Ibidem, 72-74.

be 'finished': it flows on, and takes incalculable new forms within the limitations set by human nature'.[84]

After Voltaire it is Montesquieu (1689-1755) who is important for Meinecke's research; he could relate to his views, probably because Montesquieu's conception of history is Janus-faced. Meinecke shows that Montesquieu is a thinker with an 'ambivalent train of thought': 'He is one of those borderland figures'.[85] With 'border' Meinecke means that Montesquieu should be situated between Enlightenment and historicism. These features are all reasons why Montesquieu fits Meinecke's frame of mind; he shares his interest in apparently contradictory ideas and concepts.

According to Meinecke, Montesquieu unites two movements: the rationality of natural right and the realism of empiricism. The former is stricter than the latter, for is assumes an absolute, eternal norms and values, and a God-given reason. The second movement does not renounce this, but there is also still room for the 'shortcomings' of human nature. This second movement starts from the (Machiavellian) mode of thought in which the focus is on the reality of man, and as a result practical solutions can be offered to questions concerning life.[86] On the one hand, according to Meinecke, Montesquieu appreciated this second movement's features. On the other hand, Montesquieu, in particular with regard to law, held on to absolute and timeless norms.[87]

Nonetheless, Montesquieu, in contrast to Voltaire, searched for a point or meaning in these so-called shortcomings of man: the irrational, unfathomable, the individuality of and within man. Meinecke therefore claims that Montesquieu was open to the 'irrational mental forces at work in history'[88], but he immediately adds that Montesquieu, as a result of the rationalism of his age, could not fully, internally understand the irrational.[89] Nonetheless, Meinecke asserts that: 'this kind of accommodation of reason to the irrationalities of history did not indeed reduce the stable reason of Natural Law to a really fluid state, but it did render it considerably more malleable'.[90] With a more 'fluid' and 'malleable' reason Meinecke

---

84   Ibidem, 75.

85   Ibidem, 90, 95.

86   Meinecke, *Die Entstehung des Historismus*, 130.

87   Ibidem, 130-133.

88   Meinecke, *Historism*, 106.

89   Meinecke, *Die Entstehung des Historismus*, 155-156; Meinecke's discussion of Montesquieu resembles his treatment of Hume. For example, on Hume Meinecke states: 'Den Durchbruch zum Erlebnis der ganzen Seele, zum vollen Bewußtsein ihrer Totalität und Individualität hat er nicht getan'. Though Meinecke admits that Hume, like Montesquieu, had an eye for the irrational and unfathomable in history, he considered man to be the same in each period; and the idea of conincidence was something that was not yet rationalised; it had an unknown cause. Ibidem, 195-199.

90   Meinecke, *Historism*, 141.

means an 'individualisation' of reason: 'to recognize it [reason, RK] in the thousand different forms it assumes, and see in each something unique and irreplaceable in its sheer individuality'.[91] In spite of all this, Meinecke concludes Montesquieu remained ignorant to 'the individual mind of the historical person', because in the end his 'critical intellect' triumphed over his 'powers of sympathetic intuition'.[92]

4.4   Traditionalism and historicism

As well as Montesquieu Meinecke wants us to consider the Irish writer, politician, and philosopher Edmund Burke (1729-1797) even more as an *Übergangsmensch* (a transitional man).[93] Meinecke situates Burke between the Enlightenment and romanticism. Starting from this middle position he exercised a great deal of influence on the emergence of the historicist outlook. In that sense, Burke stood between Enlightenment and historicism.[94]

Burke criticized the Enlightenment for its lack of interest in the (political) importance of history.[95] Burke's regard of history was, however, closely related to Enlightened natural law. He was convinced that history would reveal an unchanging human nature.[96] Moreover, aspects of human nature – Burke names pride, revenge, ambition, lust and the like – were the true driving forces of history.[97] According to Meinecke, Burke's focus on these partly unconscious and irrational elements of human nature was important for the rise of the historicist view. Burke's feeling for

---

91   Idem.

92   Meinecke, *Historism*, 123-124. On the notion of intuition cf the next chapter.

93   Meinecke, *Die Entstehung des Historismus*, 279. Montesquieu influenced Burke considerably. Meinecke, *Die Entstehung des Historismus*, 269; also: C. P. Courtney, *Montesquieu and Burke* (Oxford 1963).

94   This ambivalence with regard to Burke is also expressed in characterisations of Burke as 'Enlightened' and/or a conservative thinker: Wessel Krul, 'Edmund Burke en de oorsprongen van het conservatisme', *Groniek* 164 (2004) 337-348, there 341-348; also: J. C. den Hollander, 'Conservatisme en historisme', *Bijdragen en Mededelingen betreffende de Geschiedenis der Nederlanden* 102 (1987) 380-402.

95   Ankersmit, *Macht door representatie*, 24.

96   Ankersmit, *Macht door representatie*, 38-39; Meinecke, *Die Entstehung des Historismus*, 281. John Weston claims, Burke fails to take a clear stand. According to Weston, Burke thought human nature stayed the same throughout history, which also reveals this essence of humanity. At the same time, however, Weston points at Burke's idea that: "the human mind and human affairs are susceptible of infinite modifications and of combinations wholly new and unlooked-for". This point obviously does not preclude the first one, for a modification – an adaptation – can very well correspond to the view that human nature reveals itself throughout history. John C. Weston, 'Edmund Burke's View of History', *Review of Politics* 23 (1961) 203-229, there 215-216. Cited in Weston.

97   Ankersmit, *Macht door representatie*, 38.

the irrational was, in Meinecke's view, indeed very different from Enlightenment thought.[98]

Despite his interest in the more unfathomable elements of human nature, Burke remained close to the Enlightened view of history. Meinecke admits that against the background of historicism Burke's conception of history can be considered partially Enlightened and close to natural law. Nonetheless, Meinecke emphasizes that the 'static' and 'dynamic' elements coalesce in Burke's thought.[99] In this context, Meinecke refers to Burke's views on the English constitution. In the history of this constitution Burke observed a continuous change. Many historians pointed out that these changes are in fact adaptations and not really an organic, historicist growth.[100] These changes or adaptations take place because of the changing circumstances, not because of an innate evolution already present in the constitution.[101] Burke saw the state as a static organization and not as a dynamic 'organism'.[102] Meinecke partly admits this, however, he emphasizes that Burke's notion of a 'vitality of the state' did indeed contribute to the rise of the historicist outlook.[103] For that reason Meinecke characterizes Burke's conception of history as a 'revitalised traditionalism' and states that it is in fact the highest form of traditionalism. Meinecke expressed it as follows:

> the chief evidence that it [Burke's conception of history, RK] does represent the highest stage of this traditional outlook lies in its concern with the inner psychological life of man, and not merely with the faithful preservation of the institutions, customs, rights, and so on that had been handed down the centuries. This psychological life circulates through a people like a blood-stream, builds up something interconnected and organic in the body of the State and society as a whole.[104]

---

98   Meinecke, *Die Entstehung des Historismus*, 274, 280.

99   Ibidem, 279.

100   Weston, 'Edmund Burke's View', 212, 214, 217-229; Regina Wecker, *Geschichte und Geschichtsverständnis bei Edmund Burke* (Bern 1981) 58; Ankersmit, *Macht door representatie*, 41 note 40; Joseph F. Baldacchino, 'The Value-Centered Historicism of Edmund Burke', *Modern Age* 27 (1983) 139-145, there 141.

101   Wecker, *Geschichte und Geschichtsverständnis*, 58; Ankersmit, *Macht door representatie*, 41 note 40; Weston refers to Burke's idea that the constitution develops according to a natural pattern. Wecker, however, explains that Burke's idea is merely a metaphor, since Burke considered the state as an organisation, not an organism. Wecker, *Geschichte und Geschichtsverständnis*, 58.

102   Wecker, *Geschichte und Geschichtsverständnis*, 58.

103   Meinecke, *Historism*, 232.

104   Ibidem, 228.

The above quote shows again that Meinecke was mainly interested in Burke's regard for the spiritual life and the inner self, because it corresponds with the elements of historicism that Meinecke thought important. It is this 'vitality of the state' which is, according to Meinecke, Burke's 'greatest contribution to the development of the new historical outlook'.[105]

Meinecke of course sensed that certain thinkers did not fit into his idea of historicism. In his discussion of the German Movement he hints at this when he claims that all the previously discussed thinkers 'were not able as yet to appreciate the unique value of the individual'.[106] The individual could, according to Meinecke, only be valued if a fundamental change in the spiritual-intellectual life had taken place; a change that would shift the attention to the 'irrational forces in psychological life'.[107] This is how, according to him, the dualism between reason and feeling could be cancelled, and the totality of man could be realised.[108] According to Meinecke, the first thinker who broke new ground was the historian, writer, and jurist from Osnabrück Justus Möser (1720-1794): 'With Möser begins the conscious rebellion in Germany of the 'earthly-irrational', the national character and the mundaneness against the rational, abstracting and generalizing spirit', thus Meinecke.[109]

Almost all of his life Möser lived in the bishopric of Osnabrück.[110] The work that made him famous was a history of this city, entitled: *Osnabrückische Geschichte*. The fact that Möser spent nearly all his life in Osnabrück does not mean that he was unaware or uninformed about the rest of the world – Dilthey characterised him as a 'powerful autochthonous figure'.[111] Meinecke emphasizes that Möser's thought was inspired by a multitude of different thinkers, for example Leibniz, Shaftesbury, Montesquieu and Rousseau.[112] In that sense Möser was more of a 'local cosmopolitan'. It is in fact the concept of 'locality' which is central to Möser's work, and to Meinecke this is important for the rise of historicism.

---

105 Ibidem, 232.
106 Ibidem, 250.
107 Idem.
108 Idem.
109 Meinecke, 'Aphorismen', 244. 'Mit Möser beginnt in Deutschland die bewußte Auflehnung des Erdenhaft-Irrationalen, des Volkstümlichen und Bodenständigen gegen den rationalen, abstrahierenden und generalisierenden Geist (...).
110 On one occasion Möser spent some eight months in England. This stay, and also his love for the English (empiricist) writers, would prove to be important and influential on his thoughts on history and the institutions of Osnabrück. See: William J. Bossenbrook, 'Justus Möser's Approach to History' in: James Lea Cate ed., *Medieval and Historiographical Essays in Honor of James Westfall Thompson* (Chicago 1938) 397-422, there 398-399, 401-402.
111 Meinecke, *Historism*, 251.
112 Meinecke, *Die Entstehung des Historismus*, 304; Beiser, *The German Historicist*, 81-82; Bossenbrook, 'Justus Möser's Approach', 402.

In the impressive *Osnabrückische Geschichte* Möser developed in great detail his admiration of and interest in the local. This study is not only the result of his love and respect for the local and regional, it also is a reaction to the Enlightened ideals of cosmopolitanism and the threat of political centralization.[113] With regard to the Enlightenment Möser's attitude is not unambiguous.[114] Regarding the care for and focus on the 'local' – that is the historical individuality and how people experience the world in a specific time and place – Möser mainly reacts against the Enlightened ideals, in which the cosmopolitan critical reason 'liberates' itself from time and place. At the same time, he embraces the Enlightened ideal of tolerance and separation of state and church.[115] Yet, Möser emphasizes the idea that the Enlightened ideals undermine the feeling for the 'local'.

Möser had an eye for the consequences of such an undermining of the local, namely a sense of displacement. Notions like 'displacement', but also 'descend (from)', 'be rooted in', 'attached to', 'to fit in' are important for Möser and are significant for Meinecke's interpretation of his conception of history.[116] Precisely in the notions of 'descend (from)' and 'be rooted in' the dividing line between traditionalism and historicism is clear. To be 'descended from' – in the sense of Burke's idea of the historical development – refers to what Meinecke, with regard to Burke, called traditionalism. In the context of such traditionalism Möser had a love for the 'old or ancient'. 'But traditionalism is not necessarily histori[ci]sm', says Meinecke.[117] To be rooted in or to (physically) feel at home somewhere assumes an individuality – to be attached to or to be rooted in is by definition something personal and individual and thus unique and different according to time and place. Möser calls this 'historical individuality' *Lokalvernunft* ('local reason').[118] The paradox obviously lies in the idea that this notion refers to both the individual (locality) and the universal (reason); so, this is also a critique of the Enlightenment. But above all, Möser positioned himself *between* Enlightenment and historicism. To Meinecke he

---

113   Jonathan B. Knudsen, *Justus Möser and the German Enlightenment* (Cambridge 1986) 95-109; Beiser, *The German Historicist*, 67.

114   In his discussion of Möser, Meinecke also mentions elements of his view of history, which do not fit in with Meinecke's historicist outlook. He only mentions these factors in passing; Beiser elaborates in more depth on these issues. He is convinced that Meinecke underestimated the significance of natural law in Möser's thought: Beiser, *The German Historicist*, 94-97.

115   Beiser, *The German Historicist*, 65-67, 92; also: Knudsen, *Justus Möser*, 29, 95, 110; Bossenbrook, 'Justus Möser's Approach', 402, 404.

116   Meinecke, *Die Entstehung des Historismus*, 307-309, 312, 319-322; Beiser, *The German Historicist*, 66.

117   Meinecke, *Historism*, 254.

118   Beiser, *The German Historicist*, 66; Meinecke, *Die Entstehung des Historismus*, 322; Meinecke, *Historism*, 266.

is significant because of his interest in the individual – the historical individuality – which stands in opposition to the universal ideals of the Enlightenment.

The break with the Enlightenment becomes clear according to Meinecke when we consider Möser's notion of the so-called *Totaleindrücke* (total impression), which is closely related to the notion of 'local reason'. The notion 'total impression', in which Leibniz' influence is clearly articulated, refers to the idea that the whole is constructed from individualities. The concept of 'total impression' contains an empirical intuition, that is: Möser assumes that the concrete, the individual within the world contributes to the perfection of the whole.[119] In Meinecke's words: 'Only an inner love for these things and a delight in the past could produce the patience required for such a task' and: 'it fell to Möser to combine this love for the past with the new love for the human affairs arising from the increased psychological communion among men and the new prominence given to all the irrational powers of the inner man; and it was Möser who opened men's eyes to the broad totality that embraces both men and things'.[120] It will become clear that these ideas had a great impact on both Herder and Goethe.[121]

Meinecke's enthusiasm for Möser is thus mainly inspired by Möser's criticism of the Enlightenment and natural law. Meinecke claims, 'The person who interests him is not an abstract and generalized man, who is the same at all times, whose actions can be judged according to the universal standards of reason, but the concrete, historically conditioned man with his particular joys and sorrows, who must be understood as a specific person'.[122] So, Meinecke concludes Justus Möser was an 'early pioneer of historicism'.[123]

4.5   The prelude to Meinecke's historicism

In Meinecke's view, Möser, Herder, and Goethe paved the way for the historicist outlook. They formed the so-called 'individualizing movement', which Meinecke opposes with an indirect movement, the 'idealizing movement', which also contributed to historicism. The most important thinkers of this idealizing movement are according to Meinecke: Lessing, Winckelmann, Schiller and Kant.[124] It applies

---

119   Meinecke, *Die Entstehung des Historismus*, 316; Bossenbrook, 'Justus Möser's Approach', 405.

120   Meinecke, *Historism*, 258.

121   Meinecke, *Die Entstehung des Historismus*, 332. Also cf the next chapter.

122   Meinecke, *Historism*, 262.

123   Meinecke, *Historism*, 293.

124   I agree with Ernst Schulin that Meinecke's articles on Schiller are important for understanding his attitude towards historicism: Schulin, 'Das Problem', 110, 100 note 7; Frederick Kreiling is one of the few who points out that Schiller is only discussed very briefly in *Die Entstehung des Historismus*: Kreiling, *Friedrich Meinecke*, 251. Wilhelm von Humboldt is completely left out

to all thinkers discussed in *Die Entstehung des Historismus* that they struggle with the individualizing view that comes into conflict with the generalizing view: 'all of them struggle with their own individuality, and even if they cannot leave the domain of normative reasoning, they are unintentionally witnesses to the fact that the law by which they entered was precisely their own formed individuality'.[125] According to Meinecke, Schiller is an outstanding example of this, but he immediately adds that we have to make sure we distinguish between *being* an individuality and *understanding* an individuality. In Meinecke's view Schiller probably was an example of a personality instead of someone who could really fathom individualities or personalities.[126]

According to Meinecke, Schiller was dedicated to natural law even in the early years of his career. Moreover, he was constantly in conflict with the notions of the ideal and reality. This got worse when he became acquainted with Kant's philosophy.[127] However, when he came in contact with Goethe his interest in the 'individual' got stronger, even though he always kept his distance. And because of his reservations his insight into the idea of individuality, in Meinecke's view, could not really take form in Schiller, which is why he generalized the individual as something universally human: 'he willingly transformed the unique into a general form of humanity, a peculiar recurrence', says Meinecke.[128] In short, Schiller elevated the individual to the level of 'human species' to elevate this in turn to a category of the 'ideal human'.[129] Meinecke considers this a 'generalized individuality'.[130] So, Schiller individualized the ideal and idealized the individual.[131] Meinecke cannot rank Schiller among thinkers like Herder and Goethe, because Schiller falls short and is situated too much in between the generalizing and individualizing outlook.

---

from *Die Entstehung des Historismus*. It has been pointed out that Meinecke already discussed Humboldt in *Weltbürgertum und Nationalstaat*. Cf. chapter 1.

125   Meinecke, 'Schiller und der Individualitätsgedanke', 286. 'Jeder ringt nach seiner eigenen Individualität mit ihm und ist dadurch, selbst wenn er den Bezirk der normativen Denkweise nicht verlassen kann, ein ungewollter Zeuge dafür, daß das Gesetz, wonach er angetreten, eben die geprägte Form seiner eigenen Individualität war'. A year after the publication of *Die Entstehung des Historismus* the above article on Schiller was published. Another year later an article on Schiller was published entitled 'Schillers Spaziergang', which is about Schiller's poem with the same title, in which Schiller explains his philosophy of history by means of a metaphor of climbing a mountain. Friedrich Meinecke, 'Schillers Spaziergang' in: idem, *Zur Theorie und Philosophie der Geschichte* (Stuttgart 1965) 323-340.

126   Meinecke, 'Schiller', 286.

127   Meinecke, 'Schiller', 286-288, 292. Also cf the previous chapter.

128   Ibidem, 290. 'Das Besondere aber deutete er sich mit Vorliebe um in ein allgemein Menschliches, typisch Wiederkehrendes'.

129   Ibidem, 292.

130   Ibidem, 300.

131   Ibidem, 302.

Schiller, Meinecke concludes, was a *Zwischenstufe* (an intermediate stage) of the 'purely normative natural law and the individualizing way of thinking'.[132]

Herder once more

In Herder's thought the different influences of both the generalizing and the individualizing views coalesce, according to Meinecke.[133] This is in particular the result of Kant's and Georg Hamann's (1730-1788) influence on the early Herder. These two thinkers represent two wholly contradicting world views, for Kant is considered the pre-eminent philosopher of the Enlightenment and Hamann is considered the 'father' of *Sturm und Drang*.[134] Through these thinkers Herder came in contact with pietism, Neoplatonism, and the Enlightenment.[135] According to Meinecke, it was Hamann who opened Herder's eyes to the 'individual', irrational and unfathomable in and of life, man, and history. In Meinecke's words:

---

132   Ibidem, 319. Relatively recent articles on 'Schiller as an historian' give evidence of this idea of 'Schiller as an intermediate stage' and thus with Meinecke's contradiction of Enlightenment and historicism, which Meinecke partly coined. With regard to this second point, Meinecke was not the first to make this distinction: Karl Mannheim also identified a break between the Enlightenment and historicism. Cf. the conclusion of this chapter. A few short characterizations will show how influential Meinecke's idea of 'Schiller as an intermediate stage' proved to be to the 'Schiller-historiography: Daniel Fulda considers Schiller, more or less in the same way as did Meinecke, a representative of a process of transformation from the Enlightenment to historicism. Daniel Fulda, *Wissenschaft aus Kunst*, 228-229. Ulrich Muhlack, on the contrary, views Schiller as the one who completes the 'change of paradigms'. Thomas Prüfer considers Schiller a representative of a specific 'idealistischen Geschichtsauffassung', that is Schiller connects an 'empirische Geschichtsforschung' with a 'philosophisch-poetischer Geschichtsschreibung'. Cf. Horst Walter Blanke 'Vereinnahmungen: „Schiller als Historiker" in der Historiographiegeschichte der letzten 150 Jahre' in: Michael Hofmann, Jörn Rüsen and Mirjam Springer ed., *Schiller und die Geschichte* (Munich 2006) 104-123, there 116-121; Thomas Prüfer, *Die Bildung der Geschichte. Friedrich Schiller und die Anfänge der modernen Geschichtswissenschaft* (Cologne, Weimar and Vienna 2002); Wolfgang Wittkowski ed., *Friedrich Schiller. Kunst, Humanität und Politik in der späten Aufklärung. Ein Symposium* (Tübingen 1982); Otto Dann, Norbert Oellers and Ernst Osterkamp ed., *Schiller als Historiker* (Stuttgart 1995); Gerhard Fricke, 'Schiller und die geschichtliche Welt' in: idem, *Studien und Interpretationen. Ausgewählte Schriften zur deutschen Dichtung* (Frankfurt am Main 1956) 95-118.

133   Meinecke, *Die Entstehung des Historismus*, 361-365.

134   Zammito refined our image of Kant and Herder by proving Herder was first and foremost influenced by the pre-critical Kant and Herder never really diverted from this position. Kant went into another direction and came into conflict with Herder. John H. Zammito, *Kant, Herder, and The Birth of Anthropology* (Chicago and London 2002); also: John Zammito, '(Re)Discovering Johann Gottfried Herder. A Personal Memoir', *Groniek* 171 (2006) 191-214, there 193-199; Beiser, *The Fate of Reason*, 16-18; Dobbek, *J. G. Herders Weltbild*, 21-23.

135   Meinecke, *Die Entstehung des Historismus*, 361, 363.

In Hamann, Herder was confronted with the disturbing figure of an original and independent thinker of great psychological and spiritual force. Here was a man who discerned in the sensual impulses and passions, hitherto considered sinful or dangerous by the pious, a mysterious wellspring of power; who had a new and powerful sense of the God-given unity of body and soul, and so raised the value of the irrational to a new level; and who began to survey the world of history with this new fund of vitality, though his thought was still shaped by dogmas and principles firmly based upon the Bible.[136]

At the beginning of this chapter I made clear that Herder's discovery of the individual had considerable impact on the emergence of the historicist outlook. Next to his attention to the 'individual' Herder's work is also permeated with notions like 'evolution', 'development', 'entelechy', 'organic growth'. These notions are clearly references to nature, because to Herder nature and history are one and the same, since both are subject to the same laws.[137] In my discussion of 'Meinecke's Leibniz' I have already shown that Herder did not assume any form of determinism, for his conception of 'development' should be conceived of as an open entelechy.[138] Herder's metaphor of the tree, which is already present in the seed and able to develop freely, is symbolic for his conception of development in nature and history; as mentioned, it is here that the internal dialectic of historicism presents itself. Herder's idea of development undermines the idea of individuality. After all, the development of, for instance, a culture implies a succession of stages of growth or even a higher goal, whereas the principle of individuality refers to the single, the unique, and above all to the purpose in itself of every culture, period and so on.[139] Herder solved this apparent contradiction by stating that every period, culture, people, individual was both means and goal. Meinecke explains it as follows: 'It is through development that something wonderful takes place, whereby the self-same man does not remain the same'.[140] From this follows a notion which is of importance to both Herder and Goethe, namely 'Dauer im Wechsel' (continuance

---

136   Meinecke, *Historism*, 301.

137   This unity of nature (matter) and human history (spirit) proves his spinozism. Herder's spinozism is, however, dynamic. Cf. the next chapter. Beiser, *The German Historicist*, 100; Wilhelm Dobbek, J. G. *Herders Weltbild*, 55; Hugh Barr Nisbet, 'Goethes und Herders Geschichtsdenken', *Goethe Jahrbuch* 110 (1993) 115-133, there 119; Irmscher, 'Aspekte der Geschichtsphilosophie', 28, 37-47; John H. Zammito, 'Herder, Kant , Spinoza und die Ursprünge des deutschen Idealismus' in: Marion Heinz ed., *Herder und die Philosophie des deutschen Idealismus* (Amsterdam and Atlanta 1997) 107-144, there 128; Meinecke, *Die Entstehung des Historismus*, 386-388.

138   Meinecke, *Die Entstehung des Historismus*, 373.

139   Beiser, *The German Historicist*, 136.

140   Meinecke, *Historism*, 314.

within change).[141] In the next chapter I will come back to Meinecke's discussion of Goethe and his theory of the metamorphosis in relation to this concept of change and continuity. For the moment it is enough to know that 'continuance' refers to the fixed within man, culture and periods in spite of all the changes they go through.

Meinecke considers Herder – who succeeded in uniting the idea of individuality with that of development – *not* as the high point of historicism. There are two reasons for this: first, Herder's idea of a divine plan in history, which in his first writings was blended with Neoplatonism and later on developed into an idea of progress in history. Meinecke thought this incompatible with historicism.[142] The second reason is Herder's *Humanitätsideal* (ideal of humanity), which is closely related to his idea of progress; he combined both in the idea that a historical development should no longer be connected to divine providence, because its only goal is the realization of humanity; Herder elaborated on this in detail in his famous work *Ideen zur Philosophie der Geschichte der Menschheit* (1785-1791). Meinecke considered this study 'from the point of view of the development of history a retrograde step when compared with the sketch of 1774 [*Auch eine Geschichte*, RK]'.[143] So, it is Enlightenment thought that dominates Herder's later works, and that is what Meinecke, with regard to his view of historicism, cannot agree with.

Perhaps Meinecke agrees with Goethe, who also criticized the later works of Herder.[144] For it was the early Herder (particularly in *Auch eine Philosophie*) that had an enormous impact on the young Goethe. Conversely, Meinecke also considers Goethe's influence on Herder an important factor in the origin of historicism. This influence is, as will be discussed in the next chapter, clearly articulated in Goethe's use of organic metaphors with regard to history. Moreover, Herder's ideas on nature, on Spinoza, and on the concept of *coincidentia oppositorum*[145], were largely adopted by Goethe who made them his own.[146]

Commenting on Herder's feeling for the individuality and uniqueness of, for example, cultures and peoples, Meinecke admits he was indeed able to capture the specific features of these individualities, but Herder fails to actually fathom the depth of a so-called *Einzelpersönlichkeit* (the fully formed individual personality).

---

141 Gerhart Baumann, *Goethe. Dauer im Wechsel* (München 1977), 40; Hans Börnsen, *Leibniz' Substanzbegriff und Goethes Gedanke der Metamorphose* (Stuttgart 1985) 93.

142 Meinecke, *Die Entstehung des Historismus*, 382, 385, 387, 390, 398, 408-409, 412, 417; Dobbek, *J. G. Herders Weltbild*, 15-22, 54, 143; Nisbet, 'Goethes und Herders Geschichtsdenken', 133.

143 Meinecke, *Historism*, 354.

144 Nisbet, 'Goethes und Herders Geschichtsdenken', 133.

145 The view that nature, the world, life and thus history consist of polarities which can be reconciled.

146 With regard to Goethe's notion of *Natur* and his interest in Spinoza cf. the next chapter. On the meaning of *coincidentia oppositorum* regarding Herder: Dobbek, *J. G. Herders Weltbild*, 70-78.

In that sense, Herder was more interested in individualities as collectivities: the suprapersonal. The first writer who, in Meinecke's view, really succeeds in revealing the 'individual personality' is Goethe (maybe with the help of Herder): 'The man who was once instrumental in arousing the personality of Goethe appears in the end, alongside that towering growth, as no more than a withered stump'.[147] In spite of all that, Meinecke considers Herder to be one of the most important precursors (*Wegbahner*) of the historicist outlook. But eventually it is Goethe who elevates everything to a higher level. To do justice to Meinecke's admiration of Goethe, I will elaborate in detail on this in the next chapter.

The above sections show that the rise of historicism, in Meinecke's view, was instigated by those thinkers who focussed on the irrational, the individual, and the inner self. Meinecke's view or better: his invented tradition met a lot of resistance by different colleagues over time. In the next section I will discuss the most fundamental criticism with regard to *Die Entstehung des Historismus*.

4.6   Science or world view

A lot has been written on *Die Entstehung des Historismus*. It has been, and still is, considered to be a major work and it has been labeled a masterpiece, but it also has been attacked and referred to as escapist, idealistic and outdated. These contradictions can nearly always be traced back to the definitions of historicism that the critics themselves provide.[148] The most interesting account of *Die Entstehung des Historismus* – which can also clear up the later discussions on Meinecke's historicism – was written by the Italian philosopher, historian and dissident politician Benedetto Croce (1866-1952).[149] Meinecke saw the focus of the argument between the two historicists in the difference between *Wissenschaftsprinzip* and *Lebensprinzip*, that is: historicism as a scientific principle or as a principle of life. To Croce the key

---

147   Meinecke, *Historism*, 335.

148   Cf. my Introduction.

149   Croce, *Die Geschichte als Gedanke*, 107-134. Croce had undergone a profound development before he came to his definitive view of historicism as represented in this work. It would be beyond the scope of this book to discuss his whole philosophical development; on this subject see the outstanding study: Peters, *History as Thought and Action*, passim; also: Pyper, 'Meinecke, Croce, and the Individual', 319-345, 429-442. The next best known critic of Meinecke was Carlo Antoni (1896-1959) who in 1940 published his: *Dallo storicismo alla sociologica*, and that was translated into German in 1951: *Vom Historismus zur Soziologie*. Well-known is the English translation by Hayden White: Carlo Antoni, *From History to Sociology. The Transition in German Historical Thought* (Detroit 1959). Meinecke thought Antoni had completely misread his work. This he writes in a letter to Walter Goetz: Meinecke, *Ausgewählter Briefwechsel*, 214.

question, however, revolved around what 'true' historicism was. In later accounts on Meinecke's historicism – particularly those of Rüsen and Schulin – Meinecke's distinction resounds time and again. Croce's point of view and especially the value he attached to 'pure reason' can be found in these recent accounts only in Meinecke's view on all this.

Croce versus Meinecke

From 1914 onwards, Croce and Meinecke were in contact with each other.[150] In many reviews and a lively correspondence they admired and respected each other's work. On the level of content, they, in their own views, held two very different and conflicting views of historicism.[151] Particularly after the publication of *Die Entstehung des Historismus* in 1936 the debate between the two intensified. In *La storia come pensiero e come azione* (1938) Croce explains his conception of historicism and points out where his view differs from Meinecke.[152] In 1939, in a response to Croce's work, Meinecke expressed in a letter to Croce their different views – and in particular their apparent incompatibility –, as follows: 'our contrasting spiritual directions are, from the start two incompatible types that clash with each other, and yet they are mutually productive'.[153] This 'productive' inspiration that Meinecke derived from Croce's work mainly convinced him further of his own views. The question is: what are these differences between Croce and Meinecke?[154]

According to Croce the intellectual (*geistesgeschichtliche*) change – the emergence of historicism, which was also the key issue for Meinecke in *Die Entstehung des Historismus* – was brought about by 'logical reason' (*logischen Vernunft*).[155] This *logical* genesis of historicism was in Croce's view not the problem of historical writing, as was according to him the case with Meinecke's *historical* view. The historical genesis of historicism should not, according to Croce, be considered a spiritual or intellectual development of human life as a whole, but as the specific genesis of what was later referred to as the age of historicism.[156] Croce agreed with Meinecke that his-

150 Pyper, 'Meinecke, Croce, and the Individual', 11.

151 Meinecke, *Neue Briefe*, 465, 474; Meinecke, *Ausgewählter Briefwechsel*, 181, 187-188; Croce, *Die Geschichte als Gedanke*, 109 note 3. On the exchange of ideas between Meinecke and Croce: Pyper, 'Meinecke, Croce, and the Individual'.

152 I made use of the German translation: Benedetto Croce, *Die Geschichte als Gedanke und als Tat* (Bern 1944).

153 Meinecke, *Ausgewählter Briefwechsel*, 187. Letter from 2 September 1939. 'Es sind zwei, wohl von Geburt an verschiedene Typen geistiger Richtung, die in uns beiden auf einander stoßen und sich doch gegenseitig dabei befruchten'.

154 I want to thank Rik Peters for his comments on an earlier version of this section.

155 Croce, *Die Geschichte als Gedanke*, 107-110.

156 Ibidem, 111.

toricism had had a deep impact, but he could not agree with Meinecke's explanation for this revolutionary change. In Croce's view the true revolution was not generated by the individualizing movement of Möser, Herder, Goethe and Ranke – as Meinecke thought it to be – but by the idealizing movement of Kant, Fichte, Schelling and Hegel.[157] Moreover, Croce considered the Italian philosopher Giambattista Vico (1668-1744) the true precursor of historicism.[158] According to Croce, Vico criticized the Enlightenment's abstract rationalism, but he simultaneously elevated this 'rationalism of historical development' to a higher level.[159] That is, in Croce's view 'real historicism' is about encapsulating Enlightenment's rationalism and broadening it instead of rejecting or breaking with it.[160]

A crucial difference between Meinecke and Croce concerns the preference for precursors of historicism. Croce 'followed' Vico and thereby considered an *historical individuality* as an individuality of action in reality. For that reason, he disagrees with Meinecke's choice for Leibniz as (one of the) precursors. For in Croce's view the Leibnizian monad is quite the opposite of an historical individuality, because it is focussed on 'soul-substances' that are practically unrelated to history.[161] This fundamental difference between Croce and Meinecke is also expressed in both views on the (in)accessibility of the individual.

Croce considers the individual to be rational and historical. That means Croce asserts that historicism entails reality as historical. By this he means that within every period we are conscious of the fact that life and reality are nothing but history.[162] Historical consciousness and reality are thus inseparable; to Croce there is no reality outside of historical thinking. For that reason, Croce considered historicism a logical principle, a category of the mind. Historical thinking is based on the 'true ideas', the pure concepts and categories which are a prerequisite for creating and valuing history.[163] These concepts realize within the individual. That is why Croce considers the individual a rational and historical individuality. This

---

157   Ibidem, 116, 128; Meinecke, 'Zur Entstehungsgeschichte', 342; Pyper, 'Meinecke, Croce, and the Individual', 438-439; Schulin, 'Das Problem', 110; Hofer, *Geschichtsschreibung*, 393-394. From Meinecke's point of view Croce's statement could be countered by the fact that both Kant and Fichte are not involved in historicism in Croce's sense.

158   Croce, *Die Geschichte als Gedanke*, 117, 120. Meinecke also devotes some attention to Vico in his *Vorstufen*-section of *Die Entstehung des Historismus*. He, however, did not consider him to be *the* precursor of historicism, but merely one of many. Moreover, in Meinecke's view Vico was not one of the thinkers who forced a break with regard to the individualising view; that, to Meinecke, are Möser, Herder and Goethe.

159   Croce, *Die Geschichte als Gedanke*, 117.

160   Ibidem, 120.

161   Ibidem, 121-122.

162   Ibidem, 107.

163   Ibidem, 108.

rationalized individuality – in which the pure concepts and categories are realized historically (time-bound) – is the key to Croce's conception of historicism.[164] By means of Vico's doctrine of *verum et factum convertuntur* – knowable is that which was created by the knower[165] – Croce is able to claim that the categories and concepts, which are historically realized, are understandable in every time. Unlike Croce, Meinecke emphasizes the irrational elements in human life. According to him, pure reason cannot fully understand the individual; the soul and feelings are unfathomable, and Meinecke claims they even transcend the rational. Croce rejects this 'individualized ahistorical view'.[166]

History – as a logical principle – reveals itself in Croce's view in the mind of the (contemporary) individual. Since the individual is bound to the present, history should be considered as an infinite sequence, a development of 'present-moments' in, and of the mind.[167] Croce cannot agree with Meinecke's idea of a *second* 'world' – a world of ideas and values – next to the real historical world. In Meinecke's view the individualities strive for the higher, universal, whereas Croce claims the universal is active within the individual – this principle is identified as 'immanentism'.[168] Croce rejects the idea that the universal 'hovers above' reality within a realm of ideas, which can only be sensed or divined (*geahnt*) by the individual, as it is (partly) the case in Meinecke's view.[169]

In his criticism of Meinecke, Croce also focusses on the relationship between historicism and politics. He claims that 'the completion of historicism, being the legacy of the Enlightenment, was the active practical life; the new direction of freedom, no longer in an abstract and atomistic sense as in the Enlightenment, but concrete and connected with the social historical life'.[170] According to Croce, 'thought' and 'action' – the ideal world and the concrete world – are separated from each other in Meinecke's conception of the individual. For that reason he claims

---

164   Ibidem, 109-110.
165   R. G. Collingwood, *The Idea of History* (Oxford 1972; first 1945) 64. This principle counts for man, but also for God, for He is the only one, for example, who knows nature, which he created.
166   Croce, *Die Geschichte als Gedanke*, 110.
167   Peters, *History as Thought and Action*, 40-69, 78; David D. Roberts, *Benedetto Croce and the Uses of Historicism* (Berkeley, Los Angeles and London 1987) 335; Pietro Rossi, '"Historismus" und "Storicismo": zwei Denktraditionen' in: Arnold Esch and Jens Petersen, ed., *Geschichte und Geschichtswissenschaft in der Kultur Italiens und Deutschlands. Wissenschaftliches Kolloquium zum hundertjährigen Bestehen des Deutschen Historischen Instituts in Rom (24. – 25. Mai 1988)* (Tübingen 1989) 39-69, there 44.
168   Rik Peters, 'Italian Legacies', *History and Theory* 49 (2010) 115-129, there 119; Croce, *Die Geschichte als Gedanke*, 125.
169   Croce, *Die Geschichte als Gedanke*, 107; Pyper, 'Meinecke, Croce, and the Individual', 440-441. I state here on purpose that it is *partly* the case, for it is more complex that that. Cf. the next chapter.
170   Croce, *Die Geschichte als Gedanke*, 129.

his own conception of historicism – on the basis of his idea that the universal is immanent in the individual and therefore realized in time – should be a guide for true liberal politics.[171] Croce asserts that in Meinecke's historicism 'individuality' is attached to the state, not the actual individual.[172] What follows from this view is that precisely (individual) freedom is limited. Croce further claims that in Germany freedom is elevated to an ideal which is more and more connected with Germany as such; Germany as a *Menschheitsnation* (nation of all mankind).[173] This is, in Croce's view, a threat to freedom – in this context he refers to the dangers of the present (Nazism) in which the individual freedom is also restricted by the state.[174] Croce considers his own view of historicism, which is directed at freedom and the unity of theory and practice, to be more important, and more suitable for Europe's political and moral future than Meinecke's historicism.[175]

Croce's criticism encouraged Meinecke to write an 'Antikritik'.[176] In this critique he does not discuss the relationship between historicism and politics – perhaps Meinecke considered *Die Idee der Staatsräson* and his articles on the crisis of historicism sufficient with regard to political and ethical questions. He does touch upon the difference between his view of historicism and that of Croce.

According to Meinecke, the difference between the two views concerns the fact that to Meinecke it is 'not merely a scientific principle and its use, but a life principle, a new view on human life all together, from which this scientific principle developed'.[177] Meinecke considers historicism as a *Lebensproblem* – a guiding principle of life – whereas to Croce, in Meinecke's view, it is a logical scientific principle or an epistemological principle: the problem of the philosophy of mind.[178] He considers Croce's explanation of historicism – in which the historical individuality is an individuality of action and in which rationalism is *the* essential element in the historicist revolution – too narrow a view. In his own view of historicism, Meinecke emphasizes the spiritual, unfathomable 'irrationality' of the individual.[179] In criticizing Croce he also states that the idealizing movement of Kant, Fichte, and

---

171   Ibidem, 132.
172   To Croce the state is an abstraction; only the individuals and their mutual relations are real.
173   For this notion cf section 1.2.
174   Croce, *Die Geschichte als Gedanke*, 131.
175   Ibidem, 133.
176   It is found in the first pages of: Friedrich Meinecke, 'Zur Entstehungsgeschichte des Historismus und des Schleiermacherschen Individualitätsgedankens' in: idem, *Zur Theorie und Philosophie der Geschichte* (Stuttgart 1965) 341-357, there 341-344.
177   Ibidem, 341.
178   Ibidem, 341-342; Hofer, *Geschichtsschreibung*, 394-395.
179   Meinecke, 'Zur Entstehungsgeschichte', 342; Meinecke, *Die Entstehung des Historismus*, 6, 10; Hofer, *Geschichtsschreibung*, 395; Pyper, 'Meinecke, Croce, and the Individual', 448.

Hegel trivialize the idea of individuality. Moreover, Meinecke asserts that he himself situates the origin of historicism in the eighteenth century, which grounds it more profoundly than Croce – apparently Meinecke overlooked the fact that Croce considered Vico the most important precursor of 'true historicism'.[180] With these arguments, and above all the distinction between a 'life principle' and a 'scientific principle' Meinecke avoids Croce's criticism, for the fundamental discussion was, in Croce's view, not this distinction, but the question what 'true historicism' entails.

Next to his distinction Meinecke shifts the discussion in yet another way. In his criticism of Croce he relates their discussion to the one between Ranke and Hegel of a century before. The problem with this comparison is that in his criticism of Hegel Meinecke adopts *Ranke's* interpretation of Hegel. Ranke claimed Hegel was only interested in the universal; the individual only served as a stage or link in the process of obtaining knowledge of the universal. The historian on the other hand, claimed Ranke, was mainly interested in the individual 'for its own sake'.[181] Now, when Meinecke relates his discussion with Croce to that of Ranke and Hegel, he implies that Croce, like Hegel, also considered the individual a means to an end: the realization of world reason (*Weltvernunft*).[182] Meinecke here fails to acknowledge that Croce always kept his distance to Hegel. In fact, Croce wrote a piece on the shortcomings of Hegel's philosophy.[183] In particular, Croce could not agree with Hegel's idea of a goal in history.[184] It is even more important to acknowledge that it was not Hegel but Vico (who did not assume a goal in history) who formed the basis of Croce's historicism.[185] So, when Meinecke claims the contrast between Hegel and Ranke continues with Croce and him, it is all based on a Rankean-coloured image of Hegel and on an inaccurate comparison of Croce and Hegel.

In reality, as suggested by the philosopher Frederick Beiser, Hegel and Ranke did not differ that much on certain issues.[186] Hegel stated, as Beiser shows, that the universal could also be reached through the individual. The plan or goal in history could be approached, according to Hegel, by means of a detailed study of individual

---

180 Meinecke, 'Zur Entstehungsgeschichte', 342.
181 Ranke, *Über die Epochen*, 7. Cf. Ranke's well-known phrase: 'Jede Epoche ist unmittelbar zu Gott, und ihr Wert beruht gar nicht auf dem, was aus ihr hervorgeht, sondern in ihrer Existenz selbst'. Beiser, *The German Historicist*, 260.
182 Meinecke, 'Zur Entstehungsgeschichte', 343
183 Peters, *History as Thought and Action*, 46-50.
184 Rossi, '"Historismus" und "Storicismo": zwei Denktraditionen', 49: 'Croce wies effektiv den Anspruch zurück, den Geschichtsprozeß auf eine logische Entwicklung zurückzuführen, deren Epochen a priori bestimmt werden könnten, und daher lehnte er auch die Hegelsche Geschichtsphilosophie ab. Er übernahm aber gleichzeitig aus dieser die Auffassung des Geschichtsprozesses als Verwirklichung eines unendlichen Geistes'.
185 Croce, *Die Geschichte als Gedanke*, 125-126.
186 Beiser, *The German Historicist*, 261-266.

epochs. In other words, with the exception of his idea of a goal in history, Hegel's idea of individuality is not that different from Ranke's. With regard to the discussion between Meinecke and Croce it is important to note that Hegel tried to gain insight into the spirit by concentrating on different individual epochs, in which the spirit realizes itself *historically*. Ranke on the other hand was more focussed on intuitively sensing (*ahnen*) the unfathomable in and of the individual.[187] In that sense the discussion between Ranke and Hegel was indeed somewhat comparable to that between Meinecke and Croce.

The fundamental distinction between Meinecke and Croce lies, however, in the essentially different movements in which they position themselves and the different metaphysical views which they hold. Meinecke saw his ideas rooted in Leibniz, Spinoza, Herder, Goethe, Humboldt and Ranke. Meinecke thus assumes (unfathomable) monads, substances and individualities. Moreover, he also accepts a second 'world' of ideas and values 'above' the real, historical world. Croce, who was wrongly equated with the Kant-Fichte-Hegel-movement, particularly agreed with Vico and his idea of timeless pure concepts and categories which realise themselves historically through individuals. Croce started thus from a concrete historical rationalism, while Meinecke preferred a historically 'irrational' theory of understanding (*Verstehen*). In sum, both had a fundamentally incompatible world experience (*Welterfahrung*).[188] In later debates on Meinecke and Croce, the latter's logical distinction is left out and instead Meinecke's (simplified) distinction between 'Scientific principle' and 'Life principle' is referred to by the commentators.

Reprise

One of the most interesting or rather the most conspicuous account of Meinecke's historicism in recent historiography is that of the late Otto Gerhard Oexle (1939-2016). He claims Meinecke's historicism heralded a wrong development (*Fehlentwicklung*).[189] Oexle comes to this conclusion on the basis of his distinction between *Historismus I*, which indicates an 'intellectual-cultural movement' and *Historismus II*, which refers to historicism as an academic discipline. 'Historicism I' refers to the idea of the historicization of reality.[190] This means that everything in reality is temporal and relative; an insight which was already acquired in the eighteenth

---

187   Ibidem, 264-266.

188   Hofer, *Geschichtsschreibung*, 398; Hughes claims both worlds were not that different and both thinkers were essentially rooted in Dilthey: H. Stuart Hughes, *Consciousness and Society. The reorientation of European social thought 1890-1930* (Brighton 1979) 247.

189   Cf. my Introduction. Oexle, "'Historismus'", 66-68. The notion *Fehlentwicklung* was coined by Nipperdey, 'Historismus und Historismuskritik heute', 65. Also: Paul, *Het moeras van de geschiedenis*, 18-21.

190   Paul, *Het moeras van de geschiedenis*, 21.

century. Oexle's 'Historicism II' refers to a historicism based on the ideas of Humboldt and Ranke.[191] Where Oexle objects to Meinecke's historicism, particularly as laid down in *Die Entstehung des Historismus*, is that Meinecke equates both forms of historicism, and thus he reinterprets (*Umdeutung*) historicism. The outcome, in Oexle's view, was that (German) historians began to consider historicism as the intellectual-cultural movement which reached its high point in the ideas of Humboldt and Ranke.[192] As a result of Meinecke's reinterpretation, Oexle says, the 'real' problem of historicism was ignored, namely: (ethical) relativism, which is the main characteristic of the first form of Oexle's historicism. For that reason, Oexle claims Meinecke's historicism lacks the disturbing or alarming (*beunruhigende*) element of historicism of the beginning of the twentieth century.[193] Meinecke's historicism is therefore limited to an early phase, which is why Oexle sees Meinecke's historicism as closed: 'It [Meinecke's historicism, RK] was out of date in his time'.[194]

Meinecke's position towards ethical relativism will be discussed in the next chapter. For now, I will limit myself to Oexle's claim that Meinecke had caused a 'reinterpretation' of historicism. What is striking in Oexle's argument is that he distinguishes between two forms of historicism, and subsequently reproaches Meinecke of making this distinction. With regard to *Die Entstehung des Historismus* this is in fact remarkable, for the second, academic form of Oexle's interpretation of historicism is barely discussed by Meinecke. Moreover, as was clear from Meinecke's discussion with Croce, to Meinecke historicism was more than 'merely' a scientific principle. 'The science of history' merely attributes according to Meinecke 'what had already been present and active in the spirit of the modern man, as principle and direction, a means of knowledge and disposition, and had its effects beyond the academic science'.[195] To Meinecke it was more than 'a scientific principle and its application; it was a guiding principle of life as a whole, from which that scientific principle arose in the first place'.[196] Meinecke expressed this in a similar

---

191 Otto Gerhard Oexle, 'Die Geschichtswissenschaft im Zeichen des Historismus. Bemerkungen zum Standort der Geschichtsforschung' in: idem, *Geschichtswissenschaft im Zeichen des Historismus: Studien zu problemgeschichten der Moderne* (Göttingen 1996) 17-40, there 30-31; also: Fulda, *Wissenschaft aus Kunst*, 267-268.

192 Oexle, 'Die Geschichtswissenschaft', 30-31.

193 Oexle, '"Historismus"', 65.

194 Idem. 'Er [Meineckes Historismus, RK] war in seiner Gegenwart unaktuell'.

195 Meinecke, 'Zur Entstehungsgeschichte', 341. '(...) die Fachwissenschaft der Geschichte' (...) 'was schon vorher im Seelenleben des modernen Menschen als Prinzip und Richtung, Erkenntnismittel und Gesinnung gewirkt hat und weiter wirkt, über den Kreis der Wissenschaft weit hinaus'.

196 Idem. 'ein Wissenschaftsprinzip und dessen Anwendung, sondern um ein Lebensprinzip (...), eine neue Schau menschlichen Lebens überhaupt, aus der jenes Wissenschaftsprinzip erst entsprang'. Translation J.E Anderson in Meinecke, *Historism*, xvii.

way in *Die Entstehung des Historismus*: 'histori[ci]sm stands for more than simply the application of scientific methods of thought'.[197] So it seems Oexle overinterpreted (*Umdeuten*) Meinecke.

Unlike Oexle, Jörn Rüsen states that Meinecke's historicism is not 'closed' or outdated.[198] He claims Meinecke describes historicism in *Die Entstehung des Historismus* 'not so much as a tradition of scientific research and historical writing, but as a cultural phenomenon that made this research possible; thus, a cultural system of interpretation of the experience of temporal change of man and his world, which made this type of historical science possible in the first place'.[199] Rüsen considers Meinecke's interpretation of historicism a 're-theorisation', for Meinecke reminds the professional science (*Fachwissenschaft*) of its theoretical foundations.[200] Thus Rüsen tries to incorporate Meinecke's view in the professional scientific discussion. He objects to the lack of method in Meinecke's interpretation of historicism and considers this a serious weakness. To Meinecke, who considered historicism a principle of life and a world view, it is not unusual to omit an explicit scientific method. Rüsen is of course aware of this, as is clear from his description of Meinecke's position: 'A pre-rational inwardness in dealing with the human past, which reaches into the realms of religious grasping, is regarded as a necessary precondition for objective and practical effective historical knowledge'.[201] Meinecke's emphasis on the inner, the 'irrational' – Rüsen calls this 'anti-intellectualism' of empathising (*Einfühlen*)[202] – leads, according to Rüsen, to a highly subjective view of historicism, which cannot be scientifically verified and is therefore impractical.[203] Meinecke's point however was that the spiritual-irrational within man as well as events are anything but logical and cannot be explained in that way, which means that *Einfühlen* (sensing) is the only way to gain insight into the individual. Rüsen is convinced that such an historical view in the end will lead to a deep relativism.

Rüsen expressed a similar criticism of Meinecke's historicism with regard to politics. Rüsen thought that Meinecke ranked the so-called *Vernunftanspruch* (claim of reason) lower than an irrational decisionism in political actions.[204] In my discussion of Meinecke's *Staatsräson* study, however, it is clear that Meinecke avoids such decisionism. Moreover, Meinecke did not want to replace the rational with the irrational; he wanted the irrational next to the rational. The 'substitution' of the

---

197   Meinecke, *Historism*, lv.
198   This is Rüsen's view on Meinecke from the 1980's, beginning of the 1990's. Cf. my Introduction.
199   Rüsen, 'Friedrich Meineckes', 84.
200   Ibidem, 82.
201   Ibidem, 89.
202   Ibidem, 88-89.
203   Ibidem, 89.
204   Ibidem, 92.

individualizing view for the generalizing view, which Meinecke states in his definition, refers to the development from Enlightenment thinking in generalisations toward the focus on the individual and the irrational. Just as the romantics were not against Enlightened reason, so Meinecke did not want to abandon reason in favour of irrationalism. He wanted both at the same time, as an addition and a deepening of insight into individualities.

Yet Rüsen attached more value to a verifiable method, which is why he claims Meinecke's historicism is not scientific. Rüsen, for that matter, also tried to evaluate Meinecke's conception of history from a scientific point of view.[205] Meinecke however was primarily interested in an account of the development of the mind and not in a scientific discussion, which he also expressed in his criticism of Croce.

Finally, it is important to discuss (again) Ernst Schulin's view, the most nuanced critic of Meinecke. Schulin, like Rüsen, claims that the main problem of *Die Entstehung des Historismus* was the idea of individuality. In the introduction of this book I briefly stated that Schulin considered Meinecke's conception of history situated on too high or too individual a level, so that only a select group of initiates could agree with him. Schulin concludes therefore that the spiritual devotion to the individual is too dominant in Meinecke's historicism.[206] Like Rüsen, Schulin also points at Meinecke's preference for subjective experience, feeling, the voice of the inner self as opposed to 'pure reason'. So, Schulin also concludes that Meinecke's view lacks a certain scientific attitude.[207] It is clear that the accounts of Rüsen and Schulin, and to a lesser degree that of Oexle, are all similar to Meinecke's discussion with Croce – and not the other way around, for Croce had a very different view on his debate with Meinecke. In short, Rüsen and Schulin essentially adopt Meinecke's distinction of a 'principle of life' versus a 'scientific principle' to subsequently criticize Meinecke's historicism on these grounds.

Meinecke's view of historicism as a 'principle of life', as a world view, does not specifically include the logical, methodical or ethical difficulties of historicism, for

---

205 Rüsen considers historicism as: '(...) die erste Epoche in der Geschichte des historischen Denkens und der Geschichtsschreibung, in der sich die Geschichte als Fachwissenschaft ausbildet und in den Formen eines akademischen Betriebes institutionalisiert, die wir heute noch kennen: Mit der historischen Forschung wird ein Prozeß dauernden Wissenszuwachses in Gang gesetzt, auf dem die Geschichtsschreibung aufbauen muß, wenn sie glaubwürdig bleiben will. Die Historiker werden zu Fachleuten; mit dem Bewußtsein ihrer akademischen Professionalität grenzen sie sich von Dilettanten oder bloßen Literaten ab'. Jaeger and Rüsen, *Geschichte des Historismus*, 8.

206 Schulin, 'Das Problem', 100.

207 Meinecke's contemporary Johan Huizinga was criticized in a similar vein; especially his *Herfsttij der Middeleeuwen; Waning of the middle ages*, received some criticism, which is reminiscent of the criticism Meinecke received. Cf.: W. E. Krul, *Historicus tegen de tijd. Opstellen over leven en werk van J. Huizinga* (Groningen 1990) especially: 208-239.

it refers to historicism as including the whole of life, or simply the basis of life. Meinecke, for that matter, was not alone in this view of historicism. The Hungarian sociologist and philosopher Karl Mannheim (1893-1947) expressed a similar view of historicism as far back as 1924. His definition runs as follows:

> Historicism is therefore neither a mere fad nor a fashion; it is not even an in-tellectual current, but the very basis on which we construct our observations of the socio-cultural reality. It is not something artificially contrived, something like a programme, but an organically developed basic pattern, the *Weltanschauung* itself, which came into being after the religiously determined medieval picture of the world had disintegrated and when the subsequent Enlightenment, with its dominant idea of a supra-temporal Reason, had destroyed itself.[208]

Meinecke and Mannheim agree that the individualizing historicist mode of thought had caused a revolutionary change; it was a clear break with Enlightenment thought. Ernst Troeltsch held a similar view, which he had laid down in his major study two years before Mannheim's essay, namely: *Der Historismus und seine Probleme*. He defined historicism, also in contrast with the Enlightenment, as a new *Weltanschauung*. According to Troeltsch, historicism is 'the fundamental historicization of all our thinking about man, his culture and his values'.[209] So he sees historicism as a *Weltanschauung* and not as a scientific programme. Oexle, Rüsen and Schulin thus use a narrow sense of historicism compared to Meinecke, Troeltsch and Mannheim, who consider historicism to be more than a scientific principle. As far as Meinecke and Croce are concerned, there is a difference between paradigms; both views are essentially incompatible.

Conclusion

Meinecke claims the essence of historicism lies in its history, in particular in *his* interpretation of that history. It is about the emergence of a world view in which an understanding and a sensing of the individual and the unique in history are

---

208   Karl Mannheim, *Essays on the sociology of knowledge* ed. Paul Kecskemeti (New York 1952) 84-85.

209   Troeltsch, *Der Historismus und seine Probleme*, 102. Translation Beiser, *The German Historicist*, 2. At the beginning of his major study, Troeltsch states: 'Sie [historicism, RK] wurde die leitende Macht der Weltanschauungen, die den Dogmatismus der Aufklärung und der französischen Revolution ablösten'. Troeltsch, *Der Historismus und seine Probleme*, 10. On the difference or the similarities between Meinecke and Troeltsch, in particular their views of historicism and their way of coping with the so-called crisis of historicism, I will come back to in the next chapter.

essential. In his selection of precursors and Enlightenment thinkers Meinecke had found the initiatives that were important for this world view. *Die Entstehung des Historismus* is therefore not the history of the tradition of historicism, but rather a constructed account – a *narratio*[210] – of Meinecke's view of historicism. The different personalities that he discusses are all evidence of his view of historicism.

With his account of historicism as a break with the Enlightenment and as a revolution of the mind, Meinecke tried to revitalize historicism, which was under severe pressure during this period. *Die Entstehung des Historismus* is thus a self-identification of historicism.[211] In that sense, Meinecke's 'affirmative attitude' (*Bejahung*) towards historicism is an (implicit) acknowledgement of the problems and moreover a first suggestion of how to solve these problems. For Meinecke historicism is the spear that injures and heals at the same time. In one of his aphorisms Meinecke expressed it as follows:

> Historicism, which analyzes itself and tries to understand itself from its genesis, is the serpent that bites its tail. It is a sign that the stage of its naïve self-understanding, as I experienced it in my youth, is over, that doubts and problems arise, that a crisis exists, perhaps that its end is near. In this mood I wrote my book on the rise of historicism. But even if the end is near, will it be dying in the sense of being destroyed? There is no such dying in intellectual history. Perish and rise again also applies here. Historicism, even if the mature, perhaps over-ripe fruit of its present form disintegrates, will still contain seeds for new forms.[212]

In that sense, when Meinecke reverts to the history of historicism he, on the one hand, gives evidence of his view of historicism: the identity of, in this case historicism, lies in its past. On the other hand, he falls back on the history of historicism to affirm its viability for the present.[213]

---

210 On this subject: F. R. Ankersmit, *Narrative Logic. A Semantic Analysis of the Historian's Language* (The Hague 1983).

211 In that sense, Meinecke's study fits Hayden White's definition of the romantic plot: 'The Romance is fundamentally a drama of self-identification (...)' White, *Metahistory*, 8.

212 Meinecke, 'Aphorismen', 215. 'Der Historismus, der sich selbst analysiert und aus seiner Genesis zu verstehen sucht, das ist die Schlange, die sich in den Schwanz beißt. Es ist ein Zeichen, daß das Stadium seiner naiven Selbstverständlichkeit, wie ich es noch in meiner Jugend erlebte, vorüber ist, daß Zweifel und Probleme auftauchen, daß eine Krisis da ist, vielleicht sein Ende nahe ist. In dieser Stimmung schrieb ich mein Buch über die Entstehung des Historismus. Aber selbst wenn das Ende nahe ist, wird er darum sterben im Sinne von vernichtet werden? Solch Sterben gibt es in der Geistesgeschichte nicht. Stirb und werde gilt auch hier. Der Historismus enthält, auch wenn die reife, vielleicht überreife Frucht seiner jetzigen Gestalt zerfällt, Samenkörner für neue Gestaltungen'. Ernst Troeltsch had somthing similar in mind with his idea of 'Geschichte durch Geschichte zu überwinden'. Troeltsch, *Der Historismus und seine Probleme*, 772.

213 Cf. Hofer, *Geschichtsschreibung*, 376, 408.

In Meinecke's view it was Goethe who embodied the point of view in which past and present coalesce, in which pietism, Neoplatonism, the irrational and the rational, feeling and reason, the local and the universal, life and science are reconciled. He managed to reach a level in which metaphysics is connected to reality. According to Meinecke, the self-identification of historicism and the revolution that caused an awareness of the 'principle of life' of historicism, recovers in its most elevated form in the 'high peak' (*Hochgebirge*) of Goethe's thought – the subject of the next chapter.

5. Harmony regained

'Versöhnung ist mitten im Streit und alles
Getrennte findet sich wieder'[1]

Introduction

'We would not be what we are today if it wasn't for Goethe'.[2] With this statement
Meinecke starts his closing chapter of *Die Entstehung des Historismus*, which is entirely devoted to this poet, playwright and natural scientist.[3] Meinecke's claim is
telling for two reasons. Firstly, it is a reference to Goethe's immense cultural legacy
as a representative of the *Kulturnation*. Secondly, Meinecke considers Goethe the
high point of the historicist outlook, the new view on human life in which the idea
of individuality is key. 'In this change [*Umschwung*] of the mind', claims Meinecke,
'Goethe is and remains the most powerful representative'.[4] Further, even though
this is less explicit in Meinecke's statement, the statement also seems to refer to
what Goethe could still mean for Meinecke's present, the 1920's and 1930's, when
historicism faced a crisis.[5] That Goethe indeed could still play an important role is
clear at the end of Meinecke's account; he claims: 'it is Goethe who, with regard to
intellectual heights, still remains a signpost'.[6]

---

1 Friedrich Hölderlin, 'Hyperion oder der Eremit in Griechenland' in: idem, *Friedrich Hölderlin
   Sämtliche Werke und Briefe in drei Bänden*, ed. Jochen Schmidt (Frankfurt am Main 1994) Volume
   II, 9-175, there 175.
2 Meinecke, *Die Entstehung des Historismus*, 445.
3 The chapter on Goethe – which was also published as a *Sonderausgabe*: Friedrich Meinecke,
   Goethe und die Geschichte (Leipzig and Munich 1949) – covers approximately 150 pages. This
   chapter is followed by a short lecture (as a *Beigabe*) on Ranke, but that is generally considered
   to be of less importance regarding the theme and structure of the book.
4 Meinecke, 'Zur Entstehungsgeschichte des Historismus', 341.
5 A few authors (especially Hessing, Iggers and Pois) refer to the rise of Hitler (and claim
   Meinecke sets Goethe against him). Further, these and other authors point at the apparent
   relation between *Die Entstehung des Historismus* and the rise of Hitler. Cf. the previous chapter
   and my introduction.
6 Meinecke, *Die Entstehung des Historismus*, 584. 'ein Wegweiser in die Höhe muß er uns bleiben'.

When it comes to German *Kultur* and particularly historicism Goethe is the alpha and the omega for Meinecke. He is even convinced that the apotheosis of historicism was realized in and with Goethe, that is Goethe the natural scientist (*Naturforscher*). In Meinecke's view, it is therefore not an historian, but a natural scientist – not Ranke, but Goethe – who should be praised.[7] That is a striking choice, for with this choice for Goethe, Meinecke shifts the high point of historicism to a period well before its scientific foundation. Against the background of the previous chapter Meinecke's choice for Goethe dovetails with his view of historicism as a 'principle of life' instead of a scientific principle, for Meinecke based himself on Goethe's natural scientific observations, which are, as will become clear in this chapter, primarily focussed on the principles of life and nature (*Natur*).

Things get more complicated when it turns out that Goethe had a very difficult relationship with history and historical writing. Throughout his oeuvre one can find many statements of dissatisfaction with regard to history. He characterized it, for example, as a storeroom of rubbish or dustbin of the past, as gossip, foolishness and depravity, and more often than not certain periods from the past had a cadaverous smell to them.[8] The older he got, the more often he expressed himself in this manner. For instance, in 1828, four years before his death, he claimed, 'I have not grown so old as to care for world history, which is the most absurd thing; whether this or that dies, this or that people perish is all the same to me; I would be a fool to worry about that'.[9] Further, he considered historical writing the most ungrateful and dangerous profession, one that is likely to cause the writer considerable trouble.[10] These are unmistakable statements. So, how is it that Meinecke declares this anti-historical thinker to be the high point of historicism? Meinecke acknowledges Goethe's negative comments on history but – and that is the thesis of this chapter – he reconciles the negative view of history with its positive dimension.[11] Goethe had also expressed himself positively with regard to history. He claimed, for example, that he had had many experiences in which he felt the mutual connection between

---

7   Walter Goetz states that Goethe never considered the world from a standpoint of a trained professional (philosopher, natural scientist or historian): 'Ihm galt es, das Rätsel der Welt zu ergründen, die Einheit von Natur und Geschichte aufzufinden. So geht sein Suchen immer auf das Allgemeine, auf die Zusammenhänge alles Seins, auf Mensch und Weltall und auf das Göttliche in ihnen'. Walter Goetz, 'Die Entstehung des Historismus' in: idem, *Historiker in meiner Zeit. Gesammelte Aufsätze* (Keulen 1957) 351-360, there 357.

8   Alexander Demandt, 'Geschichte bei Goethe', *Merkur. Deutsche Zeitschrift für europäisches Denken* (1947) 317-327, there 317-318.

9   Cited in: Rüdiger Bubner, 'Die Gesetzlichkeit der Natur und die Willkür der Menschheitsgeschichte. Goethe vor dem Historismus', *Goethe Jahrbuch* 110 (1993) 135-145, there 141.

10   Demandt, 'Geschichte bei Goethe', 318.

11   Meinecke devotes a section to Goethe's 'Mißvergnügens an der Geschichte': Meinecke, *Die Entstehung des Historismus*, 504-525.

past and present; an experience in which the past intrudes on the present. In his major autobiographical work *Dichtung und Wahrheit* he describes that these experiences influenced his life considerably. He also used these historical events and persons for his poetry, plays and novels.[12]

Meinecke is exclusively interested in Goethe's findings as a natural scientist. His botanical studies and his *Theory of Colours* provide Meinecke with insights for his view of historicism.[13] In other words, it is not so much Goethe's view in itself that is at the core of *Die Entstehung des Historismus*, but rather Meinecke's interpretation or use of these views. Considering Goethe's problematic attitude towards history, it remains to be seen why Meinecke indeed thought this *uomo universale* was more fitting for his research than an historian. In short, what function does Goethe serve for Meinecke? And in line with this: what does this say about Meinecke's view of historicism? Further, the question remains as to why Meinecke did not explicitly discuss Goethe in his earlier studies, when he was already familiar with his range of thought at an early age.[14] Put differently, why is Goethe the key figure in Meinecke's research in exactly this period?

The literature on Meinecke's historicism leaves Goethe's ideas almost always untouched, whereas these ideas are of great importance for understanding Meinecke's conception of historicism.[15] Goethe's harmonic world view, in which different polarities are in harmony, resounds in Meinecke's thought. Goethe tried to capture this duality in many concepts – *Natur*, *Urphänomen* and the *Dämonische*, which I will discuss in detail later on. Also, his interest in Spinoza and Neoplatonism give evidence of this harmonious world view. It is precisely this interest in harmony

---

12   Demandt, 'Geschichte bei Goethe', 320-321.

13   The Dutch historian Johan Huizinga also emphasized the importance of colours in historical writing, cf: Ankersmit, *Meaning, Truth, and Reference*, 200-206.

14   Meinecke, 'Erlebtes' 52, 102.

15   Nicholas Berg is an exception in this case, but he only discusses Goethe in relationship with Meinecke's last study, *Die deutsche Katastrophe*; I will discuss this in the next chapter. Pyper only mentions Goethe and his apparent panentheism in his conclusion, without any further explanation. Pyper further claims that in his view Meinecke mainly 'follows' Goethe, whereas the opposite is in fact the case as will become clear: Meinecke adjusts Goethe to his own views. Pyper, 'Meinecke, Croce, and the Individual', 484. The well-known Dutch historian Von der Dunk has criticized Meinecke's 'sponsorship of Goethe', because Goethe's thought was pre-eminently concrete, which is why, according to Von der Dunk, it does not fit Meinecke's more or less 'hovering' view of historicism: 'Nergens is bij Goethe sprake van het soort personalisatie van historische collectiviteiten en fenomenen dat nu juist een kenmerk voor het latere historisme in de zin van Meinecke is en dat debet is aan het zweverige en vrijblijvende van het begrip'. Notwithstanding the fact that Meinecke's principle of individuality (to which Von der Dunk refers) does not only refer to collectivities, I will also make clear that Goethe's thought is not as concrete as Von der Dunk suggests. Dunk, *De glimlachende sfinx*, 46.

that characterizes Meinecke's thought; therefore it is remarkable, to say the least, that this relationship between Goethe and Meinecke has never been researched in detail, particularly since Meinecke considered Goethe's thought to be the high point of historicism.

Meinecke's familiarity with Goethe's world of ideas is well-known, but in the Meinecke historiography there is no detailed discussion on the contents of Goethe's views on, for example, nature and history and the difference with Meinecke's views on these issues.[16] The status of Goethe in *Die Entstehung des Historismus* is of course mentioned, but these are only short remarks. It has been noted, for example, that Meinecke found a solution for the crisis of historicism in Goethe, but there is no detailed account. Ernst Schulin, in his famous article on Meinecke and historicism, gave an initial impetus to Meinecke's interpretation of Goethe. He, for example, states it was not so much Goethe's contribution to historicism which gave him the prominent position in *Die Entstehung des Historismus*, but indeed what he could still mean for historicism.[17] Schulin claims that Meinecke uses Goethe as an ideal for present purposes, but Schulin does not explicitly explain this.[18] An elaborate

---

16   Jakob Hessing refers to Meineckes 'use' of Goethe, but claims it is 'Goethe the poet', which Meinecke uses for his own view of historicism, while in fact – that is the thesis of this chapter – it is Goethe the natural scientist that is key for Meinecke. Moreover, Hessing fails to discuss Goethe's concept of *Natur*. Hessing's studies are by the way characterized as highly polemical. For example, Meinecke's work is considered a *Fehlleistung* with regard to his view on his own time: Meinecke is characterised as an 'spätgeborenen Vertreter des ancien régime' (332), whereas in fact Meinecke started his historical research alsmost always from present problems: Eberhard Kessel, 'Einleitung des Herausgebers' in: Friedrich Meinecke, *Zur Theorie und Philosophie der Geschichte* Eberhard Kessel ed., (Stuttgart 1965) vii-xxxiv, especially: viii, xx. Cf on Hessing: Stefan Meineke, *Friedrich Meinecke*, 29-32. Hessing's work: Jakob Hessing, 'Friedrich Meinecke: Naturbegriff und Goethebild. Zur Problematik der konservativen Goetherezeption in Deutschland', *Jahrbuch des Instituts für Deutsche Geschichte* 12 (1983) 317-351; Jensen also mentions Goethe's concept of nature and even that it differs from Meinecke's, but because he claims that Meinecke never explicitly stated what he means by nature, Jensen leaves the issue untouched. Jensen, '"Unity in Antinomy"', 188 in particular note 2. Also see: Iggers, *The German Conception*, 220-221. Mandelkow mainly shows that, thanks to Meinecke, the relation between Goethe's view of history and nature has, since the publication of *Die Entstehung des Historismus*, barely been touched upon, for Meinecke had uncovered Goethe's negative attitude towards history. Karl Robert Mandelkow, 'Natur und Geschichte bei Goethe im Spiegel seiner wissenschaftlichen und kulturtheoretischen Rezeption' in: Peter Matussek ed., *Goethe und die Verzeitlichung der Natur* (Munich 1998) 233-258, there 246-247. Bruch claims: 'Für Meinecke selbst blieb Goethe der Fluchtpunkt', but he also does not explain why this is the case. Rüdiger vom Bruch, 'Ein Gelehrtenleben zwischen Bismarck und Adenauer' in: Gisela Bock and Daniel Schönpflug ed., *Friedrich Meinecke in seiner Zeit. Studien zu Leben und Werk* (Stuttgart 2006) 9-19, there 19.

17   Schulin, 'Das Problem', 106-107.

18   Ibidem, 104, 107.

detailed discussion of the relationship between Meinecke and Goethe is thus still lacking, until now.

Walter Hofer and before him Siegfried Kaehler thought that *Die Entstehung des Historismus* should in general be viewed as a *Vorstufe* (precursor) of Meinecke's own view of historicism or even identical to it.[19] In the previous chapter I more or less defended the same thesis, arguing that Meinecke considered historicism as his interpretation of the history of historicism. This, however, only applies to the build-up, the precursors of historicism that Meinecke discusses, because in his account of Goethe something else seems to be going on. 'Goethe' is not identical with Meinecke's historicism.[20] The previously discussed thinkers were, in essence, also not identical with Meinecke's historicism; Meinecke only took those aspects of the ideas (of, for example, Leibniz, Burke, Möser and Herder) he thought important for what he considered to be the essential elements of the rise of the historicist outlook. In the case of Goethe, it is a wholly different story. As Meinecke's account of Goethe unfolds, something striking happens. Gradually it becomes more and more difficult to determine whether it is still Goethe's relationship towards history that is being discussed or that little by little Meinecke's own conception of historicism is central, under the pretext of Goethe's range of thought. Dealing in such a way with Goethe's legacy is, by the way, not new. To this very day, there seems to be no end of publications on Goethe's life and work, and thereby the appropriations of his ideas and their use for many different goals. 'Immortal' authors such as Goethe, who have written on almost every subject, and who are open to many (mis)interpretations, are of course a goldmine (a *Fundgrube*) for new ideas for every new generation. In Meinecke's time this was no different: historians, philosophers and writers all tried to relate themselves to Goethe and used his ideas for their own purposes.[21]

---

19   Schulin also refers to this: Schulin, 'Das Problem', 107; Hofer, *Geschichtsschreibung*, 371, 373; Kaehler probably described this in a letter to Meinecke, that is, Meinecke's answer implies this: Meinecke, *Ausgewählter Briefwechsel*, 355.

20   Hofer claims Meinecke links up with Goethe: Hofer, *Geschichtsschreibung*, 411. For that matter, Hofer considered Meinecke to be the 'realiser' of Dilthey's ideas, and is he not directly related to the Goethean tradition. Hofer, *Geschichtsschreibung*, 12.

21   Meinecke's contemporaries, among others, on Goethe: Wilhelm Dilthey, 'Das Erlebnis und die Dichtung. Lessing, Goethe, Novalis, Hölderlin' in: idem, *Gesammelte Schriften* XXVI (Göttingen 2005; 1906); Thomas Mann with some lectures, essays, and a novel: 'Goethe als Repräsentant des bürgerlichen Zeitalters' (1932), 'Goethes Laufbahn als Schriftsteller' (1932): both in: Thomas Mann, *Adel des Geistes. Sechzehn Versuche zum Problem der Humanität* (Stockholm 1945) 104-145, 146-179. The novel on Goethe: *Lotte in Weimar* (1939); on Georg Simmel, Ernst Cassirer, Walter Benjamin and Karl Löwith cf: Ulrike Böge, 'Die Inbesitznahme Goethes durch die Philosophie. Goetherezeption bei deutschsprachigen Philosophen in der ersten Hälfte des 20. Jahrhunderts' (Kiel 2001) unpublished dissertation; Erich Franz, *Goethe als religiöser Denker* (Tübingen 1932); E. Menke-Glückert, *Goethe als Geschichtsphilosoph und die geschichtsphilosophische Bewegung seiner Zeit* (Leipzig 1907).

It was Schiller who already predicted that Goethe's legacy would be (mis)used in such a way: 'Goethe rightly deserved to be inherited and plundered by others', he said.[22] Meinecke also consciously distorts Goethe's ideas to fit his own views of historicism. Building on Schulin, I will make clear in this chapter that Meinecke did not so much consider Goethe as an ideal; he also subjects Goethe's views on nature and history to a reinterpretation (*Umdeutung*) in favour of his own conception of historicism.

5.1   Polarities and harmony

Meinecke divided his chapter on Goethe into a 'genetic' and a 'systematic' part. First, he chronologically describes the development of Goethe's relation to history, to subsequently deduce from this development a certain coalescence in his thought on history. In a way, Meinecke deduces a *Historik* (historiology) from Goethe's thought. It will become clear that this *Historik* is remarkably close to Meinecke's own conception of history. Furthermore, it is my contention that Meinecke deduces much more than merely a research method or the 'task of the historian' from Goethe's ideas. The systematic part is subdivided into Goethe's negative and positive relation to history. What is striking about his analysis of this negative dimension of Goethe's image of history is that Meinecke squirms out of the negative part to find something positive in it, while at the same time failing to be critical of Goethe's positive view of history. At the end of his section 'Goethes Mißvergnügen an der Geschichte' Meinecke justifies this neglect in the following (Goethean) phrase: 'Goethe can only be understood through the polarities of his thought'.[23] According to the general view this is related to Goethe's harmonious world view, in which polarities like the individual and the absolute, light and dark, law and freedom, past and present are reconciled.[24] Meinecke, however, alludes to Goethe's negative relation to the past, which could or should also turn into a positive dimension. According to Meinecke, it is an oscillation: '(...) making it possible for any dissatisfaction with history to swing over into the deepest

---

22   Cited in: Demandt, 'Geschichte bei Goethe', 319.

23   Meinecke, *Historism*, 442. The German literary critic Reich-Ranicki agrees with this statement: Marcel Reich-Ranicki, *Goethe noch einmal. Reden und Anmerkungen* (Munich and Stuttgart 2002) 78-79, 137.

24   Reinhart Koselleck, 'Goethes unzeitgemäße Geschichte', *Goethe Jahrbuch* 110 (1993) 27-39, there 33; Gerhard Schulz, 'Chaos und Ordnung in Goethes Verständnis von Kunst und Geschichte', *Goethe Jahrbuch* 110 (1993) 173-183, there 178; Werner Schultz, 'Die Bedeutung des Tragischen für das Verstehen der Geschichte bei Hegel und Goethe', *Archiv für Kulturgeschichte* 38 (1956) 92-115, there 101-102.

satisfaction'.[25] But Meinecke kept silent with regard to the possibility that the positive dimension could also swing over into a negative stance towards history. The contrast in Goethe's conception of history is, in Meinecke's view, a necessary polarity, which, he immediately adds, was not experienced by Goethe as such.[26]

The vagueness of Goethe's thought with its polarities and/or harmony has everything to do with the notion of *Natur* (nature) and the meaning he attributes to it. For a better understanding of Goethe's ideas (on history) and Meinecke's interpretation, a clarification of the concept of nature is important. Goethe shaped his idea of nature under the influence of the views of Spinoza, Kant, Herder, Schelling and Neoplatonism.[27] In this context, Spinoza is of the utmost importance, for at the end of the eighteenth century in Germany, under the influence of the so-called *Pantheismusstreit* (1785-1789), a rehabilitation of Spinoza took place. Goethe's idea of nature was partially shaped by this pantheism controversy. Moreover, this Spinoza-affair should be considered a fundamental and important intellectual event, in which thinkers such as Kant, Herder, Fichte, Schelling, Hegel and Goethe participated. The Spinoza controversy marks a transition from the Enlightenment to Romanticism and German Idealism. It is therefore not surprising that Meinecke's thought also reveals traces of this. So, before we discuss Goethe's concept of nature and Meinecke's reading of this, we should first take a closer look at this controversy.

Pantheism controversy

At the end of eighteenth-century Europe, the general attitude towards Spinoza or Spinozism was one of hostility. The dominant view on Spinoza's thought was its incompatibility with any view starting from a personal God, thus with almost all of Christianity.[28] The heart of Spinoza's doctrine is about the 'one Substance', which can also be indicated as God or Nature (*Deus siva natura*). This Substance is simultaneously the free 'self-creating' creator (*natura naturans*) as well as the sum of his creation, that is: all things in the world, which are considered as modifications (*modi*) of the Substance (*natura naturata*). These modi exist because of the Substance

---

25   Meinecke, *Historism*, 449.

26   Idem.

27   Robert J. Richards, *The Romantic Conception of Life. Science and Philosophy in the Age of Goethe* (Chicago and London 2002) 407-502; Alfred Schmidt, 'Natur' in: Bernd Witte ed., *Goethe Handbuch in vier Bänden* (Stuttgart and Weimar) Volume 4/2 755-776, there 775, 770, 774; Beiser, *German Idealism*, 361-368; David Bell shows that Goethe partly found a confirmation of his ideas in Spinoza's thought: David Bell, *Spinoza in Germany from 1670 to the Age of Goethe* (London 1984) 147-168; Franz, *Goethe als religiöser Denker*, 211; Helmut Thielicke, *Glauben und Denken in der Neuzeit. Die großen Systeme der Theologie und Religionsphilosophie* (Tübingen, 1988 first 1983) 223.

28   Pinkard, *German Philosophy*, 93.

and are therefore equally necessary, because their relation to the Substance is internal. The modi are thus only understandable from the Substance from which they emanated, 'in' which they 'are'. There is, so to speak, a hierarchy from the Substance to the modi: they are the product of the Substance; so, there is nothing outside of the Substance.[29] This doctrine proved incompatible with a belief in a personal, orthodox Christian God; hence, Spinozism's frequent equation with atheism.[30]

At the end of the eighteenth century an accusation of Spinozism could still completely destroy a person's intellectual career. That was not the initial objective of Friedrich Heinrich Jacobi (1743-1819) when he revealed in *Über die Lehre des Spinoza in Briefen an den Herrn Moses Mendelssohn* (1785) that Gotthold Ephraim Lessing (1729-1781), shortly before his passing, had confessed to him that he was a spinozist.[31] With this revelation of Lessing's alleged Spinozism Jacobi – who first and foremost wanted to direct attention to himself – unleashed a debate that has come to be known as the Pantheism Controversy. Precisely by accusing Lessing – the paragon of the Enlightenment thinker pur sang – of Spinozism, Jacobi warned against the dangers of Enlightened rationalistic philosophy. He suggested that all Enlightened thinkers, following Lessing, have to admit that they are essentially spinozists. Jacobi went even one step further in stating that (radical) rationalism, which is what Spinozism was in Jacobi's view, will lead to atheism, nihilism – a term coined by Jacobi –, and eventually fatalism.[32] After all, the core of Spinozism, Jacobi claims, is a denial of free will, providence and a personal God.[33]

The reason for Jacobi's attack on Lessing, as is clear from the title of his work, was a correspondence between Jacobi and the philosopher Moses Mendelssohn (1729-1786). After Lessing's death in 1781 Mendelssohn – a good friend of Lessing – intended to write a biography of Lessing. When Jacobi found out about these plans, he wrote Mendelssohn a letter in which he informs him of Lessing's Spinozism. Mendelssohn, who was deeply shocked by Lessing's confession, entered into a heated correspondence with Jacobi. When Mendelssohn decided he would still publish his biography of Lessing – leaving aside whether he would include Lessing's Spinozism – Jacobi felt forced to publish, as soon as possible, his recollections of his discussions with Lessing and his letters to Mendelssohn (who was unaware

---

29   Wolfgang Bartuschat, *Baruch de Spinoza* (Munich 1996). 64-68.

30   Pinkard, *German Philosophy*, 93.

31   Pinkard, *German Philosophy*, 91-93; Beiser, *German Idealism*, 361. For a detailed discussion on the pantheism controversy: Beiser, *The Fate of Reason*. On Jacobi: Frank Ankersmit, 'Jacobi. Realist, Romanticist, and Beacon for Our Time', *Common Knowledge* 14 (2008) 221-243; Dale Evarts Snow, 'F. H. Jacobi and the Development of German Idealism', *Journal of the History of Philosophy* 25 (1987) 397-415. Beiser, *The Fate of Reason*, 44-91.

32   Beiser, *German Idealism*, 362.

33   Beiser, *The Fate of Reason*, 66.

of this), and this he did in 1785.[34] The following year, in 1786, Mendelssohn died. Surrounding his death, it was rumoured that Jacobi was 'directly' responsible. Whether he really was guilty of Mendelssohn's death will remain a mystery, and is of minor concern here; it is more important that Jacobi, with his revelation of Lessing's Spinozism, had committed an attack on the Enlightenment with its (radical) rationalism.[35]

Apart from Jacobi's intention to be the center of attention by revealing Lessing's Spinozism, he also had three other things in mind with his attack on Lessing. Next to the biographical reason, he was particularly interested in the right interpretation of Spinoza, and especially the authority or reason.[36] So basically, Jacobi's target was Kant.[37] Jacobi did not abandon reason, but he was convinced it could not provide all answers to the problems of the world. Reason, according to him, was not the basis of all knowledge.[38] He claimed one had a choice between on the one hand following reason, which would eventually lead to atheism and fatalism, and on the other hand a decision to distance oneself from reason and to make a so-called *salto mortale* of faith (*Glaube*) in a personal God and freedom. Jacobi's faith is by the way not exclusively related to a belief in God, for it is also a belief in moral, political, rational, and – with reference to the *common-sense* philosophy of Thomas Reid – a belief in the existence of reality.[39] It is a choice between rational skepticism and an 'irrational' faith. That is why Jacobi points out to Kant (and Mendelssohn) that *faith* in reason is the basis, not reason itself, because it could not be the justification of itself.[40] For that matter, faith is to Jacobi what reason is to Kant and Mendelssohn: the touchstone of truth.[41]

Jacobi's warning against Spinozism was largely ignored by his contemporaries, especially by Herder, Goethe and Schelling. Indeed, many felt attracted to the philosophy of Spinoza, particularly now that an authority like Lessing had engaged himself in this philosophy.[42] Furthermore, many thinkers thought they now could

---

34   Pinkard, *German Philosophy*, 93; Beiser, *The Fate of Reason*, 72.
35   Beiser, *The Fate of Reason*, 74-75.
36   Ibidem, 47.
37   Pinkard, *German Philosophy*, 93.
38   Ibidem, 91.
39   Isaiah Berlin, *The Age of Enlightenment. The Eighteenth-Century Philosophers* (Oxford 1979) 261-265.
40   Beiser, *The Fate of Reason*, 80-81; Pinkard, *German Philosophy*, 92-93.
41   Beiser, *The Fate of Reason*, 89. Jacobi's *Gewißheit* and *Glaube* bears a resemblance with Meinecke's idea of conscience and intuition; I will come back to this in (a footnote) in the section on the crisis of historicism.
42   Beiser, *German Idealism*, 362-363; Pinkard, *German Philosophy*, 93; Michael N. Forster, 'Herder and Spinoza' in: Eckart Förster and Yitzhak Y. Melamed ed., *Spinoza and German Idealism* (Cambridge 2012) 59-84, there 60.

solve Kant's dualism by means of Spinoza's motto *hen kai pan* (one and all). Kant distinguished between a noumenal and a phenomenal reality – the object of knowledge and the knowing subject – which were in Spinoza's view both *modi* of the one Substance.[43] And Spinoza's view that God is present in everything also dovetails with Herder's, Goethe's and Schelling's pietism. To them the personal experience of God was closely related to Spinoza's idea that God is nature, for to them the personal experience of the divine was simultaneously an experience of being in contact with nature as a whole.[44] These thinkers could, however, not agree with Spinoza's rationalistic determinism. They were mainly attracted to his love of God, that is: the idea of perceiving God in everything – for that reason Novalis once characterized Spinoza as the 'God-drunken man'.[45]

A personal experience of God or the god-like, as pietism suggests, dovetails with Spinoza's so-called 'third form of knowledge'. His system starts from three forms of knowledge. The first, the *imaginatio*, refers to knowledge by association. The second form is knowledge by reason, which corrects the first form, but stays focussed on the general and therefore does not yield any knowledge of the individual. The third form, the *scientia intuitiva* – to know intuitively – provides a solution here.[46] Since the individual and the whole are closely connected to each other in Spinoza's system, or, put differently: because they are one 'in' the Substance (*hen kai pan*) it is possible, by means of intuitive knowledge, to divine (*ahnen*) the essence of the Substance. So, the *modi* originated from the Substance, and because of this they can acknowledge, feel and approach *within themselves* the higher god-like. It can be compared with the notion of conscience as a reflection (*Abglanz*) of absolute morality, as mentioned in my discussion of Meinecke's *Staatsräson* in chapter three. Compared with conscience, which merely carries a spark of the absolute in itself, intuitive knowledge will also never fully reach complete knowledge of the Substance, for it is a form of knowledge of temporal modi, which can never really fathom the infinite Substance.[47]

Goethe and Herder thus oppose Jacobi's explanation of Spinoza, for both are convinced that Spinozism is unlikely to lead to atheism; on the contrary, it will confirm the divine, for God is itself existence: nature and God are identical. Goethe agrees with Spinoza's idea that by means of intuition we can approach the essence

43   Pinkard, *German Philosophy*, 92; Ankersmit, 'Aftermaths and 'Foremaths'' (Groningen 2010) unpublished lecture.

44   Beiser, *The Fate of Reason*, 60-61.

45   Novalis, *Werke, Tagebücher und Briefe* 3 volumes (Munich 1978) volume 2, 812; Beiser, *German Idealism*, 363.

46   Bartuschat, *Baruch de Spinoza*, 89-104.

47   Ibidem, 102-104.

of things, and even catch a glimpse of the divine. And finally, the idea that the 'individual' can only be understood in relationship with the whole also appealed to him. Studying the individual self can provide insight into nature and vice versa. Like Spinoza, Goethe is aware that this (self-)insight into nature or God will never be complete. Goethe puts it in his *Maximen und Reflexionen* as follows: 'There is an unknown quality of lawfulness in the object which corresponds to the unknown quality of lawfulness in the subject'.[48] Subject and object thus have the same basis: the Substance, which cannot rationally be comprehended, only by means of divining (*ahnen*). Instead of 'Substance' Goethe rather speaks of *Gottnatur*, a notion clearly inspired by Spinoza's *Deus sive natura*.[49]

Goethe's view of nature

Goethe's notion of *Natur* is about the tensions between matter and mind. The two major themes of nature are, according to him, polarity (*Polarität*) and elevation (*Steigerung*); the first is the characteristic feature of matter, the second of the mind.[50] Both matter and mind contain similar elements and yet depend on each for their own existence. In 'Erläuterung zu dem aphoristischen Aufsatz „Die Natur"' Goethe states that as a result of the attraction and rejection between polarities a temporal elevation or unification takes place on a higher level.[51] In his *Theory of Colours*, by way of illustration, Goethe claims that an intensification of colour is the result of the clashes between light and dark; likewise with electricity it is the plus

---

48   Johann Wolfgang von Goethe, *Maxims and Reflections* transl. Elisabeth Stopp (London etc. 1998) 170. I will come back to this notion of the 'unknown' which manifests itself in the object as well as the subject. German original: 'Es ist etwas unbekanntes Gesetzliches im Objekt, welches dem unbekannten Gesetzlichen im Subjekt entspricht'. Cited in: Dietrich von Engelhardt, 'Natur und Geist, Evolution und Geschichte. Goethe in seiner Beziehung zur romantischen Naturforschung und metaphysischen Naturphilosophie' in: Peter Matussek ed., *Goethe und die Verzeitlichung der Natur* (Munich 1998) 58-74, there, 64; Richards, *The Romantic Conception*, 376-377.

49   Ranke's notion of the 'Real-Geistige' is also clearly derived from such a Spinozist or Goethean concept. Cf. chapter two on Meinecke's *raison d'état*-thinking and Meinecke, *Die Idee der Staatsräson*, 450.

50   Schmidt, 'Natur', 775; Engelhardt, 'Natur und Geist', 66; Angus Nicholls, *Goethe's Concept of the Daemonic: After the Ancients* (New York 2006) 217; also: Meinecke, *Die Entstehung des Historismus*, 559-560, 566-567.

51   Schmidt, 'Natur', 775; Johann Wolfgang von Goethe, 'Erläuterung zu dem aphoristischen Aufsatz „Die Natur"' (1828) in: idem, *Goethes Werke. Hamburger Ausgabe in 14 Bänden*, ed. Erich Trunz (Munich 1982) Volume XIII, 48-49, there 48.

and minus which converge and cause an elevation.[52] Goethe's view of nature, for that matter, starts from the idea of inhalation and exhalation. Goethe indicates this process with the notions of *Systole* and *Diastole*, which to him are realizations of the concept of polarity. With this pair of concepts (polarity and elevation) he refers to apparent contrasts or antinomies which are in reality manifestations of a higher unity, reached by means of an elevation, like the plus and minus in electricity.[53] In his *Theory of Colours* he aptly states: 'to polarize unity and to unify polarities is nature's being'.[54] I will come back to the Neoplatonic dimensions of this idea.

In contrast to the partially Spinozist and metaphysical speculation above, Goethe was primarily an empiricist and completely trusted his senses. He assumed he had a direct experience of reality, and he was convinced he could express this experience in his art, among other things. When he was introduced to Kant's transcendental philosophy, in which the subject is opposed to the object, he encountered some problems. For Kant assumed that the subject could only have knowledge of the object by means of the categories of reason, but the 'thing in itself' was essentially unfathomable.[55] In other words, the *experience* of reality was in Kant's view not reality itself.

With the concept of *Urphänomen* (primal phenomenon) Goethe tried to solve Kant's dualism. This concept of the 'primal phenomenon' is twofold. On the one hand, it refers to a mental or theoretical abstraction from which other complicated phenomenon can be derived. On the other hand, it can be interpreted as an actual existing 'archetype' from which existing physical phenomenon or biological forms can develop. In general, there is some overlap between both.[56] Goethe claimed these primal phenomenon correspond with different categories of objects, for example: plants, animals and colours.[57] In his research of botany he tried, for instance through observation of development or growth – the metamorphosis – of plants, to discover the essence which would be found in all plants, and thereby to find the so-called archetypal plant (*Urpflanze*). At the same time, Goethe thought it possible – starting from the 'primal phenomenon as abstraction' – to extrapolate to an

52   Nicholls, *Goethe's Concept of the Daemonic*, 218; Goethe, 'Erläuterung', 48; Johann Wolfgang von Goethe, 'Zur Farbenlehre' (1808) in: idem, *Goethes Werke. Hamburger Ausgabe in 14 Bänden*, ed. Erich Trunz (Munich 1982) Volume XIII, 314-523, there 488.

53   Peter Huber, 'Systole/Diastole' in: Bernd Witte ed., *Goethe Handbuch in vier Bänden* (Stuttgart and Weimar) Volume 4/2, 1034-1035, there 1034.

54   Goethe, 'Zur Farbenlehre', 488. 'Das Geeinte zu entzweien, das Entzweite zu einigen, ist das Leben der Natur'.

55   Pinkard, *German Philosophy*, 40.

56   John Erpenbeck, 'Urphänomen' in: Bernd Witte ed., *Goethe Handbuch in vier Bänden* (Stuttgart and Weimar) Volume 4/2, 1080-1082, there 1080. also: Nicholls, *Goethe's Concept of the Daemonic*, 186-187; Richards, *The Romantic Conception*, 424.

57   Nicholls, *Goethe's Concept of the Daemonic*, 187.

idea which would indicate the possibility of an archetypal plant, which cannot be realized in reality. In 1787 Goethe visited Italy, where he hoped to find the archetypal plant; in a letter to Herder he described this as follows:

> The archetypal plant will be the most wonderful creation of the world, which nature will envy. With this model and the key to it one can then invent plants infinitely, which must be consistent, that is, which – though they do not exist – could exist, and are not, for example, graphic or poetic shadows and semblances, but bear an inner truth and necessity.[58]

In that sense Goethe's concept of the primal phenomenon can be considered Neo-platonic, since on the one hand, the physical forms are free to (eternally) develop in reality, and on the other hand, it is possible to strive for an abstract idea – the unity in 'God' – which can be divined (*ahnen*), but is unfathomable in reality.[59]

Kant considered concepts like Goethe's primal phenomenon to be speculative.[60] And Schiller, who was very Kantian, claimed Goethe's archetypal plant was essentially an idea, which was therefore nowhere to be found in reality.[61] Goethe was, however, convinced that experience should complement Kant's transcendental philosophy. In his *Theory of Colours* Goethe puts it as follows: 'Every act of seeing leads to consideration, consideration to reflection, reflection to combination, and thus it may be said that in every attentive look at nature we already theorize'.[62] The

---

58 Cited in: Erpenbeck, 'Urphänomen', 1080.

59 *Urphänomene* are, for that matter, *dämonisch*; cf. section 5.3.

60 Nicholls, *Goethe's Concept of the Daemonic*, 170-172. With regard to Kant's analysis of 'teleology' Zammito shows that Kant's *Kritik der Urteilskraft* was in fact a reaction to Herder's *Ideen zur Philosophie der Geschichte der Menschheit*. John H. Zammito, 'Herder, Kant, Spinoza und die Ursprünge des deutschen Idealismus' in: Marion Heinz ed., *Herder und die Philosophie des deutschen Idealismus* (Amsterdam and Atlanta 1997) 107-144, there 110. In particular: John H. Zammito, *The Genesis of Kant's Kritik der Urteilskraft* (Chicago 1992) 151-262; F. R. Ankersmit, 'Aftermaths and 'Foremaths', (2010) unpublished lecture; Richards, *The Romantic Conception*, 396, 424, 429.

61 When Schiller and Goethe first met in 1794 this was discussed. Goethe later on referred to this in his 'Glückliches Ereignis'. Based on his memory, Goethe described how he explained to Schiller the theory of metamorphosis, and by way of illustration he drew a 'symbolic plant'. Schiller thereupon claimed: 'Das ist keine Erfahrung, das ist eine Idee'. To which Goethe replied: 'Das kann mir sehr lieb sein, daß ich Ideen habe ohne es zu wissen, und sie sogar mit Augen sehe'. Johann Wolfgang von Goethe, 'Glückliches Ereignis' (1817) in: *Goethes Werke. Hamburger Ausgabe in 14 Bänden*, ed. Erich Trunz (Munich 1982) Volume X, 538-542, there 540-541; Bell, *Spinoza in Germany*, 162; also: Meinecke, *Die Entstehung des Historismus*, 531.

62 *Goethe's Theory of Colours* transl. ed. Charles Lock Eastlake (London 1840) xx. Goethe, 'Zur Farbenlehre', 317. 'Jedes Ansehen geht über in ein Betrachten, jedes Betrachten in ein Sinnen, jedes Sinnen in ein Verknüpfen, und so kann man sagen, daß wir schon bei jedem aufmerksamen Blick in die Welt theoretisieren'.

primal phenomena (*Urphänomene*) are, in Goethe's view, not eternal and unchange-able principles, but models with their own limits, and simultaneously indicate the limits of human knowledge.[63] In other words, Goethe assumes that there are certain things that human reason is unable to understand, but unlike Kant, who distinguished the unknowable object from the knowing subject, Goethe took the view that, besides experience, man is also able to gain insight into the object by means of his intuition. Whereas Kant rejected all forms of intuition because they cleared the way for speculation, Goethe greeted Spinoza's *scientia intuitiva*.[64] Instead of 'intuition' Goethe refers to this as *Anschauen* (contemplation or intellectual intu-ition), which makes it clear that in approaching a primal phenomenon it essentially refers to a combination of physical perception and mental intuition.[65]

In attempting to go beyond Kant's dualism, not only Spinoza's 'intuitive knowl-edge' was important to Goethe; equally or even more important was Schelling's (Spinoza-based) 'intellectual intuition'. Schelling considered intellectual intuition as 'the capacity to see the universal in the particular, the infinite in the finite, and indeed to unite both in a living unity'.[66] Schelling arrived at an identity of the universal and the individual through intellectual intuition, that is: the individ-ual is part of the universal, and the universal cannot do without the individual.[67] Schelling's Absolute Idealism is, therefore, engrafted into Spinoza's pantheism. Schelling also starts from the idea that self-insight grows through knowledge of God or nature, since the 'self' is a 'mode' of the Substance. In other words, whatever the self thinks or does, eventually, it is God who acts and thinks 'through' and 'in' the self: *hen kai pan*.[68] In that sense, and this is important also for Meinecke, one can realize the highest objectivity through subjectivity. After all, the ideas which realize spirit and nature and shape the creation of nature are not the property of individualities, but belong to nature or the divine, which acts through and in these individualities.[69]

---

63   Nicholls, *Goethe's Concept of the Daemonic*, 187.

64   Beiser points at 'intellectual' and 'sensible' forms. Beiser, *German Idealism*, 299.

65   Bell, *Spinoza in Germany*, 162.

66   Cited in: Beiser, *German Idealism*, 580.

67   Ibidem, 581: 'In an intellectual intuition (...) I intuit (...) the *identity* of the universal and the particular; I see how the particular is inseparable from the whole of which it is a part, and how the whole cannot be without that particular'.

68   Beiser, *German Idealism*, 583. Again, this reminds us of Goethe's *Werther* (cf Chapter 4): 'Ich kehre in mich selbst zurück, und finde eine Welt!', but now it would be more accurate to state: I return to myself, and find the *Gottnatur*!

69   Richards, *The Romantic Conception*, 471; Franz, *Goethe als religiöser Denker*, 212-214. This idea also gives more weight to Herder's notion of *Einfühlen*, with which Meinecke for the greater part agrees. Also the notion of *Verstehen* becomes more intellectually charged. With regard to the relationship between Goethe and Schelling: Dalia Nassar, 'From a Philosophy of Self to

According to Schelling – and also Goethe – we are able, by means of intuition, to know the Substance, God or nature, in which subject and object form a unity. Again, not the (isolated) subject is central in this case – as in Kant's philosophy – but the relationship or unity between nature and subject; the subject's consciousness is after all not only *his* consciousness, but also the absolute or divine consciousness within him.[70], 358. By means of his ideas on colours, Goethe, in conversation with Eckerman, puts it as follows: 'The light is there, and the colours surround us; but if we had no light and no colours in our own eyes, we would not perceive the outward phenomena'.[71] Goethe formulated it even stronger in his *Theory of Colours*:

If the eye were not sunny,
How could we perceive light?
If God's own strength lived not in us,
How could we delight in Divine things?[72]

In a draft letter to a friend Goethe went even a step further in his ideas on the relationship between subject and object:

a. In nature is everything which is also in the subject.
y. and something more.
b. In the subject is everything which is also in nature.
z. and something more.
'b' is able to know 'a', but 'y' can only divine through 'z'. From this follows the equilibrium of the world and our circle of life to which we were consigned. The essence, which, in the highest degree of clarity, comprehended all four, has been by all people always referred to as God.[73]

a Philosophy of Nature: Goethe and the Development of Schelling's *Naturphilosophie*', *Archiv für Geschichte der Philosophie* vol 92 issue 3 (2011) 302-321.

70   Beiser, *German Idealism*

71   Johann Peter Eckermann, *Gespräche mit Goethe in den letzten Jahren seines Lebens* (Berlin and Darmstadt 1958) 102. 'Das Licht ist da, und die Farben umgeben uns; allein trügen wir kein Licht und keine Farben im eigenen Auge, so würden wir auch außer uns dergleichen nicht wahrnehmen'.

72   *Goethes Theory of Colours*, xxxix. 'Wär nicht das Auge sonnenhaft/ Wie könnten wir das Licht erblicken?/ Lebt nicht in uns des Gottes eigne Kraft,/ Wie könnt uns Göttliches entzücken?'. Goethe, 'Zur Farbenlehre' 324; Goethe here sums up Plotinus' ideas – Goethe refers to an 'alten Mystikers'. Trunz claims Goethe refers to Plotinus. Cf. 'Anmerkungen' in *Goethes Werke. Hamburger Ausgabe in 14 Bänden*, ed. Erich Trunz (Munich 1982) volume XIII, 643.

73   Cited in Andreas Anglet, *Der „ewige" Augenblick. Studien zur Struktur und Funktion eines Denkbildes bei Goethe* (Cologne, Weimar and Vienna 1991) 258. 'a. In der Natur ist alles was im Subjekt ist. / y. und etwas drüber. / b. Im Subjekt ist alles was in der Natur ist. / z. und etwas drüber. / b. kann a erkennen, aber y nur durch z geahndet werden. Hieraus entsteht das Gleichgewicht der Welt und unser Lebenskreis in den wir gewiesen sind. Das Wesen, das in höchster Klarheit alle viere zusammenfaßte, haben alle Völker von jeher *Gott* genannt'.

With this 'something more' after the 'y' Goethe refers to nature in the sense of Spinoza's Substance from which all subjects emanated. For that reason, nature is in the subject and simultaneously something more, for it is the source of *all* subjects; it transcends everything. Furthermore, all that is in nature is also in the subject, but here also 'something more', since the subject is free to develop itself independently from nature, and it has, through intuition, the capacity of striving 'back' to nature – which is what the 'something more' after the 'z' refers to. This quote makes it clear that Goethe not only managed to circumvent Kantian dualism, but also went beyond Spinoza's static idea of the Substance in which everything 'is' or in which 'being' is central; Goethe 'vitalized' this notion. After all, the subject gains more room to develop freely and independently from the Substance. It was in this case Herder – appealing to Aristotle's notion of entelechy – who pointed out to Goethe that Spinoza's system was based on 'being' and that an idea of 'becoming' was lacking.[74] Herder 'dynamized' Spinoza's Substance, and as a result, nature and spirit became both manifestations of a living force in which everything is in harmony.[75]

The above dovetails with Goethe's concept of the metamorphosis. For, everything in nature should be considered as infinite manifestations emanated from the *Gottnatur*, and which through polarity and elevation clearly gives evidence of the metamorphosis.[76] According to Meinecke, Goethe's view is about 'flux and flow' (*Fluß und Wandel*)[77] and at the same time he assumes a permanent element (*Urtypus*) which indefinitely multiplies. Essentially, according to Meinecke, the harmony and dynamism of unity and variety became central.[78] Put differently, Goethe advocates an 'Aristotelian Neoplatonic' interpreted Spinozism[79]; such a characterization of course expresses Goethe's eclecticism. This 'Aristotelian Neoplatonic Spinozism' we can simply characterize as panentheistic, that is: in a sense, the world is in God or the Substance, but in another sense, God simultaneously transcends the world

---

74   Schmidt, 'Natur', 770-771; also: Beiser, *German Idealism*, 367.

75   Beiser, *German Idealism*, 367, 371. To Herder the Aristotelian notion of entelechy is the element that connects history and nature. Ankersmit, 'Aftermaths and 'Foremaths'', unpublished lecture. Zammito shows that Herder borrows this idea of 'dynamics' from Leibniz: Herder transforms Spinoza by means of Leibniz. Zammito, *Kant and Herder. The Birth of Anthropology*, 317. also: Dirk Kemper, „Ineffabile". *Goethe und die Individualitätsproblematik der Moderne* (Munich 2004) 385-386; Dobbek, *J.G. Herders Weltbild*, 46-49.

76   Jennifer Mensch, 'Intuition and Nature in Kant and Goethe', *European Journal of Philosophy* (2009) 1-23, there 15; Bartuschat, *Baruch de Spinoza*, 99-104.

77   Meinecke, *Historism*, 399.

78   Ibidem, 399.

79   Meinecke, *Die Entstehung des Historismus*, 450, 501. Goethe's Neoplatonism, however, is also a hybrid; it was passed down by pietism. In particular Gottfried Arnold's pietism was important for Goethe. On pietism: the previous chapter, section 4.2.

and leaves room for the world to develop freely.[80] Meinecke puts it as follows: it is a matter of a 'constant remodeling of the self, while yet remaining true (...)', it is a matter of 'individuality, which was both malleable and yet permanent', (...), the primal forms of nature and humanity created by God, and the thousand changes that they undergo'.[81] Hence, Goethe succeeded in merging 'becoming' and 'being'. In this context, he refers to what he calls 'constancy in the midst of change' (*Dauer im Wechsel*).[82] To Meinecke this is the essence of Goethe's thought, which he therefore considers the high point of historicism. For, equal to Goethe's idea of nature in which reconciliation but also tensions are central, he, according to Meinecke, by means of the idea of 'constancy in the midst of change' succeeds in reconciling or even transcend both Enlightenment thought and historicism: 'Goethe's basic relation to history was, therefore, neither that of the Enlightenment, nor that of historicism, but stands in the middle, or, if one wishes, transcends both'.[83] According-ing to Meinecke, Goethe shares with the Enlightenment (history) the search for a general humanity and supratemporality, and with historicism he shares 'the mean-ing for the inner law of life, for the structure, for the manifoldness, the richness and depth of the individuality of the individual man'.[84] Goethe is able to understand both the general and the individual at the same time, for both emanated from the

---

80   Hinrichs, *Ranke und die Geschichtstheologie*, 117-118; Thielicke, *Glauben und Denken*, 177. Gregersen distinguished three forms of panentheism: a *soteriological* variant in which evil is no part of God; an *expressivistic* variant, which is mainly associated with the rise of German Idealism in which the main idea is that the world stems from God and strives back to God; finally, the third variant is the so-called *dipolar* form in which God is on the one hand 'timeless, beyond space and self-identical, while in other respects temporal, spatial, and affected by the world'. Gregersen, 'Three Varieties of Panentheism', 19-35.

81   Meinecke, *Historism*, 422. On the relationship between Goethe's metamorphosis and Burckhardt's *Verpuppungen* see my section on 'the crisis of historicism' below, the same applies to the notion of 'Dauer im Wechsel'.

82   Demandt, 'Geschichte bei Goethe', 323; Meinecke, *Historism*, 423.

83   Meinecke, 'Goethes Geschichtsauffassung', 282-283; Meinecke, *Die Entstehung des Historismus*, 503. Eberhard Kessel claims that Meinecke, on a different level, was trying to bridge a gap between German thought and Western thought, or in the very least that Meinecke's historicism study could contribute to a mutual understanding of both traditions. Eberhard Kessel, 'Einleitung des Herausgebers', xxvi. In this context it is appropriate to refer to the project *Beyond Historicism*, of which this book originally (in the Dutch version) was a part. For Meinecke's interpretation of Goethe seems to prelude a world *beyond* (the problems of) historicism. In particular the synthesis of the aforementioned project is of importance in this context, because it treats this '*Beyond*', the future, viability and/or contemporary interpretation of historicism: Frank Ankersmit, *Meaning, Truth, and Reference in Historical Representation* (Ithaca 2012).

84   Meinecke, 'Goethes Geschichtsauffassung', 283. 'den Sinn für das nach innerem Lebensgesetz Gewachsene, für die Struktur, für die Mannigfaltigkeit, den Reichtum und Tiefe der Individualität des Einzelmenschen'.

Gottnatur. This Neoplatonic thought, in which the particular is rooted in the divine, was, in Meinecke's view, the principal foundation and strongest formative force of historicism.[85]

5.2   Meinecke's Umdeutung of Goethe

In the light of Goethe's idea of nature, it is striking that Meinecke initially distinguishes between Goethe's negative and positive attitude towards history. Considering Goethe's emphasis on harmony within, and of nature it would seem more logical for Meinecke to not make this distinction. Even more so, Meinecke had claimed that this disunity of a negative and positive dimension of history was not considered a problem by Goethe himself.[86]

Meinecke's idea of the disunity in Goethe's conception of history is, according to him, related to Goethe's distinction between nature and history. Despite Meinecke's claim that Goethe in the end considered history and nature as one and the same, and that history merely is an *Ausschnitt* (section) of the *Gottnatur*, he comes to the conclusion that Goethe, on a practical level, could not find the same line in history which he found in nature.[87] According to Meinecke, most of the time Goethe rejected history, because he was not able to find the same laws[88] as he did in nature.[89] Meinecke thought Goethe tried to find in history, as in nature, some kind or *Urtypus*: 'Guided by the sense of the organic unity of all life, Goethe sought in history for similar laws as found in nature, especially in the metamorphosis of plants, that is, after ever-recurring original forms, which, however, are still distinguished by the infinite creative power of Nature to differentiate individuality'.[90] In nature, Meinecke writes, Goethe found a sound development – 'Nature does not proceed by leaps'[91] – whereas history, as a result of human actions, could be 'led astray in

---

85   Meinecke, *Die Entstehung des Historismus*, 581. Meinecke, *Historism*, 492.

86   Mandelkow, 'Natur und Geschichte bei Goethe', 246-247.

87   Meinecke, *Die Entstehung des Historismus*, 511; Rudolf Vierhaus, 'Goethe und der Historismus', *Goethe Jahrbuch* 110 (1993) 105-114, there 109, 113-114; Karl Robert Mandelkow, *Goethe in Deutschland*, 181-182; Ziegler, 'Zu Goethes Deutung', 233-234, 237, 247, 249, 263, 267.

88   Goethe's laws are not the 'fixed' positivistic laws but 'laws' which he discovered through his empirical research on (different) metamorphoses: Meinecke, *Die Entstehung des Historismus*, 504. Also: H.B. Nisbet, *Goethe and the scientific tradition* (London 1972).

89   Meinecke, 'Goethes Geschichtsauffassung', 282.

90   Meinecke, 'Goethes Geschichtsauffassung', 282; Meinecke, *Die Entstehung des Historismus*, 531. Again, the similarities with Burckhardt's concept of *Verpuppungen* is apparent; on this issue: the section on the crisis of historicism (5.5).

91   Cited in: Meinecke, *Historism*, 474. This principle of continuity reminds us of Leibniz' famous statement: *natura non facit saltus*.

a morbid manner from the true path'.[92] For that reason, Meinecke distinguishes between Goethe's view of nature in which all is in harmony, and his view of history in which Meinecke makes a second distinction between Goethe's negative and positive stance towards history.[93] The problem is that Goethe's negative view, according to Meinecke, does not tally with Goethe's harmonious world view. For that reason, Meinecke, as his argument unfolds, tries to dispose of the negative elements of Goethe's view of history by means of Goethe's view of nature. In sum, Meinecke attempts to reconcile the negative with the positive dimension, or even alter the negative into a positive.[94]

---

92 Meinecke, *Historism*, 431.

93 The German literary theorist Klaus Ziegler (1908-1978) claims in his article 'Zu Goethes Deutung der Geschichte' that an antithesis like Meinecke's account of Goethe's view of history is unsatisfactory for two reasons. On the one hand, Ziegler claims that not only nature but also history is harmonious in Goethe's thought. On the other hand, analogous to history, nature should also be considered as 'disharmonious chaos'. In other words, Ziegler claims that the negative features of Goethe's view of history do not contradict his view of nature. According to Ziegler, we should consider Goethe 'as a whole'. That is, we should simultaneously view nature and history in its polarities and harmony. In the words of Ziegler: 'Die Polarität von Positivem und Negativem ist ebenso für Goethes Auffassung der Natur wie für seine Auffassung der Geschichte konstitutiv'. Like the positive dimension, Ziegler states that, analogous to Goethe's view of nature, we can also discover *Urphänomenen* for the negative dimension in history. Ziegler concludes: '(...) daß nämlich Goethes Deutung der Geschichte mit dem Ganzen seiner Natur und Seinslehre, seines Menschen- und Gottesverständnisses in der Tat unauflöslich zusammenhängt'. Klaus Ziegler, 'Zu Goethes Deutung der Geschichte', *Deutsche Vierteljahrsschrift für Literaturwissenschaft und Geistesgeschichte* 30 (1956) 232-267, there 234, 267. For a more philosophical, but similar argument: Ludwig Landgrebe, 'Die Geschichte in Goethes Weltbild' in: Gerhard Haselbach en Günter Hartmann ed., *Beiträge zur Einheit von Bildung und Sprache im geistigen Sein. Festschrift zum 80. Geburtstag von Ernst Otto* (Berlin 1957) 371-384, in particular: 373, 381-383. Literary theorist Thomas Zabka claims Zieglers interpretation of Goethe's *Urphänomen* misses a vital point: 'In Goethes Urphänomenen soll der Gegensatz von Ordnung und Chaos (...) aufgehoben sein'. Thomas Zabka, 'Ordnung, Willkür und die "wahre Vermittlerin". Goethes ästhetische Integration von Natur- und Gesellschaftsidee' in: Peter Matussek ed., *Goethe und die Verzeitlichung der Natur* (Munich 1998) 157-177, there 158. In sum, Ziegler also remains 'caught' in Meinecke's twofold distinction of Goethe's conception of history.

94 A few years before Meinecke published *Die Entstehung des Historismus*, Ernst Cassirer published in 1932 some articles in which he also discusses Goethe's attitude towards history. Meinecke does not mention these articles. Instead of distinguishing between Goethe's positive and negative stance towards history, Cassirer claims that Goethe tried to connect the 'Productive with the historical': 'In diesem Wort liegt der eigentliche Schlüssel für Goethes Anschauung von der Geschichte. Er lehnte das Historische ab, wo es nicht ihm als bloßer Stoff aufdrängte; aber er forderte es als ein Medium und als ein notwendiges Mittel, um die Form seines eigenen Seins und seines eigenen Schaffens zu entdecken. In diesem Sinne hat Goethe das Historische gebraucht und genutzt; denn, so gefaßt, bedeutete es ihm keinen Gegensatz mehr zu den produktiven Kräften, aus denen er seine Welt gestaltete, sondern

An example of Meinecke's 'reinterpretation' (*Umdeutung*) of Goethe's negative view of history is his discussion of the 'subjective element' of historical writing. All history is biased, says Goethe, influenced by the historian who wrote it. He thought it amusing, as Meinecke shows, when historical writing was attributed a label of 'scientific objectivity'.[95] Such a label could, in Goethe's view, only be ascribed to the results of (his) study of nature and to works of art, which were shaped by the inner laws of nature.[96] At first, this does not sound less subjective than Goethe's disdain for certain forms of historical writing. Goethe, however, had a different idea of subjectivity and objectivity: for him objectivity was not a scientific, verifiable – Kantian epistemological – objectivity, but a Spinozist objectivity. That means in this case, as I have shown before, that God, nature or the Substance, from which all modi emanated, are simultaneously part of the Substance: subject and object are thus one and the same. In that sense, it is God or nature that thinks and acts through and in the subject.[97] That is why Goethe could claim that by means of his subjectivity he could realize the utmost objectivity. Put differently, surrendering to subjectivity was at the same time a surrender to the claims of the object. By pointing out this Spinozist-Goethean objectivity, Meinecke nuanced Goethe's negative view of subjectivist historical writing. Meinecke claims Goethe only considered those thoughts to be true which were to himself in relation with the *Gottnatur* considered to be fruitful: 'Only that which is fruitful is true'.[98] So, Goethe could simultaneously have an aversion to historical writing in the sense of the written tradition, and accept those traditions on the account of the 'vitality of their inner content'.[99] Goethean subjectivity, which seems extremely subjective, is, according to Meinecke, in fact the strength of Goethe's thought, and did indeed not end in subjectivism (which Goethe abhorred), because the feeling of an inner connection of his own nature with the *Gottnatur* served as a basis for truth. In other words, Meinecke tried to reinterpret (*Umdeuten*) Goethe's distaste for the subjective elements of historical writing, to – analogous to Goethe's conception of nature – transcend Goethe's 'negative relationship to history' on the level of the *Gottnatur*.[100]

---

es wurde selbst eine bildende Grund- und Urkraft, die ihm die Welt des Geistigen erschloß und in ihrem ganzen Reichtum zu eigen gab'. Ernst Cassirer, *Goethe und die geschichtliche Welt* (Hamburg 1995, First 1932) 25-26.

95   Meinecke, *Die Entstehung des Historismus*, 506-507.

96   Ibidem, 507; also: Menke-Glückert, *Goethe als Geschichtsphilosoph*, 56-76; T.J. Reed, *Goethe* (Oxford and New York 1984) 15-17, 31, 35.

97   Cf. the section above on the Pantheism Controversy.

98   Cited in Meinecke, *Historism*, 428. This is a line from Goethe's poem 'Vermächtnis' (1829), which I will come back to in the section on the crisis of historicism (5.5).

99   Meinecke, *Historism*, 428.

100   Mandelkow, 'Natur und Geschichte bei Goethe', 247.

With that Meinecke sheds a different, more favourable light on Goethe's negative view of history.

Meinecke's reinterpretation of Goethe's negative view of history is also apparent in his discussion of Goethe's aversion to 'coincidence' or 'chance' in history, and especially 'evil coincidence' in history.[101] For that matter, Goethe also abhorred coincidences in nature, as they occur in natural disasters like, for instance, volcanic eruptions. Goethe preferred a steady development (metamorphoses).[102] In nature he generally could detect this kind of development, but with regard to history, Meinecke writes, he could not: 'He had no means of recognizing and eliminating the turbulent volcanic element [in history, RK], or at least of confining it within strict limits'.[103] Goethe's aversion to chance as well as his inability to tackle this problem in history does not tally with Goethe's Spinozist and Neoplatonic, harmonious world view in which everything emanated from the *Gottnatur*.[104] Meinecke thus concludes: 'Viewed in the broad light of the history of thought, Goethe would seem to have possessed the key that would have unlocked this world as well, and to have filled it out with the concepts of primal form and metamorphosis. He had it in his pocket, and yet did not use it'.[105] Put differently, in Meinecke's view, Goethe could have overcome his aversion to chance in history via his view of nature. Here Meinecke, again, adjusts Goethe's view of history by means of his view of nature, or at least he points out Goethe had the opportunity to ground his view of history in his view of nature. Meinecke concludes his discussion of Goethe's negative relationship to history as follows: 'In a good few of the motives conditioning his negative attitude to history we have seen ferments at work that were also active in his positive attitude'.[106] This is how Meinecke tried to attenuate Goethe's negative view of history or to reconcile it with his positive view of history. Meinecke thus always considers Goethe's view of history against the background of his view of nature, which enables him to reconcile the positive with the negative view of history.

Despite Meinecke's reinterpretation, Goethe really did not know how to counter chance in history. As is clear from the historical part of his *Theory of Colours*: 'Law and chance are intertwined, but the contemplative man often confuses the two, as is especially evident in the case of biased historians, who, though mostly unconsciously,

---

101 Meinecke, 'Aphorismen', 261.

102 Meinecke, *Die Entstehung des Historismus*, 511-512, 514.

103 Meinecke, *Historism*, 431.

104 When chance is interpreted as part of the broader notion of 'Schicksal' it tallies with this view and also with the concept of *daimon*. See the next section, in particular Goethe's definition of the daemonic; and also the next chapter.

105 Meinecke, *Historism*, 433.

106 Ibidem, 442.

but still artificially, use this uncertainty to their advantage'.[107] What prevented Goethe from grasping history in the same manner as he viewed nature, was, in Meinecke's view, not his 'incapacity for historical judgement', but an 'incapacity of his inmost self'.[108] To Goethe chance in history remained incomprehensible. Meinecke also shows this, but he raises this point as evidence for Goethe's reluctant acceptance of chance. Meinecke claims that Goethe indeed acknowledged chance in his own life – but in the present, not the past. He was, Meinecke writes, always aware that chance was part of the 'inscrutable texture of the world's events'.[109] This is precisely why Goethe had an aversion to chance in history: it was unfathomable, obscure, and above all daemonic (*Dämonisch*).

5.3   Goethe and the daemonic

A concept that is closely related to chance and to which Goethe did attribute value and acknowledged its effect in history is the concept of the daemonic (*Dämonische*). Initially, this concept has a mainly positive meaning for Goethe, in the sense of an 'open force' which prompts the 'heart and soul' of humans and nature into a creative act. According to Goethe, the daemonic is part of the *Gottnatur* and thereby of man's personality. It is essentially a force which is present both 'within' and 'outside' a person – in the sense of 'pre-conscious'[110] – and thus it resembles the principle of the spinozist God who acts in and through the self (*hen kai pan*).[111] For the remainder of this chapter it is important to briefly explain Goethe's interpretation of the daemonic, for it clarifies Meinecke's use of Goethe, and it explains Meinecke's position in the 'crisis of historicism', the subject of the last section of this chapter.

Goethe essentially agrees with the classical reading of the concept of the daemonic.[112] The major classical writers who discussed the deamonic are Heraclitus,

---

107   Johann Wolfgang von Goethe, 'Materialien zur Geschichte der Farbenlehre' in: idem, *Goethes Werke. Hamburger Ausgabe in 14 Bänden*, ed. Erich Trunz (Munich 1982) Volume XIV, 7-269, there 49.

108   Meinecke, *Historism*, 436. To Goethe – consistent with his research into natural phenomena – the present, nature was far more interesting to him than the past. A few years before *Die Entstehung des Historismus* was published, the Germanist Emil Ermatinger (1873-1953) had shown Goethe considered the past as being forever hidden, unlike nature, which is always present, from which Goethe tried to distil the divine 'creative force'. Emil Ermatinger, 'Goethe und die Natur. Rede zur Goethe-feier der Universität Zürich am 22. Februar 1932', *Wege zur Dichtung. Zürcher Schriften zur Literaturwissenschaft* 13 (1932) 7-33, there 15.

109   Meinecke, *Historism*, 431.

110   In that sense the concept of the daemonic transcends time.

111   Thielicke, *Glauben und Denken*, 216; also: Dirk Kemper, *„Ineffabile". Goethe und die Individualitätsproblematik*, 443-445; Nisbet, 'Goethes und Herders Geschichtsdenken', 126.

112   Nicholls, *Goethe's Concept of the Daemonic: After the Ancients*.

Socrates and Plato. They considered the *daimon* a 'divine' voice within a person. Heraclitus described it as follows: 'Man's ethos is his daimon', that is: a man's character is his fate (ἤθος ἀνθρώπῳ δαίμων)[113], which means man's fate or destiny is implicit in his character.[114] An individual's demon is thus his 'life principle' and in that sense comparable with Aristotle's concept of entelechy.[115] The daimon, destiny or fate in relation to the character or nature of man was in Plato's view a *self-chosen* fate. He was convinced that before entering earthly life all souls chose their own fate; this took place in the *topos daimonos*, the daemonic place. At the onset of their earthly life these souls all forgot their fate because in order to reach earthly life, they had to go through the river of oblivion (Lethe). Everyone chose their own fate, but since they all forgot it on earth it seemed life was ruled by necessity.[116] The daimon is considered to be the 'inner voice' which can remind us of the knowledge we have lost: our fate, but also the eternal divine 'forms' or Ideas of the True, Good and Beautiful. This 'remembering' (*Anamnesis*) can occur – says Socrates in Plato's *Phaedrus* – when the subject contemplates the beauty and goodness of certain aspects of reality. In this contemplation the subject will remember where he emanated from: the transcendental realm.[117] For this reason Plato considered the daimon as the mediator between heaven and earth, for it made fate and knowledge of the eternal forms accessible to man.[118]

The nature of the daemonic or demon is, considering the above, twofold.[119] On the one hand it transcends history, for it gains access to fate and the eternal forms. On the other hand, this remembering of our fate and the divining of the eternal takes place within reality, within history.[120] So, the daemonic reconciles both the

---

113   Heraclitus, *Fragments. The Collected Wisdom of Heraclitus*, Transl. Brooks Haxton (Harmondsworth 2001) 27, 30. Haxton translated this quote as 'One's bearing shapes one's fate'. I chose a variation of the more common translation (which Haxton mentions in his footnote): Character is fate.

114   Cf. Walter Benjamin on the relationship between character and fate: Walter Benjamin, 'Schicksal und Charakter' in: idem, *Gesammelte Schriften II.I*, ed. Rolf Tiedemann and Hermann Schweppenhäuser (Frankfurt am Main 1977) 171-179. Benjamin mainly discusses the linguistic relationship between both concepts and the connection with the German spirit.

115   Nicholls, *Goethe's Concept of the Daemonic*, 11-12, 39. This of course also reminds us of Leibniz' windowless monads and Spinoza's modi.

116   Nicholls, *Goethe's Concept of the Daemonic*, 40. That is probably the reason why Goethe abhorred chance, for it is part of our forgotten fate.

117   Nicholls, *Goethe's Concept of the Daemonic*, 41.

118   Ibidem, 37, 41.

119   Thielicke claims it belongs to a 'no-man's-land', a *Zwischenraum*: Thielicke, *Glauben und Denken*, 213.

120   In this conception of the historical and the eternal the ideas of Goethe's metamorphoses, Burckhardt's *Verpuppungen* and Meinecke's 'horizontal-vertical' conception of history meet. I will come back to this in the next section.

eternal and historical – idea and reality. In this regard, there is a connection with political action.[121] To Goethe the interweaving of the daemonic and the political was represented by Napoleon. According to Goethe, this political genius, by means of the daemonic power or force, was able to be the captain of his soul, so to speak, and thereby controlled his fate.[122] According to reports, Napoleon, in conversation with Goethe, once said: 'politics is fate'.[123] This remark implies that (in Napoleon's view) there is no such thing as a predetermined fate as Plato conceived it, but that in fact a politician can direct fate. Goethe, in the end, could not agree with this view and even revised his opinion of Napoleon. Goethe could not consider

---

121   In the chapter on Meinecke's *Die Idee der Staatsräson* I have already discussed the principle of *Staatsräson* which presumes that the intuitive, prudent statesman, who – by rising above himself to catch a glimpse of the divine – knows (or 'remembers') what should be done at a certain, crucial moment.

122   Nicholls, *Goethe's Concept of the Daemonic*, 252. A genius transcends his own age and sets the rules for art, politics, etcetera. Goethe describes it in his 'Maximen und Reflexionen': 'Das Genie bedürfte (…) keine Regel, wäre sich selbst genug, gäbe sich selbst die Regel; da es aber nach außen wirkt, so ist es vielfach bedingt durch Stoff und Zeit, und an beiden muß es notwendig irre werden (…)'. Johann Wolfgang von Goethe, 'Maximen und Reflexionen' (posth. 1833) in: *Goethes Werke. Hamburger Ausgabe in 14 Bänden*, ed. Erich Trunz (Munich 1982) Volume XII, 528. However, Goethe claims that the daemonic personality is not by definition a genius, because daemonic personalities are: '(…) nicht immer die vorzüglichsten Menschen, weder an Geist noch an Talenten, selten durch Herzensgüte sich empfehlend; aber eine ungeheure Kraft geht von ihnen aus, und die üben eine unglaubliche Gewalt über alle Geschöpfe, ja sogar über die Elemente (…)'. When discussing the more positive elements of the daemonic – insight in one's fate, the true, good and beautiful – there is in fact a relationship between the daemonic personality with his (remembered) insight and the genius. Goethe, following Schelling, considered 'ingenuity' to be an inborn mental quality; which means that the subject's consciousness is also connected to the absolute consciousness. And given that the daemonic is part of *Natur*, this innate quality – or entelechy – can also be of a daemonic nature. Related to this, as stated earlier, is the importance of the concept of intuition. Only our intuition enables us to gain insight into the unfathomable. In this case we can perhaps characterize it as a 'daemonic intuition', because a daemonic personality has 'knowledge' of the absolute, and can use this insight or knowledge to design (his) reality. In this respect one can also think of what Goethe understood as *Urphänomen* and its effect on reality, that is: the eternal, perfect 'idea' which manifests itself in reality and designs this reality. Thielicke, *Glauben und Denken*, 213; Cited in Hermann Schmitz, 'Das Ganz-Andere. Goethe und das Ungeheure' in: Peter Matussek ed., *Goethe und die Verzeitlichung der Natur* (Munich 1998) 424; Theo Buck, 'Dämonisches' in: Bernd Witte ed., *Goethe Handbuch in vier Bänden* (Stuttgart and Weimar) Volume 4/1, 179-181, there 179; Anglet, *Der "ewige" Augenblick*, 389, 407; Anglet points out that, apart from attributing the notion of the daemonic to politics, Goethe also used it with regard to art, poetry, love, and music. With regard to music and the daemonic: Kiene Brillenburg Wurth, *The Musically Sublime. Infinity, Indeterminacy, Irresolvability* (New York 2009) 170-172.

123   Nicholls, *Goethe's Concept of the Daemonic*, 251.

him a genius who transcended (his) fate, but deemed him a politician who in fact ignored the 'inner voice' of the daimon. In this context the daimon refers to Socrates' notion of the daemonic as the inner voice – our conscience – which points at moral boundaries or which advises against something. Since Napoleon refused to listen to this inner voice he finally met with disaster.[124] Hegel later on discussed this conversation between Goethe and Napoleon in relation with his notion of the world historical individuals (*Welthistorische Individuen*). According to this idea personalities like Napoleon are in fact an instrument in the hands of fate: they appear in a short space of time on the world historical stage and make their exit almost as quickly once they accomplish their task (fate) for the benefit of the (self)realisation of the spirit.[125]

To put it briefly, the daemonic cannot be controlled by man, he can only comply with it. This does not mean, according to Goethe, that the effect of certain political actions cannot still be both productive and creative and also tragic, destructive or even catastrophic.[126] So, apart from having a mediating role between higher things and reality, the daemonic is also twofold in regard to good and evil. In this context, Goethe points out that the daemonic is morally and rationally indifferent. Because man cannot find structure or morality in the daemonic – which is part of the unfathomable *Gottnatur* – he considers it amoral, disruptive or chaotic.[127] Man is thus not rationally capable of understanding why good can result in evil and vice versa. This is, by the way, not the same as the aforementioned Hegelian cunning of reason, because the daemonic is mainly concerned with a (timeless) connection, not a purpose or goal.[128]

The twofold character of the daemonic is clearly expressed in Goethe's final 'definition'. To him the daemonic is:

> (...) not godlike, for it seemed unreasonable; not human, for it had no understanding; nor devilish, for it was beneficent; nor angelic, for it often betrayed a malicious pleasure. It resembled chance[[129]], for it evolved no consequences; it was like Providence, for it hinted at connection. All that limits us it seemed to penetrate; it seemed to sport at will with the necessary elements of our existence; it

---

124   Ibidem, 260-262.

125   Cf. chapter 3. Nicholls, *Goethe's Concept of the Daemonic*, 251, 253.

126   Buck, 'Dämonisches', 179; Anglet, *Der "ewige" Augenblick*, 390, 398-408.

127   Anglet, *Der "ewige" Augenblick*, 387. This reminds us again of the principle of *Staatsräson*.

128   Buck, 'Dämonisches', 180; Thielicke, *Glauben und Denken*, 219; Hermann Schmitz, 'Das Ganz-Andere. Goethe und das Ungeheure' in: Peter Matussek ed., *Goethe und die Verzeitlichung der Natur* (Munich 1998) 414-435, there 427-429. On the relation between Hegel and Goethe: Schultz, 'Die Bedeutung des Tragischen', 92-115.

129   Chance is also part of the forgotten fate, which is why life seems ruled by necessity, while essentially it is a self-chosen fate. That is why chance resembles the daemonic, which has the same effect.

contracted time and expanded space. In the impossible alone did it appear to find pleasure, while it rejected the possible with contempt. To this principle, which seemed to come in between all other principles to separate them, and yet to link them together, I gave the name of Daemonic, after the example of the ancients (...).[130]

Below I will come back to Goethe's idea that time and space can contract and expand. For the moment it is important to realize that the twofold character of the daemonic is the reason why Goethe considered it part of the inscrutable (*Unerforschlichen*).[131]

5.4   Harmony and dissonance

Meinecke touches only briefly on the daemonic in *Die Entstehung des Historismus*. He merely states, since Goethe considered the daemonic active in all of nature, that it should not have kept him from history, in which the daemonic is most visible.[132] In *Die Idee der Staatsräson* Meinecke also mentions the daemonic in passing, but again he does not elaborate on this issue. In essence, Meinecke agrees with Goethe's interpretation.[133] But next to Goethe's interpretation, is seems that a Christian connotation[134] crept into Meinecke's account of the daemonic. That is, Goethe thought the daemonic morally indifferent, Meinecke, however, distinguishes between good and evil, whereby the daemonic is connected to evil – as is the case in our current use of language. It is this meaning Meinecke uses mainly in *Die Idee der Staatsräson*.[135] At the end of this study he even claims that the statesman should carry both God and the state in his heart, and guard against a too powerful *Dämon*, in other words: uncontrolled power politics, brute force – evil – should be kept to the minimum.[136]

---

130   Johann Wolfgang von Goethe, *The Autobiography of Goethe. Truth and Poetry: from my own life*, Two volumes, Transl. A.J.W. Morrison (London 1881) volume II, 157; Original: Johann Wolfgang von Goethe, 'Aus meinem Leben. Dichtung und Wahrheit' in: idem, *Goethes Werke. Hamburger Ausgabe in 14 Bänden*, ed. Erich Trunz (Munich 1982) volume IX-X, there volume X, 175-176.

131   Anglet, *Der "ewige" Augenblick*, 387; Meinecke, *Die Entstehung des Historismus*, 523; Meinecke, *Historism*, 441.

132   Meinecke, *Die Entstehung des Historismus*, 523.

133   Ibidem, 523; Meinecke, *Die Idee der Staatsräson*, 505; Friedrich Meinecke, 'Von Schleicher zu Hitler. Volksgemeinschaft – nicht Volkszerreißung' in: idem, *Politische Schriften und Reden* (Darmstadt 1958) 479-482, there 482; Bock, 'Meinecke, Machiavelli und der Nationalsozialismus', 163.

134   Nicholls, *Goethe's Concept of the Daemonic*, 13-14.

135   See Chapters 3 and 4. Cf. Meinecke, *Die Idee der Staatsräson*, 43, 46, 137, 321, 487, 510; also: Meinecke, 'Kausalitäten und Werte', 80, 86-87.

136   Meinecke, *Die Idee der Staatsräson*, 510.

A similar dual meaning bears Meinecke's interpretation of the (Goethean) concept of *Natur*, which to him is closely related to the daemonic, in the Christian and Goethean sense. It therefore serves a purpose to discuss Meinecke's view on the relationship between *Natur* and history, in particular because it sheds light on his appropriation of Goethe.

History and nature

In *Weltbürgertum und Nationalstaat* Meinecke considered nature as harmonious and morally just. After the First World War, his conception of nature changed and became twofold. He linked nature with evil, with the brutish and the urges of man – his struggle with good and evil in *Die Idee der Staatsräson* is clearly proof of this.[137] Opposite to this 'evil' nature Meinecke positions *Kultur*, which he also denotes as *Natur*, but now in the sense of the Goethean *Natur* or even the *Gottnatur*; the connection with higher things: the good, the true and beautiful.[138] In Meinecke's view of nature after the First World War man opposes nature – in the sense that he struggles with evil, with the animal within himself and in nature – and at the same time nature is the source from which his personality emerges and develops, for it is part of the *Gottnatur* from which it has emanated and is able to strive for a reconciliation. Meinecke thus distinguishes between a 'human' moral nature and a 'metaphysical' amoral *Natur*, but both 'natures' are present 'within' and 'outside' man, because both are part of the *Gottnatur*. In the words of Meinecke: 'Culture and nature, we might say God and nature, are undoubtedly a unity, but a unity divided in itself'.[139]

It is remarkable that Meinecke at first considers evil as something that should be overcome, instead of viewing it as a necessary element of the (Goethean) harmony. He states for example: 'Culture enters only when man takes up the struggle against nature with all his inner powers, not only with his will and intellect; only when his acts have value in the higher sense, i.e. when he creates or seeks something good or beautiful for its own sake or for its own sake seeks the truth'.[140] Nature in the sense of an 'evil nature' seems to Meinecke more an obstacle to, than part of a harmony. This does not alter the fact that Meinecke was indeed seeking harmony. In accordance with his re-interpretation (*Umdeutung*) of Goethe, he wants to show that through a kind of metaphysics analogous to Goethe's view

---

137 Cf. chapter 3 and 4.
138 In fact, the entire *Die Entstehung des Historismus* refers to this conception of *Kultur*. In the rest of his oeuvre Meinecke also frequently makes use of this conception of *Kultur*. In particular: Meinecke, 'Kausalitäten und Werte', 80, 86; Meinecke, 'Geschichte und Gegenwart' 99.
139 Meinecke, 'Values and Causalities in History', 282.
140 Ibidem, 279.

of nature, evil, chance or the demonic (in a Christian sense) can be 'overcome' in history. A Goethean metaphysics thus enables Meinecke to view evil as part of the Goethean harmony of the *Gottnatur*.[141] With this Meinecke holds a key to the distinction between the good, true, beautiful and evil, animal-like, dark, or even the destructive in history, and he may be attributing a certain meaning to this negative pole. Particularly this last point is important to Meinecke, for all too often he encounters the interweaving of good and evil in history: '(...) as though God needed the devil to realize Himself'.[142]

Meinecke uses Goethe's harmonious world view to move between the levels of 'history' – with its 'manifest' distinction between good and evil[143] – and the (metaphysical) *Natur*, and above all Meinecke wants to unify and reconcile them; like his previous works, *Die Entstehung des Historismus* is once again a search for harmony between opposite notions or antinomies. Following Goethe, Meinecke claims that we could 'feel the antinomies of life and history not as lawlessness, but as necessary dissonances within the total harmony of the universe'.[144] This may sound dualistic, but against the background of the foregoing it is clear that it is about the Goethean principle of the *Gottnatur* in which polarities are in harmony. In this context, a comparison with music, as stated by Ranke, Burckhardt and Huizinga, might clarify this idea of polarities and harmony. When Ranke speaks of Europe in the closing pages of *Über die Epochen der neueren Geschichte*, he asserts: 'The exclusion of nationalities against each other is now no longer feasible; they all belong to the great European concert'.[145] In his posthumously published *Historische Fragmente*, Jacob Burckhardt compares the different nationalities of Europe with clocks, which, even though close at hand they are in disharmony, at a distance they ring in harmony: 'From a high and distant vantage point, such as an historian's ought to be, bells harmonize beautifully, regardless of whether they may be in disharmony when heard from close by: *Discordia concors*'.[146] In a similar vein Huizinga claims, with regard to history: 'as false music at a distance sounds pure, so does the serenity of an earlier time'.[147] These statements echo Goethe: 'And so

---

141   This reminds us of Meinecke's 'new dualism' in which he tried to reconcile monism (unity of spirit and nature) with dualism. This interpretation of dualism was in the end based upon Meinecke's panentheistic conviction. Cf. chapter 3.

142   Meinecke, 'Values and Causalities in History', 281.

143   The moral problems to which this leads will be discussed in the section on the crisis of historicism (5.5).

144   Meinecke, *Historism*, 490.

145   Ranke, *Über die Epochen*, 165.

146   Jacob Burckhardt, *Judgements on History and Historians* (London and New York 2007 first 1959) 174; Jacob Burckhardt, *Historische Fragmente*, ed. Emil Dürr (Stuttgart 1957) 192-193.

147   'Zooals valsche muziek op een afstand zuiver klinkt, zoo ook de sereniteit van een vroeger tijd'. Cited in Léon Hanssen, *Huizinga en de troost van de geschiedenis. Verbeelding en rede*

we see once again that every creature is a note, a single sound in a great harmony which must be studied as a whole, or each separate element in it is a mere dead letter'.[148]

In *Die Entstehung des Historismus* and in the articles published in the 1920's and 1930's Meinecke suggests grounding historicism, which was increasingly under attack, on a Goethean panentheistic conviction. In Meinecke's view, this construction is the only way to transcend his morally Christian-tinged conception of nature and the demonic. Meinecke is able to 'disarm' the 'evil' dimension of history against the background of a transcending 'authority' like the *Gottnatur*. In 'God' (at a distance) all individualities are in harmony, but close at hand, in our daily life or in certain periods these individualities present themselves as dissonant. But these dissonances are only an 'appearance', for it is man that is not able to position himself at a standpoint from which the eternal harmony is clearly visible.[149] Nevertheless, Meinecke, following Ranke and Burckhardt, thought historians were or even should be able to take up such an 'elevated' standpoint. As he writes in a letter to Gerhard Ritter: 'And my thinking also does not prevent me, as you know, from the demonic and tragic character of historical life, but it does not shake it upside down'.[150] It is precisely these thoughts that gave Meinecke something to hold on to when confronted with the so-called crisis of historicism.

(Amsterdam 1996) 341. Original: Leiden, University Library, Huizinga-archive, 122, envelope 'Invallen'. It is remarkable how close Huizinga's view of history is to Meinecke. Krul describes the following: '(...) he repeatedly remarked [Huizinga, RK] that history could only be understood by him as a balance of antinomies. The remarkable thing about this is not the contradiction, but the balance. And: 'The culture, he learned [Huizinga, RK] exists by the grace of opposites; it takes shape in a continuous exchange of antinomies, each of which has a limited and time-bound character; the last truth (which in his opinion was indubitably hidden) remains (...), and is never entirely accessible to the human mind'. W.E. Krul, *Historicus tegen de tijd. Opstellen over leven en werk van J. Huizinga* (Groningen 1990) 23-24, 266.

148   Cited in: Meinecke, *Historism*, 394.

149   Cf. Meinecke's previously cited statement in *Die Idee der Staatsräson*: 'Wir sehen in der Geschichte nicht Gott, sondern ahnen ihn nur in der Wolke, die ihn umgibt'. Meinecke, *Die Idee der Staatsräson*, 510. Meinecke's panentheism differs from Hegel's self-realisation of the spirit, in which reality and the development of Reason are the same; Meinecke is concerned with the harmony of reality and eternity, and this is located not only at the end of history, but also 'before', 'within' and 'after' history. Meinecke is close to Ranke in this respect, for Ranke assumed a transcendental God opposite of the world, who nevertheless could be 'discovered' through or in reality. Hegel, on the other hand, assumed a god who would unfold through history. Jaeger and Rüsen, *Geschichte des Historismus*, 36-37. On the similarities between Meinecke and Troeltsch on this subject, see the next section.

150   Meinecke, *Neue Briefe*, 377.

5.5   Meinecke and the crisis of historicism

'There was talk of a 'crisis of historicism' in the post-war period, when our political and social life had been in a turmoil of doubt and experimentation', according to Meinecke when in 1942 he looked back on the 1920's and 1930's.[151] The well-known theologian Ernst Troeltsch, a friend of Meinecke and colleague in Berlin, was one of the first writers after the First World War who expressed and analysed the deep cultural disorientation of Germany.[152] In 1922 his voluminous, but unfinished *Der Historismus und seine Probleme* was published. A major part of the argument of this book was already published in several articles from 1916 onwards, among others in 'Meinecke's' *Historische Zeitschrift*.[153]

In *Der Historismus und seine Probleme* Troeltsch defines historicism as 'the fundamental historicization of all our thinking about man, his culture and his values'.[154] First of all, this refers to the idea that everything is caught in a 'stream of becoming'. That is, everything is part of a continuous historical development; an infinite, singular and unique process. Second, the consequence of this idea is that the eternal, fixed values, from which man can derive a moral standpoint, are threatened in their very existence. For, if values are tied to time and context, they become relative against the background of a continuous historical development.[155]

The essence of Troeltsch's conception of historicism is not so much the relativism of absolute values as the idea of a 'stream of becoming', which in the end proved to be a tidal wave.[156] It turned out that history seemed to develop in a direction which ran counter to the dominant (nineteenth-century) idea of a continuous, harmonious development.[157] It is not without reason that this idea took root after the First World War. The war after all had ended the thought of a continuous and harmonious development of culture. This rupture of a harmonious development

---

151   Meinecke, 'Von der Krisis', 196.

152   For the role Otto Hintze played in this debate see: Leonard S. Smith, *The Epxert's Historian: Otto Hintze and the Nature of Modern Historical Thought* (Eugene 2017) especially chapter two.

153   Erbe, 'Das Problem des Historismus', 74.

154   Troeltsch, *Der Historismus*, 102; Translation: Beiser, *The German Historicist*, 2.

155   Erbe, 'Das Problem des Historismus', 77; Blanke, *Historiographiegeschichte*, 558.

156   Paul, *Het moeras van de geschiedenis*, 83.

157   Of course, with the exception of the (anti-historical) Neokantians, who assumed absolute, timeless values. Cf: Thomas E. Willey, *Back to Kant. The Revival of Kantianism in German Social and Historical Thought, 1860-1914* (Detroit 1978). The crisis of historicism is also related to the views on the 'flux of time' that were already manifest during and after the *fin de siècle*; one can think of, among many others, Alfred North Whitehead, Henri Bergson, Marcel Proust, and Thomas Mann. On the last three: Stephen Kern, *The Culture of Time and Space, 1880-1918* (Cambridge Mass. and London 2003, First 1983).

also made clear that pre-war values proved to be no longer viable. So, from an ethical point of view everything was called into question after the war.[158]

To Troeltsch the concept of development was central. Meinecke speaks of a 'stream of becoming', but where Troeltsch struggled with the idea of a directionless tidal wave of development, Meinecke kept a certain calmness regarding this idea of an alleged tidal wave. In a letter to the philosopher Erich Rothacker for example he claims, 'Certainly, 'development' must always have a 'goal' ahead, but the essence of this goal is its ability to shift steadily within a probably fixed margin, perhaps like the summit of an unknown mountain to which one ascends under various viewpoints, but in the end, one finds out that it is completely different from what one imagined at first'.[159] So, for Meinecke, the idea of the danger of an uncontrolled development, which also has (no longer) any direction, does not immediately lead to a feeling of crisis. This has everything to do with Meinecke's attention on the individual *within* the 'stream'.

The concept of development is of course important to Meinecke, but it is the addition of the concept of individuality which is, in his view, vital for historicism. 'Historicism means not only 'seeing everything in a stream', but also seeing new forms within the stream', Meinecke writes in a review of Karl Heussi's *Die Krisis des Historismus* – in this study Heussi, according to Meinecke, only discusses the concept of development.[160] Meinecke also disagreed with Heussi's definition of historicism. Heussi defined historicism as: 'the way of thinking of German historiography around 1900'[161], whereas Meinecke's conception of historicism (following Troeltsch) was much broader than merely a 'historical way of thinking around 1900'. Meinecke claimed it was 'the new principle of understanding life and history originating from

---

158 Colin T. Loader, 'German Historicism and Its Crisis', *The Journal of Modern History* vol.48 No. 3 (September 1976) 85-119, there 88, 94, 103, 106. Loader claims the crisis was not so much intellectual as political, especially the problem of German pluralism after the First World War: Loader, 'German Historicism', 108-112.

159 Meinecke, *Neue Briefe und Dokumente*, 357. To his colleague Kurt Breysig Meinecke wrote: '"geschichtliches Leben" bezeichnet das nie Fertige, immer weiter ins Unbekannte Strömende'. Meinecke, *Ausgewählter Briefwechsel*, 169. Time and again Meinecke discussed the idea of development with colleagues like Breysig. Meinecke respected Breysig's position, but he always stated that he could not agree with Breysig's view: 'Sie sind nun mal auf's "Entwickeln", nach meiner Auffassung auf einen verfeinerten Positivismus, eingeschworen, – ich dafür auf den "Historismus", d. h. auf das Ineins von Individualitätsgedanken und Entwicklungsgedanken, – und niemand von uns kann auf seine alten Tage aus seiner Haut heraus. Lassen wir uns gegenseitig gelten, auch wenn wir uns nicht gegenseitig ganz billigen'. Meinecke, *Ausgewählter Briefwechsel*, 166. Also: Meinecke, *Neue Briefe und Dokumente*, 367. In this context also: Meinecke's reaction to Carl Schmitt. Cf. above, section 3.2 footnote 44.

160 Meinecke, 'Von der Krisis', 197, footnote.

161 Heussi, *Die Krisis des Historismus*, 20.

the spiritual revolution of the Goethe era'.[162] It was precisely in this nineteenth-century revolution that 'becoming' and the principle of individuality became central notions; in Meinecke's words: 'one of the greatest intellectual revolutions that has ever taken place in Western thought'.[163]

The tidal wave to which the stream of becoming had turned into at the beginning of the twentieth century was not a direct threat to Meinecke. His relaxed attitude can be ascribed to his conviction that it was possible to transcend this stream. In a letter to his student Siegfried Kaehler (1885-1962) Meinecke explained his philosophy of history, whereby one could transcend the problems of historicism, as follows:

> History has a demonic character; thit is something we did not realize before, and sometimes the thought that the demiurge [[164]] who brought this remarkable world to life is itself a demonic creature, blessed, unblessed, beautiful, and terrible [[165]] at the same time, plagues me, (...) Above this demiurge, which we can at most grasp with our means of knowledge, there must still be a last, absolute, which we cannot grasp – or only in such a way that we break decisively with conventional philosophy of history, which is bound to temporality and thereby to relative standards, and build a new one out of our own experience. As Spranger once said in a conversation with me: instead of the horizontal, we need a vertical philosophy of history. For this, then, Goethe's words apply: The moment becomes eternity [[166]]. For it is not the historical life itself with its confluence of forces – that always retains that demonic character – but the individual human soul that can really de-demonize [*entdämonisieren*, RK [167]] itself momentarily in the fulfilled moment, probably sinking back again and again, and yet can ascend again and again.[168]

---

162   Meinecke, 'Von der Krisis', 202.

163   Meinecke, *Historism*, liv.

164   The Demiurge is in Plato's thought the 'craftsman of the world' Cooper, *Panentheism*, 35-36.

165   In this sense there is also a connection between the daemonic and the sublime, in which the beautiful and the terrifying intertwine: Ankersmit, *The Sublime Historical Experience*.

166   In general Goethe's phrase is: The moment *is* eternity, not *becomes*. The phrase is part of his poem *Vermächtnis* (legacy).

167   Here in the sense of to dispose of evil, and not in the sense of withdraw oneself from (daemonic) fate.

168   'Geschichte hat nun einmal, was wir uns früher nicht so klar machten, einen dämonischen Charakter, und zuweilen plagt mich der Gedanke, daß der Demiurg, der diese merkwürdige Welt ins Leben rief, selber dämonischen Wesens ist, selig-unselig, schön und schrecklich zugleich. (...) Über diesem Demiurg, den wir mit unseren Erkenntnismitteln allenfalls noch fassen können, muß ein Letztes, Absolutes noch da sein, das wir gar nicht fassen können – oder nur doch so, daß wir entschlossen brechen mit der herkömmlichen Geschichtsphilosophie, die an zeitliche Verläufe und damit an relative Maßstäbe gebunden ist, und uns eine neue aus dem eigenen Erleben aufbauen. Wie es Spranger im Gespräch mit

This quote captures Meinecke's whole thought on history, which he elaborated on in several articles from the same period.[169] On the basis of a Goethean panentheistic world view he is able to shed a comforting light on the dissonances that arise in historical life itself.[170] What Meinecke actually means by Goethe's 'eternal moment', how he puts the so-called vertical conception of history into practice, and what all this says about his position in the crisis of historicism will be discussed in this section. Moreover, the role of Burckhardt in Meinecke's conception of history becomes ever clearer. I will particularly show that Meinecke's vertical conception of history is very close to Burckhardt's notion of *Verpuppungen*. But before discussing these issues, I will first elaborate on Meinecke's conception of the notion of the (Goethean) intuition, for this is the basis of his vertical conception of history.

Intuition

Meinecke uses two notions of intuition. Following Goethe, Meinecke claims that intuition alludes to, on the one hand, a sensing or divining (*ahnen*) of the *Gottnatur*. It is the intellectual intuition, in the sense of Spinoza's *scientia intuitiva*, through which the godlike can partially be fathomed or known. This concept of intuition is directed at higher things, the 'vertical', the Neoplatonic panentheistic striving for the *Gottnatur*.[171] On the other hand, on the 'horizontal' level of history Meinecke interprets intuition in the sense of empathising with the past. According to Meinecke, an historian is able to empathise with the past or even to feel the past by means of his intuitive inwardness (*Innerlichkeit*).[172] In contrast with the intellectual

---

mir einmal ausdrückte: statt der horizontalen einen vertikale Geschichtsphilosophie. Für die gilt dann Goethes Wort: Der Augenblick wird Ewigkeit. Denn zwar nicht das geschichtliche Leben selbst mit seinem Konflux der Gewalten – das behält immer jenen dämonischen Charakter – wohl aber die einzelne Menschenseele kann sich im erfüllten Augenblicke wirklich auf Augenblicke entdämonisieren, sinkt wohl immer wieder zurück, und kann doch immer wieder aufsteigen'. Meinecke, *Ausgewählter Briefwechsel*, 342. Letter from 13 April 1931. Meinecke claims in this letter that it was Eduard Spranger who handed him the terminology of 'horizontal' and 'vertical'; however, in later essays and articles – especially in 'Geschichte und Gegenwart', in which he uses this terminlogy – Spranger is not mentioned.

169   Cf. the fourth volume of Meinecke's collected works: *Zur Theorie und Philosophie der Geschichte*.

170   Meinecke, 'Kausalitäten und Werte', 80.

171   I will come back to this in the next sub-section.

172   The focus on *Innerlichkeit* in a notion like intuition corresponds with the notion of *Bildung*, which is in Goethe's and Meinecke's view closely related to pietism. Cf.: Schneider, '"Mit Kirchengeschichte, was hab' ich zu schaffen?", 84; Menke-Glückert, *Goethe als Geschichtsphilosoph*, 11-12; Franz, *Goethe als religiöser Denker*, 38-49. *Bildung*, however, is a prerequisite for intuition. We have seen that Goethe's notion of 'personality' is closely connected to the concept of *Bildung*; a moral and spiritual development for the purpose of an inner harmony and to, subsequently, act on this basis. Personality and *Bildung* are concerned with creating an (ethical) standpoint in the world, in the present. On the basis

intuition, I would like to call this an 'historical intuition'. These two notions of intuition are obviously both present within man and can coincide in an umbrella concept of intuition, which could be identified as a 'daemonic intuition'. Or instead of this we could designate it as 'prudence', as discussed in the chapter about the principle of *raison d'état* with regard to the statesman. For the intellectual intuition gives insight into the eternal ideas or forms, according to which the statesman – by means of his intuition, his conscience – the *daimon* – knows what is right, true, good, and beautiful. The historical intuition gives insight into the state's past. Both intellectual and historical intuition give the statesman the means to act, and both show the way to the state's future. The same applies to the historian's inwardness, albeit in a contemplative-creative way; that is, the historian is able to sense the past by means of his historical intuition, and through his intellectual intuition he can divine (*ahnen*) the eternal, unique, higher, true, good and beautiful (of the values) in the past. Both yield insight into the present, and form the basis on which the historian can create his historical writing.[173]

For both the statesman and the historian intuition concerns the 'moment'. It is a moment, an instant in which the eternal and the temporal or historical connect. That is why Meinecke – in his aforementioned letter to Kaehler – speaks about the 'eternal moment'. The paradox of the notion of the eternal moment dissolves when we see that this is not a quantitative indication but a qualitative one. This could include the 'moment' of love, the sudden insight into an historical problem and its solution, the aesthetic delight in seeing a landscape, and the insight into the good, true, and beautiful.[174] In all these experiences the qualitative experience is so intense that the experience of the quantitative time – clock time – ceases to exist or is in any case no longer experienced as such. On the one hand, it is an experience whose intensity and meaning transcends the temporal, on the other hand it is also an experience that gives more grip on the relationship between past, present, and future.[175]

With regard to Goethe's 'eternal moment' – Meinecke borrows this idea from him – one can distinguish between a 'fulfilled moment' and a 'pregnant moment'.[176] In Goethe's time 'pregnant' refers to the Latin meaning of the word: *praegnans*, in

---

of this personality and *Bildung*, our intuition is concerned with sensing or fathoming other individualities in the past, and with creatively acting in the present. Cf. Chapters 1 and 3.

173   This already points at Burckhardt's conception of history, which I will discuss below.

174   Anglet, *Der „ewige" Augenblick*, 1.

175   On notions like 'moment': Heidrun Friese ed., *The Moment. Time and Rupture in modern thought* (Liverpool 2001); Thomas Müller ed., *Philosophie der Zeit. Neue analytische Ansätze* (Frankfurt am Main 2007).

176   Anglet, *Der „ewige" Augenblick*, 2.

the sense of bearing, expecting, or productive.[177] That is why the 'pregnant moment' is directed at the future. In contrast to this active moment stands a more contemplative moment, a 'fulfilled moment', that is aimed at the present, the instant, but in which eternity is also experienced.[178] For that reason, the fulfilled moment, especially with regard to history, is about the (supratemporal) experience of empathy or insight into and feeling for the eternal which can be experienced in the present. With regard to Meinecke, it is the fulfilled moment which is of interest in this context. This fulfilled moment, this contemplation, is eventually transformed by the historian into action by writing a history. In the light of a Goethean *hen kai pan*-construction we can now assert that the historical intuition – the fulfilled moment – enables the historian to write a history, from a 'personally felt' past, that is, in a Goethean sense, most objective. The historian – being 'part' of the *Gottnatur* which 'breathes' in and through him – is able to divine eternity by means of his intuition, and he can fathom the past through feeling (*Einfühlung*).[179]

The historian creates a story and for that reason Meinecke connects the concept of intuition with Goethe's idea of a creative artist – Meinecke thinks in this case of the historian and his intuition and creativity. In this context, Meinecke, particularly in his essays, implicitly and sometimes explicitly refers to the connection with Jacob Burckhardt and his idea of *Kultur*, which is mostly in agreement with Meinecke's conception of culture in which he connects intuition with the creative act of the historian, which corresponds to that of the artist. For example, Meinecke claims the following in his essay 'Kausalitäten und Werte in der Geschichte': 'This means to enter into the very souls of those who acted, to consider their works and cultural contributions in terms of their own premises and, in the last analysis, through artistic intuition to give new life to life gone by – which cannot be done without a transfusion of one's own life blood'.[180] The influence of Wilhelm Dilthey's hermeneutics on Meinecke's conception of history is quite clear in this statement.

---

177   This thought comes from Leibniz: the present is pregnant with the future. Gottfried Wilhelm Leibniz, *Monadologie of De beginselen van de wijsbegeerte* transl. F. P. M Jespers (Kampen 1991) 65.

178   Anglet, *Der „ewige" Augenblick*, 2.

179   In this respect, a statement by Huizinga, which seems very close to Meinecke on this point, could be illuminating: 'From time to time one has to rise above the restrictive compulsion of hard labor, to feel again in the light of the theory that our powers are limited, but history is universal, to once again feel the profound responsibility of the historian, who, the more clearly it becomes conscious to him that only a subjective understanding is possible, the more he will hold on to the ideal of objective truth, which rests in him'. Johan Huizinga, 'Het aesthetische bestanddeel van geschiedkundige voorstellingen' in: idem, *Verzamelde Werken* VII (Haarlem 1950) 3-28, there 26-27.

180   Meinecke, 'Values and Causalities in History', 283. This essay is considered to be Meinecke's *Historik*. Schulin, 'Meineckes Leben und Werk',126; Eberhard Kessel, 'Einleitung des Herausgebers', xxix; Hofer, *Geschichtsschreibung*, 233.

In a Diltheyan vein Meinecke assumes a symbiosis between the historian's person-ality and the individualities of the past *within* the historian, on which basis he is able to write history.[181] Meinecke's 'Aphorismen', which he noted down during his research for *Die Entstehung des Historismus*, shows a similar conviction, but he goes a step further in his conclusion:

> All knowledge and all critical research should not remain an end in itself and the final conclusion of a work, but receive only higher and lasting value, if (...) subject and object merge into *one* mental unity (...) – where the subject will still remain the source of strength of this unity, but also enters into, and merges into the objective that it represents – and thereby transcends itself.[182]

So, historical writing originates from the historian's individual creativity. The his-torian uses his intuition in order to fathom the past by which he transcends the polarities of past and present, which are one in the historian's mind. Next, the history written by the historian (subject) in its turn adds something to that same historical reality (object) that he researched thereby becoming part of that history – this is what Meinecke means by transcending (*überwindet*): the subject becomes part of the object. Also, in the sphere of historical writing the unity of subject and object – in the Spinozist meaning or even a refined version of panentheism – plays an important role for Meinecke.[183]

Meinecke considered the historian's creative process a cultural act, through which the historian – the individual – is able to break away from natural, animal-like urges. Culture is a result of the spontaneous – daemonic – inner self of the individual. As is clear from Meinecke's letter to Kaehler, within the fulfilled moment it is the single human soul (*einzelne Menschenseele*) that is able to free itself from its animal nature within to subsequently transcend himself, by which he is able to catch a glimpse of the divine, the eternal. In short, when writing or creating a history, the Goethean intellectual intuition coincides with the historical intuition. It is here that Meinecke is very close to Burckhardt's conception of history.[184]

A horizontal and a vertical conception of history

The most obvious and frequently expressed criticism of Meincke's historicism, as the previous chapter shows, is that precisely when historicism fell into a crisis, Meinecke dedicated a book on the origins of it in an affirmative attitude (*bejahender*

---

181   In the next chapter I will come back to Meinecke's hermeneutics in its relation to Dilthey.

182   Meinecke, 'Aphorismen', 239.

183   In his last study *Die deutsche Katastrophe* Meinecke takes this to the extreme. Cf. the next chapter.

184   See the last sub-section, entitled 'kindred spirits', of this chapter.

Gesinnung).[185] He was nonetheless fully aware that during the time he wrote *Die Entstehung des Historismus*, historicism had become a contested concept, but it was precisely for this reason that he wrote an affirmative history of historicism. After all, as I argued in the previous chapter, Meinecke claimed that the foundation and identity of historicism was to be found in its history. With regard to the problems surrounding historicism, Meinecke thought the historian should take to heart the unfavourable elements of historicism and preserve the favourable: 'He must pay good heed to justifiable criticism, but stand by all that is best in [it] [historicism, RK]'.[186] Meinecke thus writes his genesis of historicism in an 'affirmative attitude' and can do so because it concerns the *origins* of historicism and not what Meinecke characterizes as late historicism (*Späthistorismus*) – the problematic variation of the early twentieth century that Troeltsch had referred to.[187] Meinecke was convinced that the wounds inflicted by the (ethical) relativism of late historicism could be healed by historicism itself. The legacy of historicism was still very valuable for his

---

185 Meinecke, *Historism*, liv. On the distinction between facing a crisis and causing a crisis: Herman Paul: 'A Collapse of Trust. Reconceptualizing the Crisis of Historicism', *Journal of the Philosophy of History* 2 (2008) 63-82, there 73-74.

186 Meinecke, *Historism*, lv.

187 The fundamental criticism of Meinecke's historicism, in particular Oexle's criticism, is in large part countered by Ulrich Muhlack. He pointed out to Oexle that Meinecke's view is initially very close to Troeltsch. Oexle considers Troeltsch to be the thinker that *was* indeed aware of the crisis and came up with a solution. Muhlack, 'Gibt es ein "Zeitalter"', 212; Also: Patrick Bahners, 'Literaturkenntnis schützt vor Denkmalsturz. Völlig losgelöst: der Historismus nach Annette Wittkau', *Frankfurter Allgemeine Zeitung* (17 May 1993); Ernst Schulin, 'Das Problem', 102. Oexle, '"Historismus". Überlegungen zur Geschichte des Phänomens und des Begriffs', 57-62. The fact that Meinecke's and Troeltsch' philosophy of history resembled each other, is not only apparent from Meinecke's articles on philosophy of history, but also from *Die Entstehung des Historismus*. Muhlack, 'Gibt es ein "Zeitalter"', 212-213. Troeltsch speaks of 'der grundsätzlichen Historisierung alles unseres Denkens über den Menschen, seine Kultur und seine Werte' and Meinecke talks about a 'Denkweise des dynamischen Historismus', and also historicism is, according to him: 'eine der größten geistigen Revolutionen, die das abendländische Denken erlebt hat'. Muhlack, 'Gibt es ein "Zeitalter"', 212; Troeltsch, *Der Historismus und seine Probleme*, 102; Meinecke, 'Geschichte und Gegenwart', 92; Meinecke, *Die Entstehung des Historismus*, 1. I will come back to Troeltsch below. In addition, Muhlack also determines that Meinecke comes up with a solution, a solution that is indeed very close to Weber. Oexle considered Weber to be *the* thinker that had the right solution to the crisis of historicism. Weber, for that matter, did not use the word 'historicism' in the text Oexle based his statement on: 'Wissenschaft als Beruf'. Cf.: Schulin, 'Neue Diskussionen', 110. Both Weber and Meinecke considered 'conscience' to be the solution to the crisis. While Weber holds a completely subjective view of 'conscience', Meinecke grounds his notion of 'conscience' in a belief in a higher authority. With regard to Weber's politics and ethics in relationship with the crisis of historicism: Stefan Eich and Adam Tooze, 'The allure of dark times: Max Weber, politics, and the crisis of historicism', *History and Theory* 56 no. 2 (June 2017) 197-215.

present. But how exactly did Meinecke cope with this so-called crisis of historicism? And what role did Goethe play in this?

In the aforementioned letter to Kaehler, Meinecke speaks of the 'fulfilled moment' and a 'vertical philosophy of history'. Opposite to the 'horizontal' philosophy of history he positions a 'vertical' – Neoplatonic-panentheistic – conception of history. As already stated, the horizontal conception refers to the level of the chronological, linear history with its 'dissonances'. Although these dissonances will dissolve on a higher level, they are indeed real on the level of history. In addition, on this horizontal level one is overwhelmed with the individuality, uniqueness of everything which occurs in history. This leads to some problems on an ethical level, for if, in Meinecke's view of historicism, everything is individual, unique and singular, and is caught in a stream of becoming, on what basis can we then ground our morality? Does this not lead to a deep (ethical) relativism in which 'every historical individual entity, every institution, every idea and ideology, [is considered, RK] a temporary moment in the infinite stream of becoming?'.[188] Yes, and Meinecke even admits to this: 'Thus relativism and histori[ci]sm certainly belong together'.[189] Meinecke however does not want to leave it at that. He is convinced that historicism, 'which gives us a deeper understanding of the historical human being', cannot be equated with 'decay and weakness'.[190] In Meinecke's essay 'Geschichte und Gegenwart', which was published before *Die Entstehung des Historismus*, he asked himself: 'Does (...) historicism and the particular relativism it has created have the power to heal the wounds it has struck?'.[191] In the previous chapter it was already clear that Meinecke gave an affirmative answer to this question, for it was (the history of) historicism that would offer a solution. Meinecke therefore replaced in *Die Entstehung des Historismus* his question with a conviction: 'Histori[ci]sm must itself attempt to heal the wounds it has inflicted'.[192] But what did Meinecke exactly mean by this? To answer this question, we must first discuss the idea of 'positive relativism', which Meinecke found in Goethe's thought.

Meinecke thought ethical relativism could be considered both positively and negatively. To which side the pendulum swings depends mainly on the 'character' of the individual personality.[193] According to Meinecke, ethical relativism can both have a 'life-weakening' (*lebenschwächend*) or 'life-enhancing' (*lebensteigernd*) effect. In his essay 'Von der Krisis des Historismus' he expressed it as follows:

---

188   Meinecke, 'Geschichte und Gegenwart', 93.

189   Meinecke, *Historism*, 489.

190   Meinecke, 'Aphorismen', 232.

191   Meinecke, 'Geschichte und Gegenwart', 94. First published in 1933, but already delivered as a lecture in 1930.

192   Meinecke, *Historism*, 418.

193   Meinecke, 'Von der Krisis', 204.

It [relativism, RK] weakens life when one loses faith in unconditionally binding norms of action regarding the innumerable individual powers of life which man encounters in history, and thus, leads him (...), into an "anarchy of thinking". It enhances life if one consciously and deeply senses and acts on the right to live of one's own inwardness as well as of the surrounding and nourishing forces of life (...).[194]

In Meinecke's essays from the 1920's one can find the same train of thought. For example, Meinecke claims in 'Kausalitäten und Werte': '(...) that only weak souls of little faith could despair and quit under the burden of this relativizing histori[ci]sm'.[195] According to Meinecke, relativism will not shake the 'stronger person's' belief in an 'unknown absolute' or 'common primal divine source'.[196] Following Goethe, Meinecke argues in *Die Entstehung des Historismus*: 'Thus Goethe's relativism (...) was a much more deeply based and positive relativism, uninfected by any weakening doubts about the worth of the individual will, or by vagueness or (...) opportunism about the basic forces in history'.[197] In short, it is the 'stronger' that are able to fully think through historicism and its relativistic dimension, and they will not fall victim to a paralyzing state of relativism; on the contrary, they find a foundation. In Meinecke's view, it is about finding 'intellectual control' (*geistige*

---

194 Meinecke, 'Von der Krisis', 204. The paralysing effects of historicism were already addressed by Nietzsche in his *Unzeitgemäße Betrachtung*: 'Vom Nutzen und Nachteil der Historie für das Leben' (1874). Meinecke agrees with him only where Nietzsche claims, in the words of Meinecke: 'dem Schwachen ist sie Gift, dem Starken Nahrung'. Meinecke, 'Persönlichkeit und geschichtliche Welt', 52. This essay was first published in 1918, but in 1933 it was reissued. Apart from this statement, Meinecke disagrees with the rest of Nietzsche's thought. He despises his 'Anarchie des Subjektivismus'. Meinecke, 'Erlebtes', 105. Ernst Schulin has recently shown in an article that Nietzsche's view did not tally with the actual state of historical science, because the historians of that time were indeed very much involved in the present, and there was not yet a question of a 'lebensfremde Spezialisierung und Leerlauf'. Moreover, Nietzsche, being a student in theology and philosophy, had, according to Schulin, hardly any idea of academic historical science, especially because he resided in Basel, where he was mainly confronted with the ideas of Burckhardt, which did not correspond with his fear. In short, Nietzsche's work was, according to Schulin, indeed untimely: 'Sie gab nicht die damalige Situation wieder und entsprach nicht damaligen Bedürfnissen und Nöten'. Ernst Schulin, 'Zeitgemäße Historie um 1870. Zu Nietzsche, Burckhardt und zum „Historismus"', *Historische Zeitschrift* 281 (2005) 33-58, there 33-34.
195 Meinecke, 'Values', 283 note on page 439. This essay was written in 1925 and published in 1928.
196 Meinecke, 'Kausalitäten und Werte', 83, note 1; 'Gerade individualitätsstarke Menschen und Zeiten werden diesen Glauben immer wieder haben'. Meinecke, 'Ernst Troeltsch und das Problem', 376; Kämmerer, *Friedrich Meinecke und das Problem des Historismus*, 197.
197 Meinecke, *Historism*, 490.

Beherrschung).[198] This he found in Goethe: 'Subjectivity, once it had been set free, needed, however, to revert to an objective understanding of the world, without losing the springs of individual experience and effort in the process. Goethe's own example showed that this was perfectly possible'.[199] Again, to Meinecke it is essentially about the Goethean belief in a harmonious, panentheistic *Gottnatur*, from which all individualities emanated, and which ensures that subjectivity leads to the greatest possible objectivity, because the object works 'in' and 'through' the subject.[200] Meinecke also states this clearly in *Die Entstehung des Historismus*: 'Relativism can either lead to the profoundest depths or to the dreariest of plains, according to whether it is or is not backed in the last resort by a strong creative faith; according to whether humility and reverence in the face of the unsearchable [unfathomable, RK] are the result of a relativist outlook on the world, or are absent because of the prevailing view that human insight can discern nothing but an anarchy of values'.[201] Essentially, it thus depends on the 'character of the observer whether this overpowering drama spells out a meaningful or a meaningless history, a world of consolation or a world of despair, whether it leads to a resigned relativism or to a faithful devotion to an idea, in spite of its threatened extinction', Meinecke writes.[202] It will be clear that Meinecke starts from a positive relativism that is based on his panentheistic conviction. It remains to be seen to what 'authority' a strong personality (on his Spinozist-panentheistic foundation) should appeal to withstand the crisis of historicism.

It has become clear that the absolute or divine, the higher authority, cannot be proved or actually be fathomed, for it can only be sensed or divined (*ahnen*). In this context, Meinecke claims the following at the end of his Goethe chapter in *Die Entstehung des Historismus*: 'But when it comes to deciding what are the ultimate connecting-links between our thought and our will, the intellect loses its competence over the soul, which refuses to let itself be robbed of its share, as a particular limited individuality, in the significance of life in all its immeasurable fullness'.[203]

---

198   Ibidem, 418; Meinecke, *Die Entstehung des Historismus*, 496.

199   Meinecke, *Historism*, 490.

200   And when it comes to historiography, thanks to his individual creative *Springquell*, the historian should be able to separate himself from 'disturbing nature' and 'elevate' himself to *Kultur* or 'strive back' to the divine.

201   Meinecke, *Historism*, 490.

202   Meinecke, *Historism*, 199. In this context, Gadamer compares Meinecke's historicism with a 'Historismus zweiten Grades': 'Es ist sozusagen ein Historismus zweiten Grades, der nicht nur die geschichtliche Relativität aller Erkenntnis dem absoluten Wahrheitsanspruch entgegenstellt, sondern ihren Grund, die Geschichtlichkeit des erkennenden Subjektes, denkt und deshalb geschichtliche Relativität nicht mehr als Einschränkung der Wahrheit ansehen kann'. Gadamer, *Wahrheit und Methode*, 500.

203   Meinecke, *Historism*, 491.

Put differently, Goethean 'intuition' is the core of Meinecke's thought. With regard to morality and the crisis of historicism, Meinecke considers the authority which is based on intuition to be central, that is: conscience.[204]

To Meinecke conscience is the intersection of the individual and the absolute, of the moment and eternity, but also of the present and the past. Analogous to Goethe's concept of nature, in which subject and object are in harmony, and in which the subject, by means of its intuition is able to sense (*ahnen*) the object, Meinecke employs a similar construction on the level of morality.[205] As stated in the chapter on the reason of state: conscience is subjective. To Meinecke, it is also a reflection (*Abglanz*) of absolute morality or the divine. That means, the subjective conscience bears a spark of the absolute or 'eternal good'. According to Meinecke, conscience itself is not divine, it is only akin to God and in that sense it is a 'relative absolute'.[206] In Meinecke's view it is 'based on' a panentheistic belief in a 'primal divine source'.[207] This bears a resemblance to the aforementioned concept of the

---

204 Friedrich Heinrich Jacobi states that *Glaube* gives the highest form of certainty (*Gewißheit*), for it is located in the domain where notions such as truth and falsehood have no function. It is – like intuition – on a higher level than *Verstand*. Like intuition, it is an immediate flash (of intuition), an inner vision. In this regard, there is a relationship between Meinecke's conception of intuition and conscience and Jacobi's *Glaube* and *Gewißheit*. There is another remarkable similarity between Meinecke and Jacobi: both consider the notion of 'individuality' (the core of Meineckes historicism) as the decisive standard for ontology. And because it is God who knows all the individualities, it is the humanities – which come closest to a kind of (infinite) position – that can be seen as the authority on ontology. On Jacobi's *Glaube, Gewißheit*, and ontology: Ankersmit, 'Jacobi. Realist, Romanticist'.

205 Also: Reinbert A. Krol, 'Friedrich Meinecke: Panentheism and the Crisis of Historicism', *Journal of the Philosophy of History* 4 (2010) 195-209.

206 Meinecke, *Ausgewählter Briefwechsel*, 362; Meinecke, 'Geschichte und Gegenwart', 100; Meinecke, 'Deutung eines Rankewortes', 133. Apart from being relative absolute, conscience is also relative because there are a meriad of personalities with a conscience. Following Troeltsch, Meinecke calls this *Wertrelativität*. According to Troeltsch, this does not mean: '(...) Relativismus, Anarchie, Zufall, Willkür, sondern bedeutet das stets bewegliche und neuschöpferische, darum nie zeitlos und universal zu bestimmende Ineinander des Faktischen und Seinsollenden'. Cited in Meinecke, 'Ernst Troeltsch und das Problem', 376. Meinecke explains that *Wertrelativität* is about: 'Individualität im historischen Sinne, jeweils eigenartige, an sich wertvolle *Ausprägung eines unbekannten Absoluten* – denn ein solches wird dem Glauben als schöpferischer Grund aller Werte gelten – im Relativen und zeitlich-naturhaft Gebunden'. Meinecke, 'Kausalitäten und Werte', 83. My italics. So, it is not about the idea that values change and therefore become relative and make us feel that there are no absolute values, no, it is about the fact that each individual has to decide for himself (by means of his conscience) what he has to do, how he has to act: 'Es handle sich (...) um eine Aufgabe und Pflicht, die nüchterne und praktische Weltorientierung, klares Wollen und scharfen Blick verlange'. Meinecke, 'Ernst Troeltsch und das Problem', 376.

207 Jensen claims Meinecke's preference for 'contiguous polarities' prevents him from reconciling the absolute with the individual, the vertical with the horizontal. Jensen, 'Unity in Antinomy',

daimon, the intermediary between heaven and earth. For it is the *daimon* within, the inner voice, that is able to remind us of the knowledge we have lost; or what we were previously conscious of. Thanks to our conscience, which, as daimon, serves as the mediator between the divine (good, true, beautiful) and human action, man is able to do good. In his essay 'Geschichte und Gegenwart' Meinecke expressed it as follows: 'All eternal values of history ultimately come from people acting on the basis of their conscience'.[208] It is in this statement that Meinecke connects the individual with the absolute, the present with the past, and the moment with eternity. And it is Goethe's harmonious panentheistic world view which enables him to withstand the crisis of historicism.

The historian's task, in Meinecke's view, is to recover these aforementioned 'eternal values', in order to find by means of his conscience, a fixed point within the 'continuous stream of becoming'. But how does this actually work? Meinecke mentions a few, to him unsatisfactory ways to find a fixed point in the 'stream': an 'escape into the past' whereby a certain part of the past is glorified, or an 'escape into the future' in which a future ideal is aimed at. According to Meinecke, both are inadequate, for they both are still situated 'within' the stream of becoming.[209] He therefore suggests a vertical – (Neo)platonic, or better: panentheistic – conception of history. In this context, Meinecke cites a few lines from Goethe's poem 'Vermächtnis' (1828) (Legacy), of which the most important line is: 'Der Augenblick ist Ewigkeit' (the moment is eternity).[210] I already mentioned that within the fulfilled moment time and space can contract and expand. I also linked this to the daemonic in which the temporal and eternal are closely connected. As well as Goethe's lines, Meinecke also quotes Ranke's famous expression: 'Jede Epoche ist unmittelbar zu Gott' (Every epoch is immediate to God). According to Meinecke, both these expressions manage to overcome the horizontal *Werdeströme* and force us to search for the 'God-like' in history within the moment. That is, vertically, upwards, to 'God' is what we should focus our attention on. We should search for the eternal and fixed within the change of every individuality. With regard to morality and values, we could, by means of our conscience (as daemon) approximate the eternal, good,

---

232, 182. Jensen, however, fails to discuss in detail the influence of poets and thinkers like Plato, Aristotle, Plotinus, Novalis, Herder, Goethe and Schiller on the development of Meinecke's conception of history. Also: Gerald Straus, 'Meinecke, *Historismus*, and the Cult of the Irrational', 113.

208   Meinecke, 'Geschichte und Gegenwart, 101. 'Alle Ewigkeitswerte der Geschichte stammen letzten Endes aus den Gewissensentscheidungen der handelnden Menschen'.

209   Meinecke, 'Geschichte und Gegenwart, 96-98.

210   This reminds of the ending of Faust when the choir sings: 'Alles Vergängliche ist nur ein Gleichnis'.

true and beautiful. So we should, in Meinecke's view, not focus our attention on a personal God but on an abstract panentheistic point: the Goethean *Gottnatur*.[211]

The above makes clear that Meinecke's 'vertical' conception of history can exist merely by the grace of the 'horizontal' conception. After all, contemplating the good, true, and beautiful is only possible from 'within' a finite position, since man is simply bound to or part of the horizontal, chronological history.[212] The resemblance with Goethe's conception of nature is striking for Goethe too started his search in reality – the development, the metamorphosis – to reach a fixed point – the eternal: the primal plant (*Urpflanze*) or primal phenomenon (*Urphänomene*). Similar to Goethe who, by means of empirical research, tried to find the primal plants or forms, or even to theoretically extrapolate his findings, Meinecke's vertical conception of history was only possible *after* the horizontal history had passed. So, by contemplating the manifold individualities *within* history, we can catch a glimpse of the transcendent, the good, true, and beautiful, 'above' history.[213] It is this idea, Meinecke claims, in which Enlightenment thinking and historicism come together and it was in Goethe where this occurred in its highest form. Meinecke considers Goethe's view of nature the completion of historicism as well as its transcendence.[214]

Kindred spirits

Meinecke was not alone in his panentheistic 'vertical' conception of history. Many German but also Dutch theologians advocated a similar view.[215] The theologian Richard Rothe (1799-1867) for example presumed a 'Seinsfülle' (fullness of being) which was embedded in God. That is, God is 'present' in his creation, but also leaves room for an individual development of that creation; a thought that is compatible with Meinecke's panentheism.[216]

211   Meinecke, 'Geschichte und Gegenwart', 101; Meinecke, 'Deutung eines Rankewortes', 136.

212   For many years this has been misunderstood; as if Meinecke was withdrawing into some sort of mystic Kultur and/or extreme subjectivity, for example: Pachter, 'Masters of Cultural History II: Friedrich Meinecke', 42.

213   Meinecke, 'Geschichte und Gegenwart', 98. Borowski fails to see that Goethe was an empiricist and therefore concludes that Meinecke and Goethe are merely interested in the 'mystical'. Borowski, 'Friedrich Meinecke's *Historism*', 177-178.

214   Meinecke, *Die Entstehung des Historismus*, 581; Ankersmit, *De sublieme historische ervaring*, 166. The above argument has a Hegelian ring to it; cf Chapter 4 for Meinecke's attitude towards Hegel.

215   Dutch examples of, in particular, theologians who are very close to Meinecke's and Troeltsch's views are discussed in: Paul, *Het moeras van de geschiedenis*, 57-76.

216   Thielicke, *Glauben und Denken*, 519.

In view of Meinecke's historicism it is, again, Ernst Troeltsch – by the way an admirer of Rothe – who is important in this context. Already before Meinecke's *Die Entstehung des Historismus*, Troeltsch had expounded a similar *Weltanschauung*. It is impossible to do justice here to the extremely complex work of Troeltsch, but based on the authoritative study of Jacob Klapwijk and the thorough work of Walter Wyman, I will briefly discuss Troeltsch's *Weltanschauung* and the affinity with Meinecke's panentheistic philosophy of history.[217]

Initially, Troeltsch tried to ground Christianity in an historical basis. That is he wanted to understand the Christian belief in an evolutionary sense. When this evolution turned out to be not as harmonious as he had hoped for – the afore-mentioned stream of becoming that proved to be a tidal wave – a deep feeling of crisis emerged. Next, Troeltsch shifted his research more in the direction of finding objective values in culture instead of obtaining absolute theological knowledge. The consequential historicization of historicism and the faltering of apparently unchanging values caused Troeltsch to adapt his image of God.[218] He changed it from a personal God to a panentheistic abstraction.[219] He considered God and the world as separate entities, although the world had indeed emerged from or was part of God: 'It is the process of emanation and remanation', Troeltsch writes.[220] This view, which clearly reminds us of Goethe and Meinecke, is also present in Troeltsche's *magnum opus* on historicism.[221]

---

217   Troeltsch's thought is not unambiguous, however according to Wyman, one can identify a fundamental principle that is present already in his first publications. Walter E. Wyman Jr., *The Concept of Glaubenslehre. Ernst Troeltsch and the Theological Heritage of Schleiermacher* (Chico 1983) xii-xiii; also: Gérard Raulet, 'Strategien des Historismus' in: Wolfgang Bialas and Gérard Raulet ed., *Die Historismusdebatte in der Weimarer Republik* (Frankfurt am Main 1996) 7-38, there 20-21. Jacob Klapwijk, *Tussen historisme en relativisme. Een studie over de dynamiek van het historisme en de wijsgerige ontwikkelingsgang van Ernst Troeltsch* (Assen 1970).

218   Wyman, *The Concept of* Glaubenslehre, 173; Klapwijk, *Tussen historisme en relativisme*, 119; Hartmut Ruddies, '"Geschichte durch Geschichte überwinden". Historismuskonzept und Gegenwartsdeutung bei Ernst Troeltsch' in: Wolfgang Bialas and Gérard Raulet ed., *Die Historismusdebatte in der Weimarer Republik* (Frankfurt am Main 1996) 198-217, there 201, 206; Iggers, *The German Conception*, 178, 188; Hofer, *Geschichtsschreibung*, 332-335.

219   Well over ten years before his study on historicism Troeltsch had already elaborated on his panentheistic view in: Ernst Troeltsch, 'Wesen der Religion und der Religionswissenschaft' in: idem, *Zur religiösen Lage, Religionsphilosophie und Ethik* (Tübingen 1913) 452-499, there 496-497; Ernst Troeltsch, 'Die Zukunftsmöglichkeiten des Christentums im Verhältnis zur modernen Philosophie' in: idem, *Zur religiösen Lage, Religionsphilosophie und Ethik* (Tübingen 1913) 837-862, there 843-844; but also in lectures which were published posthumously: Troeltsch, *Glaubenslehre*, 328, 381; also: Wyman, *The Concept of* Glaubenslehre, 43-45, 189-193; Klapwijk, *Tussen historisme en relativisme*, 83, 132, 237-238.

220   Ernst Troeltsch, *Glaubenslehre. Nach Heidelberger Vorlesungen aus den Jahren 1911 und 1912* (Munich and Leipzig 1925) 381.

221   Wyman, *The Concept of* Glaubenslehre, 189-192.

In *Der Historismus und seine Probleme* Troeltsch attempted, by means of a panentheistic view, to offer a solution for the (ethical) relativism which was the result of the 'the fundamental historicization of all our thinking about man, his culture and his values'.[222] According to Troeltsch, the (individual) values and the absolute were connected within the historical process. Thus, the absolute is both immanent and transcendent. The historical process is in Troeltsch's view a transformation of the divine, the god-like, the infinite to the finite world, but at the same time it is also a process from the finite to the divine; individualities striving for the divine.[223] According to Troeltsch, it was possible to sense or approximate the absolute from the finite world: '(...) we retain the possibility of grasping divine life in relative truth and relative ideal'.[224] That is essentially what Troeltsch meant by his concept of a *Kultursynthese* (cultural synthesis), for he was convinced that knowledge and values could be deduced from the past, and that they, despite being connected to a specific historical context, can in fact reflect or reveal a hidden, absolute truth beyond or behind history, which could be of importance to the future.[225] In Troeltsch's words: 'Such a philosophy of history requires a constructive constriction of the given and a granting of faith to a divine idea which reveals itself in the given'.[226] Troeltsch based or even projected his views not on Goethe, but on Leibniz' philosophy. In particular, Leibniz' monadology reflected Troeltsch's thought, for Troeltsch accepts the principle of the monad, which, by means of intuition, has, to a certain extent, access to the absolute or divine consciousness from which it emanated. However, Troeltsch rejected the idea of Leibniz' windowless monads.[227]

In many ways Troeltsch's philosophy of history corresponds to Meinecke. A major difference however is the way in which both handle the feeling of crisis, which could emerge as a result of the continuous historicization of life and thought. As mentioned previously, Meinecke was barely thrown off balance. Troeltsch, on the

---

222 Troeltsch, *Der Historismus und seine Probleme*, 102. The general idea is that Troeltsch wanted to 'overcome' historicism. Klapwijk has shown that this is an error: 'In reality, Troeltsch has recognized historicism as a fundamental achievement of Western culture, which is irreversible and whose consequences must only be courageously faced. He does, however, reject historicism, where it occurs in all kinds of negative forms'. Klapwijk, *Tussen historisme en relativisme*, 47.

223 Wyman, *The Concept of Glaubenslehre*, 190-191; Klapwijk, *Tussen historisme en relativisme*, 126-127, 132, 230, 237-238, 402.

224 Troeltsch, *Der Historismus und seine Probleme*, 184.

225 Ruddies, '"Geschichte durch Geschichte überwinden"', 210; Iggers, *The German Conception*, 188.

226 Troeltsch, *Der Historismus und seine Probleme*, 692. 'Solche Geschichtsphilosophie verlangt eine konstruktive Zusammendrängung des Gegebenen und einen Zuschuß des Glaubens an eine im Gegebenen sich offenbarende göttliche Idee'.

227 Klapwijk, *Tussen historisme en relativisme*, 397-400; Ruddies, '"Geschichte durch Geschichte überwinden"', 214-215. On Leibniz' monadology see section 4.2

other hand, fell into a deep crisis and struggled heavily with the collapse of the absolute.[228] Meinecke claims that: 'In this struggle for a fixed point (...) the theologian in him [Troeltsch, RK] was never quite involved in the philosopher of history'.[229] Troeltsch experienced an abyss in the 'deep antinomies of life which historicism had uncovered'[230], whereas Meinecke's *Weltanschauung* gave way to relatively firm ground. Supported by the thought that everything was essentially in harmony, Meinecke accepted the dissonances of life, whereas Troeltsch also questioned a belief in such a harmony.

Well before Troeltsch and Meinecke, it was in particular Jacob Burckhardt who (like Meinecke later on), analogous to Goethe's concept of metamorphosis sought for the 'transcendent' in history.[231] It was Burckhardt's conception of history that Meinecke agreed with more and more, especially during the last decades of his life; in particular during and after the Second World War Burckhardt's view of history became important to Meinecke. In the next chapter I will come back to this issue.

Burckhardt considered history as a 'miraculous process of pupations and new, eternally new revelations of the spirit'.[232] Burckhardt too searched for the eternal in history, or rather: history consisted, in his view, of pupations (*Verpuppungen*) of the eternal (spirit).[233] By contemplating the highest forms of these 'metamorphoses' of the eternal in history – art and literature, according to Burckhardt – it was possible to catch a glimpse of the good, true and beautiful. It is not necessary, as Lionel Gossman in his study on Burckhardt shows, to penetrate to what might be 'behind' these pupations: 'Indeed, to do that one would have to tear aside the last veil of phenomena and submit to the blinding light of the divine itself', thus Gossman.[234]

---

228   Klapwijk, *Tussen historisme en relativisme*, 447-449.

229   Meinecke, 'Ernst Troeltsch und das Problem des Historismus', 369.

230   Ibidem, 378.

231   On the idea that 'Burckhardt' was only possible after 'Ranke' see: Ankersmit, *De sublieme historische ervaring*, 163-176.

232   Jacob Burckhardt, *Briefe*, Fritz Kaphan ed. (Leipzig sa) 60. Burckhardt wrote this in his letter (19 June 1842) to Karl Fresenius.

233   Schnädelbach gives a clear explanation: 'Die Teilnahme allen Geschehens am Unvergänglichen ist nicht mehr die Rankesche "Unmittelbarkeit zu Gott", sondern nach Burckhardt als *innergeschichtliches* Verhältnis zu sehen. Damit stellt Burckhardt uns vor die Aufgabe, die Kontinuität und Unvergänglichkeit des Menschengeistes mit seiner historischen Wandelbarkeit und Bedingtheit zusammenzudenken (...)'. This does not mean that Burckhardt was suddenly closer to Hegel: '(...) denn "Geist" soll ja nicht mehr der Grundbegriff einer philosophischen Geschichtsspekulation, sondern der empirischen Geschichtsbetrachtung sein'. Schnädelbach, *Geschichtsphilosophie nach Hegel*, 57.

234   Lionel Gossman, *Basel in the Age of Burckhardt. A Study in Unseasonable Ideas* (Chicago 2000) 248. This is, again, exactly what Goethe describes in the second part of *Faust* in which Faust awakens and cannot look directly into the sun – a (neo) Platonic symbol for the divine – but can only see the *Abglanz* of the divine as reflection of on the landscape. Goethe puts it in

To Burckhardt, says Gossman, this would yield an insight which would be completely inaccessible and incomprehensible to man. That is why Burckhardt, in the end, focussed his attention on the real, phenomenal world, in order to subsequently contemplate the eternal, as Gossman has shown.[235]

It is clear that Burckhardt's concept of *Verpuppungen* is closely related to Meinecke's vertical conception of history. After all, like Goethe and Burckhardt Meinecke also recognizes transcending values in the development of history. Goethe refers to this as the notion of 'Dauer im Wechsel'.[236] To Meinecke, 'Dauer im Wechsel' refers to points in the horizontal history that point 'upwards'. These are points that can only be considered from a standpoint afterwards, a position that can only be taken from a Goethean perspective; in short, a view that starts from harmony at the 'beginning' and at the 'end' of history. Meinecke therefore agrees with Goethe's *Systole* and *Diastole*, inhaling and exhaling, the contraction and relaxation of the heart, emanating from 'God' or Nature and 'striving back'; the principle of a panentheistic *Weltanschauung*.

Conclusion

At the time of *Weltbürgertum und Nationalstaat* Meinecke's conception of history was coloured by his monism and idealism. After his struggle with the principle of *raison d'état*, during which his views shifted but did not get bogged down in a pronounced pessimistic dualism, Meinecke regained balance in *Die Entstehung des Historismus*. In this work his harmonious view of history is less idealistic than the monistic *Weltbürgertum und Nationalstaat* and the dualistic elements with which he wrestled in *Die Idee der Staatsräson*; he now sticks more firmly to his panentheistic view. In spite of the idea that this panentheistic view seems at first sight to be merely a metaphysical construction, I have made it clear that Meinecke's point of departure

---

his 'Versuch einer Witterungslehre' in the following way: 'Das Wahre, mit dem Göttlichen identisch, läßt sich niemals von uns direkt erkennen, wir schauen es nur im Abglanz, im Beispiel, Symbol, in einzelnen und verwandten Erscheinungen; wir werden es gewahr als unbegreifliches Leben und können dem Wunsch nicht entsagen, es dennoch zu begreifen. Dieses gilt von allen Phänomenen der faßlichen Welt (. . .)'. Johann Wolfgang von Goethe, 'Versuch einer Witterungslehre' in: idem, *Goethes Werke. Hamburger Ausgabe in 14 Bänden*, ed. Erich Trunz (Munich 1982) volume XIII, 305-313, there 305. also: Erich Trunz, 'Das Vergängliche als Gleichnis in Goethes Dichtung' in: idem, *Ein Tag aus Goethes Leben* (Munich 2006, first 1999) 167-187, there 167.

235   Gossman, *Basel in the Age of Burckhardt*, 248. On the connection between Burckhardt's *Querschnitt* and the historical experience: Ankersmit, *De sublieme historische ervaring*, 170-176.

236   Meinecke, *Die Entstehung des Historismus*, 503.

was always the *experience* of the diversity *within* the world. It was also this thought that brought him to Goethe's philosophy of nature.

In *Die Entstehung des Historismus*, Meinecke shows that Goethe did not manage to achieve in his conception of history what he had achieved in his view of nature. Next, Meinecke elaborates precisely on what Goethe, according to him, failed to do, namely: to formulate a Goethean-based conception of history. In that sense, Meinecke turns Goethe, the natural scientist, into an historicist *avant la lettre*. If, according to him, Goethe personifies the apex of historicism, then in view of the above, it says more about Meinecke's idea of historicism than about Goethe's relationship with the past. It is clear that Meinecke's account of Goethe is meant to give more weight to his own ideas about historicism. Goethe's philosophy of nature confirms his philosophy of history, but he transfers it into a field that Goethe had a difficult relationship with, to say the least. In other words, this is an *Umdeutung* of Goethe. That is to say, Meinecke internalizes Goethe's natural scientific theories and then converts them into his own idea of historicism. Another reason for Meinecke to shape his historicism in this way is the fact that Goethe reconciles the 'Enlightenment' and 'historicism', and even transcends both. It is a harmonious conception of history, which is elevated above all dissonances, and at the same time it does acknowledge these dissonances. It is, in other words, an all-embracing conception of the past, in accordance with life as a whole. Because it is equal to life itself, we cannot escape historicism as Meinecke understands it.

On the basis of his panentheistic conviction Meinecke was able to consider history consisting of an ineffable amount of individualities, who are essentially all unfathomable and at the same time in a continuous development. This multitude of individualities, but also the idea of a continuous development did not cause a feeling of crisis in Meinecke. Indeed, he wrote about the 'stream of becoming' and the principle of individuality in an affirmative attitude (*bejahender Gesinnung*) in *Die Entstehung des Historismus*. And it was exactly this 'revolution' in thought that Meinecke affirmed. And even when this 'stream of becoming' overflowed, Meinecke, unlike Troeltsch, was not driven to despair.

When everything is subject to continuous change and it is in fact impossible to identify change as such, since it dissolves in the flow of becoming, it is not that strange that Meinecke (like Burckhardt and Troeltsch), in following Goethe, found the immutable or fixed in history in the transcendent – the divine or the good, true and beautiful. That is not to say that individual periods, morality and values became relative. On the contrary, every period expresses the individual, unique values which refer to the transcendent. In short, every period has its absolute. That is why Meinecke suggests a vertical conception of history, which is related to the horizontal development of history. In this vertical view it is our conscience, which – on the basis of a panentheistic view and in connection with the Goethean idea of the *daimon*, the Spinozist or Schellingian intuition – resists ethical relativism. The 'abyss' is always close by; however it depends on our individual personality whether someone in the end is likely to resist or succumb. It is certain that Meinecke had indeed regained his harmony in Goethe.

Next to Meinecke's appropriation of Goethe and his re-interpretation (*Umdeutung*), Goethe also represents the cultural nation (*Kulturnation*). Meinecke, who was familiar with the oeuvre of this *uomo universal* early on, falls back on Goethe's humanism precisely when *Kultur* was under a lot of pressure as a result of political radicalization. Given that Meinecke in *Die Entstehung des Historismus* was mainly interested in the development of world views and did not intend to discuss any politics in this study, his choice for Goethe as the symbol of (German) culture is understandable, for Goethe lived well before the unification of Germany. Goethe, in that sense, was not related to any political ideology. In addition, we have seen at the beginning of the previous chapter that Meinecke intended to research the roots of historicism early on. The aftermath of the First World War, the 'crisis of historicism', the unstable Weimar Republic, the rise of National Socialism were all additional reasons for him to bury himself in his research on historicism. As well as regarding Goethe the apex of historicism, Meinecke also considered him to be a guide in the unstable period between the two wars. In that sense, *Die Entstehung des Historismus* was, apart from being an affirmation of the historicist tradition, also an affirmation of the cultural legacy of Goethe and thereby of Germany as a whole. It was this legacy, which would become increasingly important to Meinecke in the years to come. In his defence of *Kultur*, he not only falls back on Goethe, but also, and more often, on Jacob Burckhardt, one of the main figures of the next chapter.

6. The authority of the personality

> 'Eben die intensivste vita contemplativa ist
> es, die oft die vita activa am stärksten be-
> fruchtet'[1]

Introduction

During his research on the origins of historicism, Meinecke was still very concerned with the political situation in Germany. In the early 1930s he expressed his concerns with regard to the rise of National Socialism and Hitler's coup in many different newspaper articles.[2] Immediately after the Second World War, Meinecke reminds his audience of this:

> In the spring of 1933, I was the last to publicly warn for Hitler in the press, just two days before the Reichstag fire. Then the terror came over us, and we, who from the outset felt something satanic in the dazzling beginnings of the work of Hitler, hostile to the spirit of Western Christian culture, we had to remain silent.[3]

---

1   Meinecke, *Die Entstehung des Historismus*, 369.

2   Cf. among others: Friedrich Meinecke, 'Staatsräson', in: idem: *Politische Schriften und Reden* (Darmstadt 1958) 467-470, there 469; Friedrich Meinecke, 'Von Schleicher zu Hitler', in: idem, *Politische Schriften und Reden* (Darmstadt 1958) 479-482, there 481. Furthermore, Meinecke expressed his horror of Hitler in many different letters.

3   Meinecke, 'Zur Selbstbesinnung', in: idem, *Politische Schriften und Reden* (Darmstadt 1958) 484-486, there 484. 'Ich war im Frühjahr 1933 der letzte, der, zwei Tage vor dem Reichstagsbrand, in der Presse öffentlich vor Hitler gewarnt hat. Dann kam der Terror über uns, und wir, die wir in dem blendend begonnenen Werke Hitlers von vornherein etwas Satanisches, dem Geiste christlich-abendländischer Kultur Feindliches spürten, mußten fortan schweigen'. In the preface of *Die deutsche Katastrophe* (1946) Meinecke states it in a similar way: 'Ich habe von vornherein die Machtergreifung Hitlers als den Beginn eines allergrößten Unglücks für Deutschland angesehen (...)'. Friedrich Meinecke, 'Die deutsche Katastrophe. Betrachtungen und Erinnerungen' in: idem, *Autobiographische Schriften* (Stuttgart 1969) 323-445, there 324.

Indeed, Meinecke had warned about Hitler.[4] In one of the last articles before Hitler's take-over, Meinecke claims that once the Third Reich would be realized, 'Then farewell freedom of personality; the age of bourgeois German culture with its synthesis of state and spirit would definitely be over'.[5] The synthesis between power and culture, which Meinecke had hoped to be realized from his first studies onwards, would be further away than ever before with the appointment of Hitler.

Despite Meinecke's aversion to Hitler, he also expressed his admiration for the victories and swift expansion of the army at the outbreak of the war. In 1940 he wrote his son-in-law: 'I am also full of admiration for what our army accomplishes and it goes far beyond all expectation. That it was possible to build such an army in a few years is the greatest positive achievement of the Third Reich'; and Meinecke immediately adds: 'but [I, RK] do not forget for a moment what has happened to the negative'.[6]

---

4   In recent studies and the re-issue of *Die deutsche Katastrophe* this is emphasized: Bernd Sösemann, ed., *Friedrich Meinecke, Die deutsche Katastrophe. Betrachtungen und Erinnerungen. Edition und internationale Rezeption* (Berlin 2018) 463, 466-468; Bock, 'Friedrich Meinecke und seine Briefe', 10; Meineke, 'Parteien und Parlamentarismus im Urteil von Friedrich Meinecke', 51-93; Wehrs, 'Demokratie durch Diktatur?', 95-118; Wippermann, '"Deutsche Katastrophe"', 177-191; And also older studies confirm this: Wolfgang Wippermann, 'Friedrich Meineckes Die deutsche Katastrophe. Ein Versuch zur deutschen Vergangenheitsbewältigung' in: Michael Erbe ed., *Friedrich Meinecke heute. Bericht über ein Gedenk-Colloquium zu seinem 25. Todestag am 5. und 6. April 1979* (Berlin 1981) 101-121, there 106 ff; Walther Hofer, '"Keine Fahnenflucht vor der Schlacht". Friedrich Meineckes Warnungen vor dem Nationalsozialismus' in: Catherine Bosshart-Pfluger ed., *Nation und Nationalismus in Europa. Kulturelle Konstruktion von Identitäten. Festschrift für Urs Altermatt* (Frauenfeld, Stuttgart and Vienna 2002) 323-345.

5   'Dann ade Freiheit der Persönlichkeit, und das Zeitalter der bürgerlichen deutschen Kultur mit der Synthese von Staat und Geist wäre endgültig geschlossen'. Meinecke, 'Staatsräson', 467.

6   Meinecke, *Ausgewählter Briefwechsel*, 192, also: 180, 193-194, 364. In his professional capacity as the editor-in-chief of the *Historische Zeitschrift* Meinecke experienced firsthand the pressures of the political and ideological circumstances. In 1933 Meinecke was forced (in his view) to fire the Jewish editor Hedwig Hintze – wife of the historian Otto Hintze, who was a colleague and a good friend of Meinecke's from 1889 onwards. On this history: Peter Th. Walther, 'Die Zerstörung eines Projektes. Hedwig Hintze, Otto Hintze und Friedrich Meinecke nach 1933' in: Gisela Bock and Daniel Schönpflug ed., *Friedrich Meinecke in seiner Zeit. Studien zu Leben und Werk* (Stuttgart 2006) 119-143. On Hedwig Hintze also: Friedrich Meinecke, *Akademischer Lehrer und emigrierte Schüler*, 81-92. Within the restrictions (from 1933 onwards) Meinecke tried to hold a steady course both politically and scientifically for the *Historische Zeitschrift*. Walther, 'Die Zerstörung eines Projektes', 122; G. Eckert en G. Walther, 'Die Geschichte der Frühneuzeitforschung in der Historische Zeitschrift', *Historische Zeitschrift* 289 (2009) 149-197, there 160. Due to this course, which was partly prompted by the circumstances, Meinecke came into conflict with Johan Huizinga. On this: Willem Otterspeer, *Huizinga voor de afgrond. Het incident-Von Leers aan de Leidse universiteit in 1933* (Utrecht 1984) 23-24.

Meinecke's initial admiration for the successes of National Socialism appears at first to be similar to his enthusiasm around 1914.[7] The difference is that around 1914 Meinecke thought the promise of a synthesis between power and culture – between Bismarck and Goethe – to be realistic, whereas at the beginning of the Second World War he only speaks of the military and political victories.[8] Apparently, he could not find any *Kultur* in (the violence of) National Socialism. His enthusiasm quickly made way for a deep aversion to the war. From 1941, as his letters prove, he considered the excessive militarism a threat to *Kultur*, and thereby to Germany as a whole. In a letter to his daughter and son-in-law Meinecke writes: 'The Germany I love has perished, and how much of any life values will and can be maintained in the new Germany, I do not know'.[9] After the war Meinecke managed to turn the disillusion which is expressed in these words into a hopeful argument.

Immediately after the war *Die deutsche Katastrophe* (1946) is published, in which Meinecke – by now 83 years old – discusses the political fall of Germany. In this study he tries to find and explain the roots of the twelve catastrophic years (1933-1945).[10] Thus, Meinecke does not focus so much on the twelve years of the Third Reich, but on the previous history of more or less 120 years: 'Out of the abundance of experiences only certain problems of inner and more permanent significance will be selected', Meinecke writes.[11] In *Die deutsche Katastrophe* Meinecke asks himself where it all went wrong in German history, and next, what should be done after the destruction of the Second World War. Meinecke's frequent use of the notion of *Schicksal* (fate) reveals that at least some of the causes are, to him, inexplicable.

The notion of *Schicksal* has a long history and is, especially in Germany, a notion with many layers. The classical explanation of this notion assumes a twofold meaning. Fate is something that both *happens* to man and is *caused* by man.[12] It is this double meaning which is also used by Meinecke.[13] He moves between the

---

7   Jensen, '"Unity in Antinomy"', 201. I elaborated on this in Chapter 2.

8   On the promise of the *Reformzeit* which had its effect well into Meinecke's age: Schönpflug, 'Revolution und "Erhebung"', 28.

9   'Das Deutschland, das ich liebe, ist vergangen, und wieviel des Lebenswerten auch im neuen Deutschland sich halten wird und kann, weiß ich nicht'. Meinecke, *Neue Briefe*, 407. Bock, 'Friedrich Meinecke und seine Briefe', 11. Also: Meinecke, *Ausgewählter Briefwechsel*, 401: 'Es wird einem wirr vor den Augen, wenn man zuerst hinblickt, – aber allmählich sieht man dann klarer die ungeheure Tragik des heutigen Geschehens'.

10   Meinecke, 'Die deutsche Katastrophe', 323.

11   Friedrich Meinecke, *The German Catastrophe. Reflections and Recollections* transl. Sidney Fay (Boston 1963, first 1950) xi; Nicolas Berg, *Der Holocaust und die westdeutschen Historiker. Erforschung und Erinnerung* (Göttingen 2003) 69-70.

12   M. Kranz, 'Schicksal' in: Joachim Ritter ed., *Historisches Wörterbuch der Philosophie* 13 vol (1971-2007) volume 8, 1275-1289, there 1275-1278.

13   Martin Heidegger has a similar conception of fate; he distinguishes between *Schicksal* (fate) and *Geschick* (destiny). The latter refers mainly to the fate of a certain group or community –

idea that a catastrophe had happened to Germany and that Germany had caused it itself. Meinecke agrees with Herder's and Goethe's conception of *Schicksal*, which is mainly derived from the classical interpretation of the word. Characteristic for the nineteenth-century conception of *Schicksal* is the dichotomy between a notion of 'fatal' fate and a notion in which there is room for one's own responsibility.[14] Most of the time the latter is closely linked on an individual level to the aforementioned concept of the daemon, which gives insight into fate or reminds us of our fate.[15] Goethe in particular considered the *Schicksalsbegriff* to be based on a close connection between *Geschick* (destiny) and the individual's personality. In his view, man's character or nature and fate are one and the same: a unity.[16] Fate in this sense is thus part of our own actions.[17] Between the two world wars this notion of *Schicksal* had undergone a change, and its connotation became more normative.[18] That is to say, it acquired the meaning of a task, something that has to be accomplished. To the National Socialists it meant, for example, that the German idea (or myth) of superiority had to be fulfilled. After the Second World War Meinecke seems to hold on to the Goethean idea of fate, but it becomes more difficult for him to view Germany's fate (caused by or happened to) as something that is not normative – in the sense of a demonic evil that had befallen Germany.[19] Nevertheless, Goethe and the *Goethezeit* act as a (therapeutic) counterweight for Meinecke.

Despite his notion of *Schicksal*, Meinecke finds his initial answers to the question of the causes for the German catastrophe in Burckhardt.[20] Meinecke's interest in Burckhardt was already aroused well before the Second World War, but at that time he considered Burckhardt's conception of history as an alternative to the German conception of history.[21] According to Meinecke, the fact that Burckhardt

---

in the sense of 'to send' (schicken). cf: Georg Steiner, *Heidegger* (Stanford Terrace 1978) 108; Kranz, 'Schicksal', 1286-1287. Maubach jumps to conclusions when claiming that the use of the notion Schicksal means that no one is responsible anymore: Maubach, '"Wie es dazu kommen konnte"', 174, 172.

14   Kranz, 'Schicksal', 1284.

15   See the previous chapter, section 5.3.

16   Klaudia Hilgers, 'Schicksal' in: Bernd Witte ed., *Goethe Handbuch in vier Bänden* (Stuttgart and Weimar) Volume 4/2, 941-943, there 941.

17   See for example the explanation of Heraclitus (a man's character is his fate) in section 5.3 of the previous chapter.

18   Kranz, 'Schicksal', 1286.

19   In this period the notion of *Dämonisch* is often associated with evil. Jost Hermand, *Kultur im Wiederaufbau. Die Bundesrepublik Deutschland 1945-1965* (Munich 1986) 51.

20   Meinecke, 'Die deutsche Katastrophe', 325.

21   Initially, it was thought that Meinecke late in his career buried himself in Burckhardt's oeuvre. This has been refuted by Herkless among others: J. L. Herkless, 'Meinecke and the Ranke-Burckhardt Problem', *History and Theory* 9 (1970) 290-321, there 291-292. Two reviews, of 1906 and 1928, prove that Meinecke considered Burckhardt's conception of

was settled in Basel played an important role in this case. Because of this, says Meinecke, Burckhardt had failed to experience the problematic nature of the relationship between state and nation, something that was central to German historical writing: 'Our historical thinking has by and large been developed by the struggle for state and nation', Meinecke writes.[22] That is why he, initially, placed Burckhardt's conception of history outside the general (German) conception of history: 'In this way he created his own area of work and interests in a sovereign manner and thus separated himself from the general direction of historical studies'.[23] Despite Burckhardt's special position (*Sonderstellung*), Meinecke always sympathized with his alternative view of history.[24]

During the Second World War Burckhardt's conception of history became more and more important to Meinecke.[25] It is not Burckhardt's pronounced pessimism which Meinecke finds interesting – Meinecke, among other things, speaks of hope of renewal in *Die deutsche Katastrophe* – it is Burckhardt's consideration of *Kultur* and his aversion to power politics which Meinecke finds interesting. To Meinecke, during and after the Second World War, Burckhardt's conception of history is a welcome addition to the German, mainly Rankean conception of history. Meinecke's correspondence in this period gives also evidence of this: 'I often deal with this question [the relationship between power politics and *Kultur*, RK] with Burckhardt, who perhaps had a more profound understanding of this issue than even Ranke

---

history to be a viable alternative to the German conception. In the review of 1906 Meinecke discusses Burckhardt's *Weltgeschichtliche Betrachtungen*: Friedrich Meinecke, 'Jacob Burckhardt, die deutsche Geschichtsschreibung und der nationale Staat', in: idem, *Zur Geschichte der Geschichtsschreibung* (Munich 1968) 83-87; In his 1928-review Carl Neumann's work on Burckhardt is central: Friedrich Meinecke 'Carl Neumann über Jacob Burckhardt' in: idem, *Zur Geschichte der Geschichtsschreibung*, 88-92. Gossman, *Basel in the Age of Burckhardt*, 578 note 24 and note 36.

22   Meinecke, 'Jacob Burckhardt', 85.
23   Meinecke, 'Carl Neumann', 92; Meinecke, 'Jacob Burckhardt', 87.
24   Meinecke also praised Burckhardt's *Weltgeschichtliche Betrachtungen*. Meinecke, 'Jacob Burckhardt' 83-84 and 87.
25   In chapter 2 and 3 I have made clear that Meinecke, after the First World War, in *Die Idee der Staatsräson*, expressed a certain amount of criticism with regard to Ranke's views. In 1942, in an extensive article entitled 'Deutung eines Rankewortes' Meinecke asked himself a burning question: to what extent is Ranke's famous expression 'Jede Epoche ist unmittelbar zu Gott' still viable (if at all). He concludes that, albeit in a secularized fashion (may be even a panentheistic one) that this expression is indeed still valuable. Friedrich Meinecke, 'Deutung eines Rankewortes' in: idem, *Zur Theorie und Philosophie der Geschichte* (Stuttgart 1965) 117-139. Bruch claims that Meinecke's (alleged) shift from Ranke to Burckhardt occurs already in *Die Idee der Staatsräson*, Bruch, 'Ein Gelehrtenleben zwischen Bismarck und Adenauer' 17; Mark W. Clark, *Beyond Catastrophe. German Intellectuals and Cultural Renewal after World War II 1945-1955* (New York 2006) 39-40.

ever had', Meinecke writes in a letter to the historian Willy Andreas.[26] And in a letter to his friend and colleague Kaehler, Meinecke claims that Burckhardt is to him, 'now, more and more the most important man of the nineteenth century'.[27] This does not mean, however – as I will show below – that Meinecke turns his back on Ranke.[28]

In *Die deutsche Katastrophe* Burckhardt's influence is clearly visible on many different levels. Meinecke agrees with Burckhardt's views about the causes of political decline. Furthermore, as mentioned at the end of the previous chapter, there is a relationship between Burckhardt's notion of *Verpuppungen* and Meinecke's ideas on the 'horizontal' and 'vertical' conception of history. Meinecke's interest in Burckhardt should not only be understood against the background of the Second World War; it should also inform the perspective of his autobiographical works, for during his writing of *Die deutsche Katastrophe* Meinecke was also writing his memoirs; *Die deutsche Katastrophe* is not without reason incorporated in his *Autobiographische Schriften*, for it essentially is an autobiographical work.[29] After Meinecke fled in 1945 from Berlin to Wässerndorf and then to Göttingen (because of the unsafe situation in Berlin) he wrote *Die deutsche Katastrophe* practically without access to sources or his books, and for the most part on the basis of his own recollections.[30] Of course, that does not make it an autobiographical work. It indeed becomes autobiographical when Meinecke lets his own individual past merge with Germany's past as a whole. This way, as I will make clear in this chapter, Meinecke pressed his philosophy of history to its ultimate conclusion.

The German historian Nicholas Berg has already pointed out that Meinecke in *Die deutsche Katastrophe* reconciled history with autobiography.[31] In Berg's view

---

26   Über diese Frage setze ich mich jetzt oft mit Burckhardt auseinander, der doch vielleicht tiefer und schärfer als selbst Ranke in diese Problematik hineingeleuchtet hat'. Meinecke, *Neue Briefe und Dokumente*, 423.

27   Ibidem, 420, also: 417-418, 436; and: Meinecke, *Ausgewählter Briefwechsel*, 360, 420.

28   Gossman even claims that Meinecke's focus in *Die deutsche Katastrophe* is very 'un-Burckhardtian', because this study is in his view on the lessons that a state can learn, not on the evil that the state has inflicted upon its people. Gossman, *Basel in the Age of Burckhardt*, 450.

29   Meinecke's memoirs were originally published in two volumes: *Erlebtes 1861-1901* in 1941 and *Strassburg, Freiburg, Berlin. Erinnerungen* in 1949; the latter was already written between 1943-1944. Both are incorporated in *Autobiographische Schriften*.

30   Meinecke, 'Die deutsche Katastrophe', 324; Ernst Schulin, 'Friedrich Meinecke', 53; Bock, 'Meinecke, Machiavelli und der Nationalsozialismus', 146. Meinecke (and with him nearly 80 refugees) stayed in Wässerndorf in the castle with the same name. Meinecke, *Neue Briefe und Dokumente*, 73.

31   Berg, *Der Holocaust*, 84-85, 90-79; There is, however, no general consensus on this matter: on the one hand Meinecke's work is regarded as a scientific study, on the other an autobiographical sketch. Those who consider it more than an autobiographical sketch are

Meinecke's study acquires more authority and authenticity because of the auto-biographical form of the work; he speaks of an *Autobiographisierung* of history.[32] Berg claims this reflects a change in Meinecke's conception of history. Berg does not compare this with the rest of Meinecke's oeuvre.[33] *Die deutsche Katastrophe* is in fact not so much a change or shift, as Berg claims, but a deepening of Meinecke's conception of history. A deepening, which Meinecke, a few years after the publica-tion of *Die deutsche Katastrophe*, further developed in a lecture entitled 'Ranke und Burckhardt'. In this lecture Meinecke positions himself between these two eminent nineteenth-century historians. Most critics saw and still see in this lecture another shift in Meinecke's views.[34] They claim that Meinecke is more inclined to agree with Burckhardt's ideas *at the expense of* Ranke's views on politics and history. In this chapter I will make clear there was no shift in Meinecke's thought, instead I suggest Meinecke reconciled Ranke's with Burckhardt's views.

6.1   A controversial study

Die deutsche Katastrophe is generally not considered to be part of Meinecke's major works, but it is one of his best-known, most translated and maybe the most contro-versial book of his oeuvre.[35] It is especially controversial because Meinecke barely

---

among others: Wippermann, 'Friedrich Meineckes *Die deutsche Katastrophe*', 102; Berg, *Der Holocaust*, 84-85, 93; William O. Shanahan, 'Germany: Realities of Power and Reconstruction', *The Review of Politics* 10 (1948) 502-509, there 506; Schulin, 'Meineckes Leben und Werk, 130; Schulin, 'Friedrich Meinecke und seine Stellung', 29, 46; Winfried Schulze, *Deutsche Geschichtswissenschaft nach 1945* (Munich 1989) 53-54; Hugo Frey and Stefan Jordan, 'Inside-Out: The Purposes of Form in Friedrich Meinecke's and Robert Aron's Explanations of National Disaster' in: Stefan Berger and Chris Lorenz eds., *Nationalizing the Past. Historians as Nation Builders in Modern Europe* (Basingstoke etc 2010) 282-297, there 282, 287-288. Frey and Jordan isolate Meinecke's study from his intellectual development; Meinecke reflected on Germany's past throughout his life; one cannot understand Meinecke's last study in isolation from the rest. So, the argument that Meinecke's study is sometimes 'contradictory' has everything to do with his love for contrasting ideas. Moreover, Frey and Jordan base some of their arguments on the very questionable book written by Pois; also see my introduction.

32   Berg, *Der Holocaust*, 84.

33   Idem.

34   The most important ones are: Iggers, *The German Conception*, 226; Schulin, 'Meineckes Leben und Werk, 130, Hofer, *Geschichtsschreibung*, 231; Jensen, '"Unity in Antinomy"', 203, 217-224; Exceptions are: Gossman, *Basel in the Age of Burckhardt*, 445, 450. Gossman does mention a *shift*, but at the same time he is convinced Meinecke did not really free himself from Ranke; J. L. Herkless, 'Meinecke and the Ranke-Burckhardt Problem', 319-321.

35   S.D. Stirk, 'Review: The German Catastrophe through German Eyes', *International Journal* 2 (1947) 343-353, there 350. Stirk claims that 'the aged Professor [Meinecke, RK] is living too much in the Prussian past to be a completely reliable guide and prophet for the new

discusses the Holocaust, he characterizes Hitler's coup as 'accident' or 'chance', and he argues for a return to the *Kultur* of the *Goethezeit*.[36] The idea of 'chance' and the *Goethezeit* I will discuss in the third and fourth section of this chapter.

The Holocaust is almost absent from *Die deutsche Katastrophe*, because to Meinecke, as shown by the historian Mark Clark, the Holocaust was part of the overall destruction caused by Nazism. In the words of Clark: 'The initial emphasis on the general causes of the catastrophe and the rise of Nazism was not just legitimate; it was crucial'.[37] Yet, Meinecke states in *Die deutsche Katastrophe*: 'But in the gas chambers of the concentration camps the last breath of the Christian feeling for humanity and of the Christian culture of the West was finally extinguished'.[38] 'Humanity' and 'culture' died in the gas chambers, but Meinecke does not explicitly mention the persecution of the Jews.[39] From the 1960's onwards this is a frequent point of criticism. Clark has shown that this criticism of the 1960's may indeed have done an injustice to Meinecke's research on the causes of the rise of National Socialism and the subsequent catastrophe. For, as Clark states: 'Yet only through the initial groundwork of Meinecke and others was the later, more nuanced, understanding of how the Holocaust emerged from German history possible'.[40]

Another explanation for the absence of a discussion on the Holocaust in *Die deutsche Katastrophe* could be Meinecke's moderate anti-Semitism. Meinecke was brought up in a social environment in which a certain amount of anti-Semitism was not unusual, as he described in his memoirs.[41] By his own account, from the end of his student days onwards until the beginning of the 1890's he was a 'blunt anti-Semite'.[42] When, in 1893, he took over the editor's position of the *Historische Zeitschrift*, and frequently had to work together with Jews, he changed his views. Yet, he found it difficult to completely distance himself from the prejudices of the cultural environment which he was part of.[43] There is, of course, a big difference

Germany of the future'; Paddock, 'Rethinking Friedrich Meinecke's historicism', 106-107; Schulin, 'Meineckes Leben und Werk', 129; Frits Boterman, *Duitse dichters en denkers*, 211-212, 275; Gay, *Weimar Culture*, 70-72.

36   Iggers, *The German Conception*, 223, 228; Berg, *Der Holocaust*, 82. Wippermann, '"Deutsche Katastrophe"', 189-190; Gossman, *Basel in the Age of Burckhardt*, 450; Boterman, *Duitse dichters en denkers*, 211-212.

37   Clark, *Beyond Catastrophe*, 42.

38   Meinecke, *The German Catastrophe*, 84.

39   Nicolas Berg claims that Meinecke in expressing himself in this way, he implicitly states that Germany's *Kultur* was destroyed instead of the Jews: Berg, *Der Holocaust*, 82.

40   Clark, *Beyond Catastrophe*, 42.

41   Meinecke, 'Straßburg – Freiburg – Berlin', 150-152; also cf.: Sösemann, ed., *Friedrich Meinecke, Die deutsche Katastrophe. Betrachtungen und Erinnerungen. Edition und internationale Rezeption*, 470-476.

42   Meinecke, 'Straßburg – Freiburg – Berlin', 150.

43   Ibidem, 150-152.

between Meinecke's moderate anti-Semitism and the murderous anti-Semitism of National Socialism – the goal of the latter being extirpation of the Jews. As the German historian Golo Mann phrased it: 'Hitler's anti-Semitism had very little to do with the average German anti-Semitism'.[44] *Die deutsche Katastrophe* is also not completely free of Meinecke's moderate anti-Semitism. He, for example, claims: 'They [the Jews, RK] contributed much to that gradual depreciation and discrediting of the liberal world of ideas that set in after the end of the nineteenth century'.[45] Meinecke immediately adds – and here we see his twofold standpoint: 'The fact that besides their negative and disintegrating influence they also achieved a great deal that was positive in the cultural and economic life of Germany was forgotten by the mass of those who now attacked the damage done by the Jewish character. Out of the anti-Semitic feeling it was possible for an anti-liberal and an anti-humanitarian feeling to develop easily – the first steps toward National Socialism'.[46]

A last point of discussion with regard to *Die deutsche Katastrophe* is the title of the book.[47] Does it refer to a catastrophe which was caused by Germany or does it refer to a catastrophe which had happened to Germany? Meinecke's argument gradually makes it clear that – in accordance with his preference for antinomies and contrasts, and the equation with the notion of *Schicksal* – the title refers to both.

Meinecke does not pass judgements and does not explicitly discuss the question of guilt; he only wants to determine how the German *Macht* of Bismarck and the *Kultur* of Goethe degenerated into the destruction of Hitler. In this regard, Meinecke shows how German history was lead astray by certain unfortunate turns.[48]

---

44 'Hitlers Judenhaß hat sehr wenig mit dem durchschnittlichen deutschen Antisemitismus zu tun'. Cited in: Wippermann, '"Deutsche Katastrophe"', 190. On Meinecke's moderate anti-Semitism: Wehrs, 'Demokratie durch Diktatur?', 111-112. There are roughly four 'movements' within anti-Semitism: religious, economic, racial and political. On this: Gavin I. Langmuir, *History, religion and Antisemitism* (Los Angeles 1990); W. Kampmann, *Deutsche und Jude. Studien zur Geschichte des deutschen Judentums* (Heidelberg 1963).

45 Meinecke, *The German Catastrophe*, 15.

46 Idem.

47 Boterman states Meinecke does speak of the German catastrophe but fails to discuss the Jewish catastrophe. Boterman, *Duitse dichters*, 211-212; Wippermann, '"Deutsche Katastrophe"', 177; Iggers assumes Meinecke is referring to a catastrophe which has happened to Germany: Iggers, *The German Conception*, 224.

48 Obviously, it is still a question what indeed is the 'right' path or development, but it is clear that Germany's road to a national unity happened in a different manner, and later than, for example, in France or England. For the nineteenth and twentieth century there are a few years, according to Meinecke, when Germany might have gone astray. For example, the year 1848 could be a starting point when the Frankfurter Parliament revolution fails, or the 1866 *Brüderkrieg* against Austria, which caused the German League to be cancelled and which opened the way to unification under the leadership of Prussia. In 1871, Germany was unified

6.2 A catastrophic synthesis

In *Die deutsche Katastrophe*, Meinecke focuses his attention on the two major movements of the nineteenth century: nationalism and socialism. His argument revolves around the question of how the worst characteristics of both these movements resulted in National Socialism.[49] Even though Germany had a different political development from the rest of Europe, Meinecke emphasized that Hitler's movement was not a particular German phenomenon, for it had its counterparts in neighbouring countries, like Mussolini's Italy. Furthermore, Meinecke claims that, historically speaking, the problem of the decline of *Kultur* already started with the French Revolution and the (British) Industrial Revolution. This is where Meinecke takes up Burckhardt's argument, who claimed that the Enlightenment and the ideals of the French Revolution bear the seed of 'false pursuit' of an unreachable happiness, which in turn changed into tyranny and a striving for wealth.[50] Meinecke's claim that the degeneration of the European and particularly the German culture began with the Enlightenment dovetails with his argument in *Die Entstehung des Historismus* in which he sets the Enlightenment in opposition to historicism. Historicism had caused a revolutionary change, but he denies at the same time that the Enlightenment was 'relieved' by historicism; to Meinecke it remained next to historicism.[51] He asserts that the negative characteristics of the Enlightenment, which had been realized in both nationalism and socialism, were combined in National Socialism.[52]

What Meinecke does not consider is the relationship between historicism – in which the principle of individuality might easily lead to ethical relativism – and

---

under the militarist regime of the Prussian Junker Bismarck; further one can think of 1914 – the beginning of the First World War or 1933 when Hitler was elected chancellor, and 1939 – the beginning of the Second World War. Meinecke, 'Die deutsche Katastrophe', 379-380. Also: Maubach, '"Wie es dazu kommen konnte"', 143-189.

49   Meinecke, 'Die deutsche Katastrophe', 334-335.

50   Ibidem, 325. Burckhardt, for that matter, saw these kinds of processes come into existence much earlier in history.

51   See the first section of chapter 4.

52   Paddock, who discusses *Weltbürgertum und Nationalstaat* and *Die deutsche Katastrophe* as if they were written at the same time, blamed Meinecke for not continuing the argument of *Weltbürgertum und Nationalstaat* in his work after the Second World War. Moreover, Paddock claims that Meinecke underestimated the 'irrational' dimension of historicism as a cause for National Socialism. He also claims that Meinecke in *Die deutsche Katastrophe* still emphasized the role of the state, whereas Meinecke explicitly changed his focus to *Kultur*, which might be more important than power. Paddock, 'Rethinking Friedrich Meinecke's historicism', 104-106; also: Iggers, *The German Conception*, 224, 226. Iggers also claims Meinecke became pessimistic after the war, while many different lectures of Meinecke from around this time show that he was in fact hopeful about the future, or, at least urged Germany to recover.

National Socialism, which, according to some scholars, became possible because of a relativism of values. Yet, Nazism itself was far from relativistic, for it advocated a range of extreme values and ideals. Moreover, Nazism was concerned with the collective and the masses and not with the individual and individuality as historicism was. To explain away Nazism on the basis of a historicist conception of history is indeed evidence of value relativism. The question whether value relativism as a consequence of historicism had caused the emergence of Nazism was probably not relevant to Meinecke, who, as a result of his panentheistic philosophy of history, was focussed on the good, true and beautiful, which could only be divined by the individual.[53]

Despite Meinecke's emphasis on the European character of the two major movements and their synthesis in Germany and also Italy, he concentrates mainly on the specific German characteristics of National Socialism, and on why this catastrophe happened in Germany. First of all, in Meinecke's view, we have to understand that in Germany nationalism rose almost half a century earlier than socialism.[54] The middle-class emerged before the proletariat, because in Germany the 'economic-technical revolution' set in later than in the rest of Western Europe.[55] In Germany nationalism or a national feeling was first mentioned during the Wars of Liberation (1813-1815). And it is precisely this period, according to Meinecke, in which fundamental changes took place within what he calls the character or identity of German man (deutschen Menschentums).[56] This notion of 'German man' reminds us of the characterization of Germany as 'Nation of all mankind' as discussed in the chapter on Weltbürgertum und Nationalstaat.[57] In that chapter it became clear that it was the Romantics of 1800, Humboldt and Goethe who considered Germany the cultural-humanistic educator of all nations; it was the so-called 'universal calling' of the German nation.[58] So, on the one hand it was about a universal-cultural ideal, and on the other an idea of a cultural-political German unity.[59]

To forge the different German states into a unity, it was important to create a synthesis between the cultural side, the spirit (Geist or Kultur) of the German lands,

---

53   On this see the introduction. The Dutch historian Von der Dunk claimed that with regard to Meinecke the relationship between ethical relativism and historicism does not lead to a value relativism, because Meinecke holds on to a universal, elementary morality. Von der Dunk, De glimlachende sfinx. Kernvragen in de geschiedenis (Amsterdam 2011) 47. Von der Dunk, moreover, favours a rehabilitation of historicism: Von der Dunk, De glimlachende sfinx, 55.

54   Meinecke, 'Die deutsche Katastrophe', 332.

55   Meinecke The German Catastrophe, 8.

56   Meinecke, 'Die deutsche Katastrophe', 332; Meinecke The German Catastrophe, 8.

57   Cf. chapter 1.

58   Meinecke, Weltbürgertum und Nationalstaat, 66; cf. chapter 1.

59   Cf. chapter 2.

and the political side, the power (*Macht* of mainly Prussia). The first initiatives to make such a synthesis possible took place in the so-called *Reformzeit* (1807-1815).[60] After the battle at Jena in 1806 Prussia was defeated by Napoleon. As a result of this defeat Prussia was forced to rethink its future, for if it wanted to triumph over France in the future, it had to reform in many different areas. According to Meinecke, a foundation was laid in this *Reformzeit* for an elevation (*Erhebung*) of personalities, the state and the nation. That means a start was made by, among others, generals like Scharnhorst, Gneisenau and Boyen to reform the army. Karl Freiherr vom Stein and Karl August Fürst von Hardenberg suggested reforms for the institutions of the state. And Wilhelm von Humboldt opposed the ideals of the Enlightenment with the idea of *Bildung*.[61] In this period a synthesis between power and culture was possible in Meinecke's view, but this process was cut short in 1819 with the so-called *Karlsbader Beschlüsse*, which were instigated by Metternich and aimed at mainly liberal and national sentiments. Meinecke considers 1819 the year in which power politics triumphed over culture (*Kultur*). Particularly the forced resignation of Humboldt and Boyen was considered by Meinecke the moment at which the ideal of a synthesis between culture and power was abandoned.[62] As I have shown in my discussion of *Weltbürgertum und Nationalstaat*, with Bismarck this synthesis seemed to be realized after all; however, Meinecke later on concluded that culture was again overrun by power politics.[63] After two world wars Meinecke paints a picture in *Die deutsche Katastrophe* of the fate of the degeneration of the German *Menschentum*. It is a development from 'spirit' to matter, from culture to power politics and militarism, from humanistic, liberal ideas and ideals of the Goethezeit to the brute reality of war and destruction of Hitler.[64] It seemed the synthesis Meinecke hoped for had dissolved to one side: power politics. It is this fate that Meinecke tries to comprehend in *Die deutsche Katastrophe*.

The development just sketched has, according to Meinecke, in part to do with what he considers the so-called *kulturfähige* (capable of culture) and *kulturwidrige* (hostile to culture) souls of nationalism and socialism.[65] Meinecke claims it was the elements hostile to culture of both movements which eventually merged into

---

60   On this topic, see Meinecke's short study of 1906: Friedrich Meinecke, *Das Zeitalter der deutschen Erhebung (1795-1815)* (Göttingen 1963).

61   Nipperdey, *Deutsche Geschichte 1800-1866. Bürgerwelt und starker Staat* (Munich 1984) 57-58.

62   Meinecke, 'Die deutsche Katastrophe', 335; Nipperdey, *Deutsche Geschichte 1800-1866*, 277-278; Schönpflug claims that the unfulfilled ideals of the *Reformzeit* turn into a promise whose influence is noticeable well into Meinecke's time. Schönpflug, 'Revolution und "Erhebung"', 28.

63   Meinecke, 'Die deutsche Katastrophe', 337-339.

64   Ibidem, 332-333, 338.

65   Ibidem, 334-335; Meinecke, *The German Catastrophe*, 10.

National Socialism. This already started with Bismarck's Prussia in which nationalism became a powerful militarism and left almost no room for spirit or culture.[66] Another development which resulted from the negative elements of socialism was, according to Meinecke, the attention to the masses. Politics was more and more directed at the masses, and these masses in turn got more 'powerful' because they came in contact with power politics. In the words of Meinecke: 'From being an aristocratic affair, Machiavellism became a bourgeois affair, and finally became mass Machiavellism'.[67] In Meinecke's view, the rise of this mass Machiavellism happened at the expense of the individual; Meinecke fully endorsed the *Bildungsideal* of Goethe and Humboldt and rejected everything associated with the masses.[68] Above all, it was the culture of the nation that suffered most, since mass Machiavellism was only aimed at benefits, direct satisfaction, and power: '(...) and its political thinking remained sound only so long as it kept in contact with the whole of the cultural life of the nation', Meinecke writes.[69] Another dimension of nationalism (hostile to culture) led, in Meinecke's view, to ethical indifference. In this context Meinecke cites the philosopher Friedrich Paulsen, who clearly explains the effect of extreme nationalism: 'Nationalism, pushed to an extreme (...) destroys moral and even logical consciousness. Just and unjust, good and bad, true and false, all lose their meaning'.[70] Meinecke claims Paulsen describes here already the ethics which would later on be dominant in National Socialism.

The German man (*deutsches Menschentum*) was gradually replaced with what Meinecke refers to as *Hitlerismus* or *Hitlermenschentum* (Hitlerism). In this regard he was again close to Burckhardt, who also preferred humans who were cultivated through individual *Bildung* within a society, instead of a trained or drilled 'man in the crowd'.[71] What was lost with Hitlerism was the man of the *Goethezeit*: 'The man of Goethe's day was a man of free individuality. He was at the same time a "humane" man, who recognized his duty toward the community to be "noble, helpful, and good" and carried out his duty accordingly', Meinecke writes.[72] Reason and spirit were substituted for the so-called technological-utilitarist spirit. Meinecke clarifies this as follows: 'The decisive factor in the development of this type was that a definitely rational concept acquired absolute dominion over all irrational elements in men'.[73] This is, according to Meinecke, already noticeable in Frederick William

---

66   Ibidem, 335-337.
67   Meinecke, *The German Catastrophe*, 52.
68   Meinecke, 'Die deutsche Katastrophe', 345.
69   Meinecke, *The German Catastrophe*, 42.
70   Cited in Meinecke, *The German Catastrophe*, 23-24.
71   Gossman, *Basel in the Age of Burckhardt*, 334, 366.
72   Meinecke, *The German Catastrophe*, 26.
73   Ibidem, 39.

I's militarism.[74] The army and power are about calculation and technique: 'For it was there [on the battle field and in training, RK] that the man was drilled, that is, made into the kind of being who was to learn how to sacrifice his life blindly for a goal not set by himself', Meinecke writes.[75] The warm blood of culture, the individual *Bildung* and the politics 'capable of culture' (*kulturfähig*) are poisoned here by the cold, calculating and culturally hostile system of war. This shift of the mind, according to Meinecke, made Hitlerism possible. And this way a gulf between the 'irrational' and the 'rational' in man emerged: the inexplicable spiritual, unfathomable man was exposed to a cold and calculating power and mass politics – and the latter overwhelmed the former. In Meinecke's words:

> The calculating intellect was excessively exaggerated on the one hand, and the emotional desire for power, wealth, security, and so forth on the other hand. As a result the acting willpower was driven onto dangerous ground. Whatever could be calculated and achieved technically, if it brought wealth and power, seemed justified – in fact, even morally justified, if it served the welfare of one's own country.[76]

This idea of Meinecke does not even come close to the radicalization of evil brought about by Hitler. And neither does it give a solid explanation for this radicalization.[77] In essence, Meinecke falls back on the theme of *Die Idee der Staatsräson*, for it is clear that he still had his mind on a principle of *raison d'état* which included ethics; he wants a *Staatsräson* capable of culture in which power and spirit coalesce. This explains why in *Die deutsche Katastrophe* Meinecke describes Hitler's power politics as satanic.[78] For this of course refers to a calculating and determined destruction, instead of following the state's interest and keeping it (and the country as a whole) sound, but it particularly shows that Meinecke did not want to confront the realities of the evil spread by Hitler, even though it was answered on a massive scale by both the military and ordinary people.

Meinecke concludes that of both movements – nationalism and socialism – the worst elements eventually coincided in National Socialism. Instead of an evolutionary socialism and a 'culturally sensitive' Prussia, an excessively militaristic Prussia – the mass Machiavellism – in combination with a raw mass socialism emerged.[79] The life-long ideal that Meinecke had in mind – the synthesis between spirit and

---

74   Meinecke, 'Die deutsche Katastrophe', 361-363.

75   Meinecke, *The German Catastrophe*, 40.

76   Ibidem, 51.

77   Cf. Sösemann, ed., *Friedrich Meinecke, Die deutsche Katastrophe. Betrachtungen und Erinnerungen. Edition und internationale Rezeption*, 476.

78   Meinecke, 'Die deutsche Katastrophe', 337.

79   Meinecke, 'Die deutsche Katastrophe', 365-381; also: Berg, *Der Holocaust*, 74-75.

power – seemed now, after the Second World War, increasingly more difficult to realise, but the demand for it became more and more imperative. In this context, it is important to see what the effect of the German fate (*Schicksal*) was on Meinecke's historicism. How did he consider this fate, and did he manage to 'transcend' it like he managed to resist the 'crisis of historicism'?

6.3   Nature, chance, fate

What is striking about Meinecke's account of the two movements is his use of metaphors of nature. When discussing these movements, Meinecke uses terms like 'waves of the ages' (*Wellen des Zeitalters*) and related notions like 'flash of lightning' (*Wetterleuchten*), 'spring of National Socialism' (*Quellgewässern des Nationalsozialis-mus*), stormy waves (*sturmische Welle*), 'strong channel' (*starkes Rinnsal*), 'hot blast of the Hitler movement' (*Sturmwinden der Hitler-Bewegung*).[80] In short, Meinecke characterizes National Socialism as a natural disaster, a catastrophe. Even though Meinecke does not discuss the question of guilt, he seems to make clear by way of these metaphors that Germany was overtaken by this catastrophe. The metaphor of the waves or currents (movements) implies, for example, that when caught in the current, it is difficult to determine when it accelerates or when the revolution actually takes place. In the case of political movements, it is not easy to draw the line between 'going with the flow' and 'being swept away by the current'.[81] The same applies to Meinecke's ideas on the change from *kulturfähig* (capable of culture) to *kulturwidrig* (hostile to culture). It is impossible to define the moment at which the catastrophe takes shape. Meinecke could only afterwards indicate where the stream possibly had accelerated and/or began to bend away.[82]

---

80   Meinecke, *The German Catastrophe*, 15, 21, 33, 43, 44; Meinecke, 'Die deutsche Katastrophe', 339, 345, 357, 368, 369; also: Berg, *Der Holocaust*, 65-66.

81   Or even causing the stream; that is essentially the theme of the unintended consequences of intentional actions. Cf. Chapter 2.

82   In a review of *Die deutsche Katastrophe* the acceleration of a stream is discussed, or better: the question is at which moment a revolution has indeed become a revolution or is experienced in this way. The reviewer, Beyerhaus, discusses what is in between 'movement' and 'rest or standstill', namely: the essence which is 'outside' of time. Plato, who is mentioned in the review, denotes this essence in *Parmenides* as the 'sudden'. It implies that from this essence an acceleration or turn can take place whose outcome is uncertain: '(...) aus dem Sein zum Vergehen und aus dem Nichtsein zum Werden', Beyerhaus writes. The 'sudden' with its 'hovering' and unpredictable character and uncertain outcome is thus located 'outside' time, for it is 'not yet'. The same latent, unfathomable character of the 'sudden' is also present in a historical crisis. 'There is something brewing', but it is not yet clear what will happen. The 'moment' at which, for example, the hostile (*kulturwidrige*) sides of both movements will coalesce and run into the tidal wave of National Socialism can be viewed

In *Die deutsche Katastrophe* Meinecke characterizes Hitler's appointment by Hindenburg as chance.[83] Initially, this reminds us of Goethe's view on chance, who compared it with sudden, abrupt events in nature, like volcanic eruptions. These are, in his view, exceptions and moreover unsound developments. At first glance, Meinecke's characterization of chance in *Die deutsche Katastrophe* dovetails with this idea.[84] For, it seems Meinecke, by means of his metaphors of nature and his idea of 'chance', tries to place the rise of National Socialism outside of the 'normal', 'sound' German history. The era of National Socialism was in this sense an unsound leap, a burst, which tore Germany away from the gradual, 'right', *kulturfähig* development that Meinecke had in mind. For Meinecke it was a way to brand National Socialism as 'un-German'. So, in his view, Germany, from Romanticism onwards, did not adhere to the wrong values, it was Hitler who executed them in an un-German way. This way, Meinecke is able to legitimate Germany's past as well as his own – his continuous search for a synthesis between culture and power.[85] In other words, the outcome could also have been good; however, chance caused a different turn.[86]

Chance, secondly, reminds us of what I discussed at the beginning of this chapter with regard to the notion of *Schicksal*, fate. 'Chance' occurs mainly on the level of individual events (like Hitler's appointment by Hindenburg). 'Fate' covers the whole of such events and, for that reason, leads to the idea that everything in history inevitably had to pass in this way. This obviously fits with the theme of the daemonic and fate, as discussed in the previous chapter. Plato, as mentioned before, claimed everyone chose their own fate, but forgot this fate as soon as they were born into this world, since everyone went before their birth to 'the river of

---

as such a sudden turn. Gisbert Beyerhaus, 'Notwendigkeit und Freiheit in der deutschen Katastrophe. Gedanken zu Friedrich Meineckes jüngstem Buch', *Historische Zeitschrift* 169 (1949) 73-87, there 85-86. Meinecke claims something similar in *Geschichte des deutsch-englischen Bündnisproblems*: 'Auf ein volles Glas können noch manche Tropfen aufgeschüttet werden, ohne daß es überläuft. (...) Nun aber gehen die Wirkungen des letzten Tropfens, der sie auslöste, weit über seine mutmaßliche Bedeutung hinaus, und es kann im weiteren Fortgange dieser Wirkungen zu ungeahnten Katastrophen kommen'. Friedrich Meinecke, *Geschichte des deutsch-englischen Bündnisproblems 1890-1901* (Munich and Vienna 1972; first 1927) 5. In a similar vein, Meinecke discusses the appointment of Hitler as Chancellor in January 1933.

83   Berg gives an overview of (negative) reactions to this idea of Meinecke. Berg, *Der Holocaust*, 95 note 113.

84   Nicolas Berg claims – as far as I know he is the only one to do so – that Meinecke in *Die deutsche Katastrophe* makes use of Goethe's historical and natural concepts: Berg, *Der Holocaust*, 71-72. In my research I will show that Goethe is not only important in Meinecke's last study, but throughout his oeuvre.

85   Cf. Berg, *Der Holocaust*, 75.

86   Meinecke thus avoids to discuss the question of guild. He was, by the way, not alone in this. Hermand, *Kultur im Wiederaufbau*, 42-52.

oblivion'. For that reason, it will appear as though everything is ruled by necessity. By aiming at higher things, the *daimon*, the divine voice within the individual, will bring us back in contact with our fate, and our lost knowledge of the divine good, true, and beautiful. In line with this idea, Meinecke's attitude to fate is twofold. Man will never fully understand or fathom the meaning of all (awful) events, but – and this is important to Meinecke – there is a possibility of obtaining insight into our fate and our own part in it. Not for nothing, Meinecke claims at the beginning of *Die deutsche Katastrophe* that he wrote this work as preliminary to future historians to finally 'understand more profoundly our fate'.[87] So, fate (*Schicksal*) and the possible perspective on this, is essentially at the heart of *Die deutsche Katastrophe*.[88]

To fathom fate is, as Meinecke also states in an article of 1942, the hardest but also the most important task of the historian: 'not identifying progress or regression, but fathoming fate is the highest task of history'.[89] Historicism, Meinecke writes, plays a central role in this, as he already made clear in *Die Entstehung des Historismus*: 'Perfect histori[ci]sm also implies the capacity to be resigned and requires respect for destiny'.[90] But can we summon respect for the atrocities of the Second World War? According to Meinecke, Ranke could soothe his conscience with a 'providence of God', for: 'His historical thinking thus ended in a religious mystery, with which a genuine historical sense of destiny and understanding of

---

87   Meinecke, *The German Catastrophe*, xi.
88   *Schicksal* (with conscience at its core) is according to Hofer the factor in which everything is merged. Hofer, however, fails to explain on what (philosophical) foundation this 'conscience' is based or how a (vertical) perspective is even possible. 'Schicksal ist der Gedanke, der für Meinecke zugleich das tiefste seines eigenen Lebens und das Geheimnis der Geschichte in sich schließt': Hofer, *Geschichtsschreibung und Weltanschauung*, 228; also: 317, 126-128, 224-231; Berg even distinguishes five dramatic acts in Meinecke's *Katastrophe*-study: (1) a golden age (*Goethezeit*) of ideals (*Kultur*), (2) the possibility of a reconciliation between spirit and power, (3) the rupture between *kulturfähig/ kulturwidrig*, (4) both hostile dimensions of the movements develop further, and finally (5) an *Aufhebung* in the idea of the tragedy of history. Meinecke compared this history of Germany with Schiller's *Demetrius*. In this context Berg cites a letter in which Meinecke claims: 'Rein und edel fängt er an, und als Verbrecher endet er!!'. Berg, *Der Holocaust*, 75; the letter: Meinecke, *Ausgewählter Briefwechsel*, 521.
89   Friedrich Meinecke, 'Gedanken über Welt- und Universalgeschichte', in: idem, *Zur Theorie und Philosophie der Geschichte* (Stuttgart 1965) 140-149, there 144. 'Nicht Fortschritte oder Rückschritte festzustellen, sondern Schicksal zu ergründen, ist die höchste Aufgabe der Historie'.
90   Meinecke, *Historism*, 279; Meinecke, *Die Entstehung des Historismus*, 337: 'Der volle Historismus schließt auch die Fähigkeit zur Resignation in sich und verlangt Respekt vor dem Schicksal'.

destiny was perfectly compatible'.[91] Now that this 'idea of providence' had become uncertain, Meinecke wanted to substitute it with something else.

Although Meinecke thought the religious dimension of Ranke's conception of history to be untenable, he himself also held on to a religious, metaphysical idea: 'what matters is becoming aware of the fact that there are two worlds that demand a final unity with each other even though inside of each things are diverse, – the world of reality and a transcendent world', thus Meinecke.[92] This of course reminds us of Meinecke's horizontal and vertical conception of history, which he used before to withstand ethical relativism.[93] In this way he is able to relativise this idea of fate (*Schicksal*) in history, since fate is only part of the real world, the horizontal history. To transcend this or to gain insight (by means of the *daimon*) into this, we need, according to Meinecke, a vertical conception of history:

> The real world is the horizontal level of the course of historical events. The supernatural world, inherent in the soul and conscience of man and urging the inner re-experiencing and purifying of what has actually been experienced, that is to say urging man towards the sensation of destiny and value, that is gained in the vertical direction by looking up to the highest guiding stars of life.[94]

Meinecke does not want to replace the horizontal view with the vertical, for this is in fact impossible, since we will always be caught in the horizontal history. Meinecke wants to make clear that even in the most horrendous periods in history we still have the possibility to focus our attention on the higher values, which are located in the 'eternal world'. In 1942 he claims in an article:

> In the real world one can be defeated and destroyed. Nevertheless, in the realm of the transcendent world, something of substance lives on with an eternal value. (...) So that even in the most gruesome abysses of world history the idea cannot go

---

91   Meinecke, 'Gedanken über Welt- und Universalgeschichte', 144. 'Sein historisches Denken endete also in einem religiösen Geheimnis, mit dem dann ein echtes historisches Schicksalsgefühl und Schicksalverstehen durchaus vereinbar war'.

92   Meinecke, 'Gedanken über Welt- und Universalgeschichte', 146-147: 'Es ist das Bewußtwerden, daß es zwei Welten gibt, die eine letzte Einheit unter sich wohl verlangen, aber in denen es verschieden hergeht, – die Welt der Wirklichkeit und eine überwirkliche Welt'.

93   Cf. Chapter 5.

94   Meinecke, 'Gedanken über Welt- und Universalgeschichte', 147: 'Die wirkliche Welt, das ist die horizontale Linie der geschichtlichen Verläufe. Die überwirkliche Welt, in Seele und Gewissen der Menschen beheimatet und zum innerlichen Nacherleben und Reinigen des wirklich Erlebten, also zur Schicksals- und Wertempfindung und Wertbeurteilung drängend, wird in vertikaler Richtung gewonnen durch den Aufblick zu den höchsten Leitsternen des Lebens'.

away that there must be a final, to us unknowable solution of this tragic conflict, a higher unity of real and supra-real worlds.[95]

Meinecke lets go of Ranke's *personal* God, but holds on to his idea of 'Every Epoch is immediate to God' because we should still consider history against the background of eternity. Meinecke substitutes Ranke's idea of providence with a 'secular Christianity'.[96] Meinecke substitutes a personal God for the Greek *theion* (the divine) in which everything is in unity. In 1943 he writes to Kaehler: 'In spite of everything, in the end I remain faithful to my θειον [*theion*], and it is for the character and testing of the character to preserve that belief (...)'.[97] This 'secularized' divine unity or harmony one can only divine or sense (*ahnen*).

Man or the individual, and especially the historian should always keep on aspiring to create higher values and focus, by means of his 'vertical' conception of history, on higher values in and of the past: 'Vertical inquiry asks the structure to be tested: in which individual form have you produced cultural values of the true, the good, the beautiful and the holy?', Meinecke writes.[98] It is the divine unity which, by means of the infinite amount of individualities, brings forth the 'cultural values of humanity' (*Kulturwerte der Menschheit*), for, as Meinecke claims: 'Everything God-related in man has an individual character'.[99] Here, Meinecke clearly echoes his panentheistic conviction, according to which all individualities carry the divine within, since they emanated from the divine. According to Meinecke, the same applies to general, collective individualities such as eras: 'In every epoch, in every individual structure of history, mental forces are stirring which rise above dull nature and mere selfishness into a higher world'.[100] Applying this to the history

---

95   Meinecke, 'Gedanken über Welt- und Universalgeschichte', 147: 'In der wirklichen Welt kann man unterliegen und vernichtet werden. In der überwirklichen Welt lebt trotzdem ein Etwas davon mit Ewigkeitsgehalt weiter. (...) So daß auch in den schaurigsten Abgründen der Weltgeschichte die Ahnung nicht untergehen kann, daß es eine letzte, für uns unerkennbare Lösung dieses tragischen Zwiespalts, eine höhere Einheit wirklicher und überwirklicher Welt geben muß'.

96   Meinecke, 'Deutung eines Rankewortes', 137.

97   Meinecke, *Neue Briefe und Dokumente*, 420: 'Ich bleibe trotz allem letzten Endes gläubig, an mein θειον [theion], und es ist Sache und Erprobung des Charakters, diesen Glauben zu bewahren (...)'.

98   Friedrich Meinecke, 'Irrwege in unserer Geschichte?'in: idem, *Zur Theorie und Philosophie der Geschichte* (Stuttgart 1965) 205-211, there 209: 'In welcher individuellen Gestalt, so fragt nun die vertikale Betrachtung das zu prüfende Gebilde, hast du Kulturwerte des Wahren, Guten, Schönen und Heiligen hervorgebracht?'.

99   Meinecke, 'Irrwege in unserer Geschichte?', 209: 'Alles Gottverwandte im Menschen trägt einen individuellen Charakter'.

100 Meinecke, 'Geschichte und Gegenwart', 99: 'In jeder Epoche, in jedem individuellen Gebilde der Geschichte regen sich seelische Kräfte, die über die dumpfe Natur und den bloßen Egoismus emporstreben in eine höhere Welt'.

of Germany and that of Prussia in particular, Meinecke concludes in one of his last articles:

> It is to be a stern question; it should not be tempered by what the horizontal view has recognized as fateful and as the wrong path in Prussian history. And yet, in the midst of its tragedy, features of true culture appear everywhere, blended into the vast majority of human imperfections, infirmities, and vices.[101]

Thus Meinecke manages, by means of his vertical conception of history, to elevate himself above the horizontal *Schicksal* that every history bears within. Evil, nature, the catastrophe can be overcome by aiming at higher things, at culture (*Kultur*). That way, Meinecke is able to resign himself to fate (*Schicksal*); a resignation which is only possible on the basis of his panentheistic conviction.[102] This remains, even after the Second World War, the foundation of Meinecke's historicism. He holds on to the idea that the highest cultural values (*Kuturwerte*) are of eternal value: 'This treasure trove of historical values (...) may at the same time be regarded as crisis-proof and as untouchable by all other fates that will strike us'.[103] Thanks to the vertical conception of history, the highest cultural values from the past can give us something to hold on to in the present; they are guides for man who searches for higher things and they comfort one in times of crisis: 'All that is good, great, and glorious, which it [the past, RK] contains, will then truly appear to the observer, pure of his own desire, immediate to God'.[104] After the Second World War Meinecke reverts to the *Kultur* of the *Goethezeit*. The last part of *Die deutsche Katastrophe*, in which the influence of Burckhardt's conception of history is clearly visible, is proof of that.

---

101   Meinecke, 'Irrwege in unserer Geschichte?', 209-210. 'Es soll eine strenge Frage sein, es soll von dem, was die horizontale Betrachtung als verhängnisvoll und als Irrweg in der preußischen Geschichte erkannt hat, nichts gemildert werden. Und doch erscheinen nun inmitten ihrer Tragödie überall auch Züge von echter Kultur, eingesprengt in die überwiegende Masse menschlicher Unvollkommenheiten, Gebrechen und Laster'. Berg claims Meinecke is discussing the Second World War in this fragment, while Meinecke is clearly referring to the era of Bismarck. Meinecke wants to make clear the twofold character of that epoch: on the one hand militarism, and on the other a revival of idealism and growth in the arts. cf: Berg, *Der Holocaust*, 89; Meinecke, 'Irrwege in unserer Geschichte?', 210.

102   Berg claims that with a vertical view one can reach a fixed point: 'Das Festhalten der fliehenden Zeit (...) verwandelt sich in zeitenthobene Beichte und in Kunst'; however, Berg does not mention Meinecke's religious – panentheistic – foundation. Berg, *Der Holocaust*, 88.

103   Meinecke, 'Irrwege in unserer Geschichte?', 210. 'Dieser Schatz geschichtlicher Werte (...) darf zugleich krisenfest gelten, als unantastbar durch alle weiteren Schicksale, die uns treffen werden'.

104   Ibidem, 211. 'Alles Gute, Große und Herrliche, das sie [the past, RK] enthält wird dann dem Betrachtenden, rein von eigenem Begehren, wahrhaft unmittelbar zu Gott erscheinen'.

6.4 Kultur as cure

'And may my jottings, limited as their value can only be today, contribute to the beginning of a new existence – to be sure a bowed down, but spiritually purer, existence. May they strengthen the determination to turn what remains of our own strength toward preserving what remains of Germany's people and culture', Meinecke writes in his preface to *Die deutsche Katastrophe*.[105] Apart from fathoming the roots of the German catastrophe, Meinecke also wanted to contribute to saving the *Kultur*. It is as if he returns to his conclusion in *Weltbürgertum und Nationalstaat*, in which he claimed that culture in the end was the motor of the German national unification. Now that the German state had fallen, Meinecke asked himself at the end of *Die deutsche Katastrophe*: 'Shall we succeed in saving the German spirit?'.[106] So, even after the Second World War Meinecke considers *Kultur* the basis for a reconstruction (of the state). It is striking that Meinecke does not mention the word 'state' once in the preface to *Die deutsche Katastrophe*. Does Meinecke abandon his 'state thinking' in this study? And, in line with his increased interest in Burckhardt, does his attention now completely shift to *Kultur*?

To Meinecke the *Goethezeit* is the cultural point of departure. In my discussion on the so-called *Reformzeit* it became clear that Meinecke considered the period around 1800 as the culmination of *Kultur*, and at the same time he considered it the period in which the possibility of reconciliation between *Kultur* and *Macht* was within reach. Now that Nazi Germany had caused a rupture on a political as well as a cultural level, he saw no alternative for a recovery of Germany than to revert to the high culture of the *Goethezeit*. How Meinecke pictured this concretely, is discussed in the last pages of Meinecke's last chapter of *Die deutsche Katastrophe*. He sketches a 'wishful picture' (*Wunschbild*), that is: the creation of so-called Goethe Communities (*Goethegemeinden*). This community of 'like-minded friends of culture'[107] should have gathered on a weekly basis to listen to the great creations of the German spirit: the poetry of the Romantics, Schiller, Goethe, Mörike and Rilke with a musical background of for example Beethoven.[108] And, if possible, these meetings should in Meinecke's view have taken place in a church, for 'the religious basis of our poetry justifies, yes demands, its being made clear by a symbolic procedure of this kind'.[109] During and especially after Meinecke's life, the general response to his suggestion was one of disapproval. After the atrocities of the Second World War,

---

105 Meinecke, *The German Catastrophe*, xiii.
106 Ibidem, 121.
107 Ibidem, 120.
108 Idem.
109 Idem.

such a proposal was considered naïve and unrealistic.[110] For this reason *Die deutsche Katastrophe* was also controversial. However, Meinecke was certainly not alone in suggesting this 'wishful picture', and on further consideration it was not a complete uncommon suggestion, as will become clear. We should therefore reconsider Meinecke's 'wishful picture' in the context of the postwar situation.

The section on the Goethe Communities has the status of an epilogue and is thus not a thorough and detailed idea. The excessive attention critics and commentators have given to Meinecke's 'wishful picture' has taken it completely out of its context, something which at the same time pushed the core of his work into the background.[111] Apart from this, Meinecke's call for Goethe Communities was not a strange idea at all at that time. If only because of the lack of libraries, bookstores and publishing houses – a lot of them were burned, damaged or destroyed – hardly any books were available. Furthermore, Germany's intellectual upper class was considerably diminished. In a cultural-intellectual respect the end of the war was to Meinecke in fact *Stunde Null*. That was the main underlying thought of Meinecke's Goethe Communities. In a letter to Siegfried Kaehler at the end of 1946 he expresses it as follows:

> People ridiculed my Goethe Communities, but I was dead serious about it. Reinhold Schneider is said to have recently claimed that it now depends on about 2000

---

110 Eugen Kogon, 'Beginn der Geschichtsrevision', *Frankfurter Hefte. Zeitschrift für Kultur und Politik* 8 (1946) 776-779, there 779; Shanahan, 'Germany: Realities of Power and Reconstruction', 506-507. Also: Clark, *Beyond Catastrophe*, 38, 114; Boterman, *Duitse dichters en denkers*, 275; Paddock, 'Rethinking Friedrich Meinecke's historicism', 106; Nicholas Berg is more nuanced: Berg, *Der Holocaust*, 71, 77. And Ritter in: Meinecke, *Akademischer Lehrer*, 26-27; also: Schulze, *Deutsche Geschichtswissenschaft nach 1945*, 55; Wippermann claimed it was indeed a naïve suggestion: Wippermann, '"Deutsche Katastrophe"', 90-191: 'Sein allgemeines Plädoyer für eine Rückkehr zum Humanitätsideal der Goethezeit war nach dem Holocaust, und wegen des Holocaust, bestenfalls naiv – auch wenn Thomas Mann ihm 1949 hierin „lebhaft" zustimme'. Peter Gay states it was a strange suggestion, because Meinecke focussed on *unpolitical* writers: 'Goethe's politics was apathy, Schiller's tyrannicide; neither was a mode calculated to prepare men for parliamentary compromises (...)'. Gay, *Weimar Culture*, 71-72. However, Meinecke was concerned with *Kultur* of a period that was still unrelated to national (power) politics. cf. the previous chapter.

111 Meinecke's Wunschbild covers only two pages. Berg, *Der Holocaust*, 71; Meinecke, *Akademischer Lehrer*, 27; Schulze, *Deutsche Geschichtswissenschaft nach 1945*, 55; Meineke, *Friedrich Meinecke*, 32. On the other hand however, we should consider the last chapter of *Die deutsche Katastrophe* as the apex of his argument, as Sösemann also claims: Sösemann, ed., *Friedrich Meinecke, Die deutsche Katastrophe. Betrachtungen und Erinnerungen. Edition und internationale Rezeption*, 477.

to 3000 Germans living today, whether the German culture, the German spirit can be preserved. Precisely this idea underlies my final chapter.[112]

Meinecke's Goethe Communities are thus not only intellectual-historical, but also of a very practical nature. It was a concrete suggestion to save culture (*Kultur*).

Considering Meinecke's oeuvre as a whole it is of course not strange at all that he focusses on the ideal of *Bildung* of the *Goethezeit*. After all, it was his main subject of research from his first studies onwards.[113] Further, we should keep in mind that the Second World War was the first war Meinecke actually experienced firsthand. He actually became a refugee, and it is clear from his letters that he found comfort and hope in Goethe's poetry. Goethe had already served this role for Meinecke during the First World War. Eberhard Kessel expressed the importance of Goethe for Meinecke as follows in his introduction to Meinecke's *Autobiographische Schriften*: 'Goethe was to him [Meinecke, RK] a living possession, from which he drew strength and gained inner elevation, an immense wealth for his own existence'.[114] This is also articulated in his brief consideration: 'Lebenströster. Betrachtungen über zwei Goethesche Gedichte', which he wrote during the composition of *Die deutsche Katastrophe*.[115]

Next to Meinecke's own arguments for realizing the Goethe Communities, his idea of 'communities' was a very practical one, considering the situation of Germany; besides the four Allied authorities, there was no coordinating power anymore.[116] The community or *Kommune* had become an important place for cultural

112 Meinecke, *Ausgewählter Briefwechsel*, 508. 'Man hat gelächelt über meine Goethegemeinden, aber mir war es bitterer Ernst damit. Reinhold Schneider soll neulich gesagt haben, es käme jetzt auf etwa 2000 bis 3000 heute lebende Deutsche an, ob die deutsche Kultur, der deutsche Geist sich würde erhalten lassen. Genau dieser Gedanke liegt auch meinem Schlußkapitel zu Grunde'. Perhaps Meinecke thought also about the many scholars in exile. Horst Möller, *Exodus der Kultur. Schriftsteller, Wissenschaftler und Künstler in der Emigration nach 1933* (Munich 1984); Hermand, *Kultur im Wiederaufbau*, 94-101; Glaser, *Deutsche Kultur*, 97-101. And cf.: Meinecke, *Akademischer Lehrer und emigrierte Schüler*.

113 Clark claims Meinecke only started to concentrate on the *Goethezeit* during the time around *Die deutsche Katastrophe*: Clarck, *Beyond Catastrophe*, 39; Bruch also states the *Goethezeit* became central to Meinecke when the synthesis failed. Bruch, 'Ein Gelehrtenleben zwischen Bismarck und Adenauer', 12.

114 Kessel, 'Einleitung', XVIII. 'Goethe war ihm [Meinecke, RK] ein lebendiger Besitz, aus dem er Kraft schöpfte und innere Erhebung gewann, ein unermeßlicher Reichtum für das eigene Dasein'.

115 Friedrich Meinecke, 'Lebenströster. Betrachtungen über zwei Goethesche Gedichte' in: idem, *Autobiographische Schriften* (Stuttgart 1969) 492-508; on the importance of Goethe in this period: Meinecke, *Ausgewählter Briefwechsel*, 207, 237-239, 592, 600, 624.

116 Maubach states that this interest in culture was no longer viable for Meinecke, and moreover not practical in the post-war years: Maubach, '"Wie es dazu kommen konnte"', 163, 174.

development (*Kulturentfaltung*) or cultural transfer (*Kulturvermittlung*).[117] In addition, Meinecke was also not alone in his attention to Goethe. After the collapse of the Third Reich many works were published in which Goethe's humanism as well as its viability was discussed and propagated.[118] For example in Albert Erich Brinkmann's work *Geist im Wandel* Goethe is considered the ideal for a humanistic 're-individualisation'.[119]

Next to Goethean humanism a frequent reference to classic and renaissance humanism was also apparent after the war.[120] With regard to classic humanism, the so-called 'Dritter Humanismus' (third humanism) of in particular Eduard Spranger and Werner Jaeger is of great importance.[121] They propagated – already in the 1920's and 1930's – the education, the cultural *Bildung* of the individual.[122] With

---

117   Frank-Lothar Kroll, 'Kultur, Bildung und Wissenschaft im geteilten Deutschland 1949-1989', *Archiv für Kulturgeschichte* 85 (2003) 119-142, there 122-123. It is about urban initiatives in the field of music, theatre, libraries and the like. Meinecke's 'Gemeinden'-idea can, for that matter, also be viewed as a variation on the general practice of conferences, study weeks and the like that were held across Europe during the 1930's until the 1950's. Or it relates also to the general practice of the formation of several *Gesellschaften* on many subjects and in many intellectual circles around 1900 (or even in the nineteenth century).

118   Cf. the chapter 'Die Goethe-Renaissance nach 1945' in Karl Robert Mandelkow, *Goethe in Deutschland*, 135-152; also: Berg, *Der Holocaust*, 81: 'Die vielfältigen Vereinnahmungen Goethes belegen weniger die umfassende Bildung der damaligen Epigonen, sondern umgekehrt deren völlige Verunsicherung, die einzig mit der Anrufung des Klassikers nichts falsch zu machen glaubte'. Hermand, *Kultur im Wiederaufbau*, 70-71. Moreover, 1949 was the 200 anniversary of Goethe's birth, which in turn led to a series of new publications, especially: *Goethe. Unesco's Homage on the occasion of the two hundredth anniversary of his birth* (Paris 1949) in which several distinguished European writers contributed a piece on Goethe. Meinecke himself is missing, but Croce is present. Goethe is portrayed in the preface as 'a great European'; further the viability of his legacy is unquestioned in this volume, and: 'It is proud to salute Goethe's memory and to recognize in it an imperishable testimony (...) of harmonizing and reconciling'. Also: Thomas Mann, 'Ansprache im Goethejahr. Gehalten am 25. Juli 1949 in der Paulskirche zu Frankfurt am Main' in: N.N. *Zum Thema: Goethe* (Frankfurt am Main 1982) 63-82; Hermann Glaser, *Deutsche Kultur. Ein historischer Überblick von 1945 bis zur Gegenwart* (Munich 1997) 107-109; Clark, *Beyond Catastrophe*, 114; Hans Mayer ed., *Goethe im zwanzigsten Jahrhundert. Spiegelungen und Deutungen* (Frankfurt am Main 1987).

119   Hermand, *Kultur im Wiederaufbau*, 71.

120   Ibidem 70.

121   It is a 'third humanism' in succession to the (first) Renaissance and the second 'Neuhumanismus' of 1750 onwards. Eduard Spranger, *Der gegenwärtige Stand der Geisteswissenschaften und die Schule* (Leipzig and Berlin 1922) 5, 8.

122   A key notion of Spranger and Jaeger is the Greek *Paideia*. Related to the German *Bildung*, this notion indicates a personal education: 'Unser deutsches Wort Bildung bezeichnet das Wesen der Erziehung am anschaulichsten im griechischen, platonischen Sinne'. Werner Jaeger, *Paideia. Die Formung des griechischen Menschen* (Berlin and Leipzig 1936) Volume I, 12-13; Barbara Stiewe, *Der "Dritte Humanismus". Aspekte deutscher Griechenrezeption vom George-Kreis*

regard to cultural *Bildung* the so-called 'Gruppe 47', which was established one year after Meinecke's *Die deutsche Katastrophe*, is also important in this context. According to this group art and literature offered comfort and gave the 'individual personality' something to hold on to. At first, they focussed on the individual and its place within the masses. They held meetings in remote, rural areas, and eventually became politically active.[123] One last symptom of this boom of cultural initiatives and promotion of cultural *Bildung*, is the immense flourishing of new journals, newspapers and magazines with striking names like: *Aufbau, Besinnung, Einheit, Ende und Anfang, Frischer Wind, Geist und Tat, Neue Ordnung* and *Zeitwende*.[124] In short, Meinecke's proposal to read the 'imperishable' *Kultur* of the great poets and thinkers in communion in order to create a new foundation for a renewed, humanistic *Kulturnation* against the background of the political and social situation, and the many cultural initiatives, was no singular suggestion.

Meinecke's emphasis on *Kultur* and his suggestion to restore contact with the ideals of the earlier *Kulturnation* dovetails with his attention to Burckhardt.[125] His view that the state should ensure peace and freedom, in order for the *Kultur* to develop and thrive, resounds more and more in Meinecke's works. In *Die deutsche Katastrophe* he states for example: 'Spiritual life and the striving for spiritual values are their own justification and work most deeply where their movements can be most free from political tendencies'.[126] But he adds: 'Indeed they work most deeply and beneficially by themselves when they go their own ways spontaneously and unregulated'.[127] So Meinecke does not abandon his political thought entirely. His critique of Burckhardt's view on the German political unification and the role attributed to power is also proof of that. Burckhardt was indeed convinced that power and the state flourished in Germany at the expense of culture. In *Die deutsche Katastrophe* Meinecke criticizes Burckhardt on this account:

> Her [Germany's, RK] former striving for unity and strength was not merely, as Burckhardt saw it in his *Reflections on History*, a blind striving of the masses to whom culture meant nothing. Rather was it borne along, as Burckhardt was not quite fully able to understand, by that great idea of an inner union of spirit and power, by humanity and nationality. Great cultural values emerged for us from it.

---

 bis zum Nationalsozialismus (Berlin and New York 2011); Hermand, *Kultur im Wiederaufbau*, 67-70; Kroll, 'Kultur, Bildung und Wissenschaft', 126.

123   Boterman, *Duitse dichters en denkers*, 271-290.
124   Glaser, *Deutsche Kultur*, 157.
125   Also: Clark, *Beyond Catastrophe*, 39.
126   Meinecke, *The German Catastrophe*, 117.
127   Idem.

But this union, as we must make clear to ourselves, was disrupted through our own fault.[128]

So Meinecke still had not abandoned all hope for a synthesis between *Macht* and *Geist*. That is also clear when Meinecke writes a few lines later: 'Our conception of power must first be purified from the filth which came into it during the Third Reich before it can again be capable of forming a union with spirit and culture'.[129] On the one hand, Meinecke distances himself from *Macht* by focussing completely on *Kultur*. On the other hand, he holds on to the Rankean tradition of political history, in which the state is central.[130] But it is *Kultur* in particular which Meinecke considers to be important for a recovery of politics. Thus, he does not abandon all hope for a synthesis, but there is room for some nuance within (the ideal of) this synthesis. In that sense, *Die deutsche Katastrophe* represents a shift of emphasis in Meinecke's thought regarding the synthesis; a shift that was caused by the Second World War.[131]

Despite Meinecke's critique of Burckhardt's view of power, Meinecke does endorse his attitude with regard to *Kultur*. When Meinecke proposed the Goethe Communities, he was not merely referring to the ideas of Goethe, but to the whole generation of German poets, philosophers, military leaders, musicians and statesmen between 1790-1815; the liberal humanism of, among others, Humboldt, Boyen and Beethoven. This generation served to Meinecke as a spiritual guide for his defeated Germany; the Spirit of past excellence as a spiritual guide in times of decline.[132] This fits with Burckhardt's description in his *Weltgeschichtliche Betrachtungen* on the means to cushion the decline of civilization, which according to him will take place regardless. Burckhardt claims: 'In the face of such historical powers [the decline, RK], the contemporary individual tends to feel utterly helpless (...)'.[133] The remedy

---

128   Meinecke, *The German Catastrophe*, 109.

129   Idem.

130   In this context there is another important point of criticism with regard to Meinecke's recommendation of 'Goethegemeinden'; why did Meinecke not directly after the war draw the attention to the fact that Goethe, a diplomat of a small German town, had no admiration for power whatsoever? Many thanks to Wessel Krul, who drew my attention to this. Cf. Gay, *Weimar Culture*, 71-72.

131   Gossman claims that Meinecke did not change his thoughts after the Second World War: '(...) Meinecke remained essentially faithful to the position that he had always occupied'. Gossman, *Basel in the Age of Burckhardt*, 450.

132   Shanahan, 'Germany: Realities of Power and Reconstruction', 507.

133   Jacob Burckhardt, *Weltgeschichtliche Betrachtungen*, ed. Rudolf Marx (Stuttgart 1978) 8. 'Gegenüber von solchen geschichtlichen Mächten pflegt sich das zeitgenössische Individuum in völliger Ohnmacht zu fühlen (...)'.

for this spiritual impotence is, according to Burckhardt, contemplation, or better: *Anschauung* (intuition) of history.[134] In the words of Burckhardt:

> History is and remains poetry to me (…), which can be mastered by intuition. (…) Of course I also include the spiritual, e.g. the historical, which results from the impression of the sources. – What I build up historically is not the result of criticism and speculation, but of the imagination, which wants to fill in the gaps of intuition.[135]

Burckhardt considered history, like poetry and art, to consist of a metamorphosis (*Verpuppungen*) of the eternal Spirit.[136] The individual is able, in Burckhardt's view, to grasp the many metamorphoses of the Spirit through intuition (*Anschauung*). By means of a 'dialogue' between the historian and the sources of the past – through subjective experience – the highest forms of these *Verpuppungen* of the Spirit of the eternal in history can be intuited or divined, which enables an (aesthetic) insight into the past.[137] Burckhardt distinguishes himself from Ranke in this, for his idea of *Anschauung* is not a philosophy of history, and he certainly did not pretend to show what essentially happened ('wie es eigentlich gewesen') in the past.[138] Burckhardt's aforementioned quote also makes clear that he positioned himself opposite of Hegel, for he does not engage in speculations like Hegel did in his philosophy of the self-realization of the spirit. Burckhardt rejects such a view: 'Accordingly, I cannot believe in a position a priori; that is a matter of the world spirit, not the man of history'.[139]

Through contemplation we can rise above the daily noise, which is, according to Burckhardt (and Meinecke), the historian's fundamental need: 'Our contemplation is not only a right and a duty, but at the same time a great need; it is our freedom in the midst of the awareness of the enormous universal restriction and the stream of necessities', thus Burckhardt.[140] That is to say, Burckhardt stresses the significance of history and reminds us that it is our duty to preserve and honour it: 'And now

---

134   John R. Hinde, *Jacob Burckhardt and the Crisis of Modernity* (Montreal 2000) 201.

135   Burckhardt, *Briefe*, 57, 60. From the letters to Willibald Beyschlag (14 June 1842) and Karl Fresenius (19 June 1842). 'Die Geschichte ist und bleibt mir Poesie (…), die durch Anschauung bemeistert werden kann. (…) Ich rechne zur Anschauung natürlich auch die geistige, z.B. die historische, welche aus dem Eindruck der Quellen hervorgeht. – Was ich historisch aufbaue, ist nicht Resultat der Kritik und Spekulation, sondern der Phantasie, welche die Lücken der Anschauung ausfüllen will'.

136   Gossman, *Basel in the Age of Burckhardt*, 248.

137   Hinde, *Jacob Burckhardt and the Crisis*, 203.

138   Ibidem, 205.

139   Burckhardt, *Briefe*, 57. 'An einen Standpunkt a priori kann ich demnach gar nicht glauben; das ist die Sache des Weltgeistes, nicht des Geschichtsmenschen'.

140   Burckhardt, *Weltgeschichtliche Betrachtungen*, 11. 'Unsere Kontemplation ist aber nicht nur ein Recht und eine Pflicht, sondern zugleich ein hohes Bedürfnis; sie ist unsere Freiheit

we also commemorate the greatness of our commitment to the past as a spiritual continuum, which is one of our highest spiritual possessions'.[141] In his frequently-cited, well-known statement Burckhardt expressed himself even more aptly:

> The mind must transform the memory into its possession. What used to be jubilation and distress must now become knowledge, as in the life of the individual.[142]

The similarities with Meinecke's vertical conception of history are once more striking, for in the vertical conception of history the historian can also, as a counterweight to the (destructive) 'horizontal' historical forces, transcend time in his mind.[143] According to Burckhardt, and Meinecke following in his footsteps, it is beneficial to call to mind our spiritual heritage in dark times, because it offers insight, faith, and satisfaction.[144] Next to this, Burckhardt's views on the subjective experience – the *Anschauung* –, and his ideas on our (own) past, which comes to an understanding (of itself), also fit in with Meinecke's views during and after the Second World War. It is not a coincidence that Meinecke chose an autobiographical form for *Die deutsche Katastrophe*, for in this manner he was able to fuse his own past with that of Germany as a whole.

6.5   Apotheosis

During the Second World War Meinecke's first volume of his memoirs was published, entitled *Erlebtes 1862-1901*. In this autobiography he portrays his youth in the small village of Salzwedel, his student years in Berlin – where he became a *Privatdozent* (unsalaried university lecturer) – and where he, under supervision of

---

mitten im Bewußtsein der enormen allgemeinen Gebundenheit und des Stromes der Notwendigkeiten'.

141   Ibidem, 9. 'Und nun gedenken wir auch der Größe unserer Verpflichtung gegen die Vergangenheit als ein geistiges Kontinuum, welches mit zu unserem höchsten geistigen Besitz gehört'.

142   Ibidem, 10. 'Der Geist muß die Erinnerung in seinen Besitz verwandeln. Was einst Jubel und Jammer war, muß nun Erkenntnis werden, wie eigentlich auch im Leben des einzelnen'.

143   Richard F. Sigurdson, 'Jacob Burckhardt: The Cultural Historian as Political Thinker', *The Review of Politics* 52 (1990) 417-440, there 436.

144   Cf. Goethe's idea of a select company of noble intellectual spirits, which are joined together over time and space. This is expressed in several ways in one of Goethe's last poems entitled Legacy (*Vermächtnis*), with phrases like: 'Geselle dich zur kleinsten Schar', which refers to such a noble company of like-minded spirits. Even more clear are phrases like: 'Vernunft sei überall zugegen, / Wo Leben sich des Lebens freut. / Dann ist Vergangenheit beständig, / Das Künftige voraus lebendig, / Der Augenblick ist Ewigkeit'. Meinecke refers to this poem on many occasions in his articles on philosophy of history, his letters and in *Die Entstehung des Historismus*.

his mentor Heinrich von Sybel, began his archival research and started writing his *Habilitationsschrift* (the biography of Boyen). The story ends when Meinecke, 39 years of age, obtained a chair in Modern History at the University of Strasbourg.

Meinecke's autobiography reads like a *Bildunsgroman* – a story of learning and personal growth, and particularly of the discovery of a vocation.[145] A famous example of an autobiographical *Bildungsroman* is of course Goethe's *Dichtung und Wahrheit*, which Meinecke considered the high point of Goethe's historical thought and historical writing in general. According to Meinecke, Goethe saw himself in this autobiography 'as the carrier of a development which needed a whole epoch as its milieu in order to be thoroughly understood'.[146] Comparable to *Erlebtes*, Meinecke portrays, in his second volume of his memoirs *Straßburg – Freiburg – Berlin. Erinnerungen* his intellectual development as a prospective historian in the same manner.[147]

In 1945 Meinecke began writing *Die deutsche Katastrophe*. Similar to his auto-biography, this study also clearly shows the entanglement of the individual and the epoch. Although the emphasis in *Die deutsche Katastrophe* is on the analysis of the origins of National Socialism, Meinecke's personal authority and the fact that he witnessed part of the history he describes, attributes more meaning to his argument.[148] In his preface Meinecke states he is well aware of the sketchy character of this study and he even makes excuses for the fact that he had few

---

145   Jeremy D. Popkin, *History, Historians and Autobiography* (Chicago and London 2005) 79, 121-123; Boterman, *Duitse dichters en denkers*, 211.

146   Meinecke, *Historism*, 414. Croce takes it a step further: '(…) daß nicht nur die Selbstbiographie Geschichte ist, sondern daß auch jede echte Geschichtserkenntnis Selbstbiographie ist, Geschichte der Welt, insofern sie in einem von uns lebendig, mit unserem Leben, unseren Notwendigkeiten, unseren Idealen verknüpft ist'. Benedetto Croce, *Goethe. Studien zu seinem Werk* (Düsseldorf 1949) 246. This fits in with Dilthey, who will be discussed below. Cassirer on the other hand is quite critical of Goethe's *Dichtung und Wahrheit* as a historical or autobiographical account: 'Goethes "Dichtung und Wahrheit" enthält keine Peripetien und keine Katastrophen, keine dramatischen Höhepunkte. Sie will das Leben in seiner Kontinuität und in seiner inneren Ganzheit vor uns hinstellen – als ein stetiges Werden, das doch von einem Einheitspunkt ausgeht und immer wieder zu ihm zurückkehrt'. Cassirer, *Goethe und die geschichtliche Welt*, 16.

147   Popkin, *History, Historians and Autobiography*, 122.

148   Berg, *Der Holocaust*, 82-85. In the introduction I already explained Berg's argument that Meinecke's work gains more authority and authenticity because of its autobiographical form. Berg, *Der Holocaust*, 93: 'Das Buch sei wissenschaftlich so gewichtig, weil es mit Persönlichkeit versehen ist, aus dem „persönlichen Bericht" sei ein „Aufriß der Zeitgeschichte" geworden – die Verschmelzung von beidem machte es zum „höchsten, was ein zeitgeschichtliches Werk erreichen" könne'. Berg claims, as mentioned before, that it reflects a shift in Meinecke's conception of history, but Berg fails to compare this shift with the rest of Meinecke's oeuvre. It is not so much a *change* in his views, but a *deepening* of Meinecke's conception of history.

sources at his disposal and that he was forced to rely on his memory.[149] But at the same time he is convinced that his work has something that all later works will lack, that is: 'the essence of the atmosphere of the period in which our fate was fulfilled, and of which one must be aware in order fully to understand this fate'.[150] The similarities between Meinecke's aforementioned characterization of Goethe's *Dichtung und Wahrheit* are striking; Meinecke claims that his individual experience of a part of the period, which he discusses in *Die deutsche Katastrophe*, contributes to a better understanding of this period, which is also the context of his own development. In other words, in *Die deutsche Katastrophe* Meinecke reconciles history with autobiography.

Given Meinecke's claim that *Die deutsche Katastrophe* carries 'the essence of the atmosphere of the period', we should situate this work (in part) in the history which it describes. But it is of course a description of the history based on his memory of that history. It is in these memories in which Meinecke saw his past coalesce with Germany's political and cultural past.[151] The German catastrophe – and also *Die deutsche Katastrophe* – yields insight into German history and at the same time into Meinecke's own individual development, since both are tied together. By tying the general history to his own experience, he elevates both to a higher level. It is a reconciliation of the two, 'because both of them (the reality as it was and its observer) meet on a common higher level', according to Meinecke in his admiration of Goethe.[152] Meinecke considers this form of historical writing, which was in his view perfected in *Dichtung und Wahrheit*, the highest form of historicism.[153] With *Die deutsche Katastrophe*, Meinecke follows in Goethe's footsteps.

In his postwar works Meinecke drives his philosophy of history to extreme consequences: he reconciles Goethe, Ranke, Burckhardt and Dilthey. In this context we should begin with Dilthey's views on autobiography, for they dovetail with Meinecke's idea of his own autobiographical works, and in particular with *Die deutsche Katastrophe*.[154] In a review of Dilthey's biography of Schleiermacher Meinecke states: 'He [Dilthey, RK] is not a writer and philosopher to everyone's taste

---

149 Also: Berg, *Holocaust*, 83-84.

150 Meinecke, *The German Catastrophe*, xii.

151 Here subject and object merge, as Meinecke had written before: 'Alles Wissen und alles kritische Forschen darf nicht Selbstzweck und letzter Abschluß der Arbeit bleiben, sondern empfängt nur höheren und dauernden Wert, wenn (...) Subjekt und Objekt in *ein* geistiges Gebilde (...) verschmelzen – wo das Subjekt dann wohl die Kraftquelle dieses Gebildes bleibt, aber auf- und eingeht in das Objektive, das es darstellt, – und sich dadurch selbst überwindet'. Meinecke, 'Aphorismen', 239. Also see the previous chapter.

152 Meinecke, *Historism*, 487.

153 Cf. the previous chapter.

154 Dilthey did not fully specify his ideas on autobiography.

(…) To love him, one must already bring inclinations which are related to his (…)'.[155] That these 'inclinations' were not altogether foreign to Meinecke is clear when one considers the similarities between Meinecke's autobiographical work and Dilthey's ideas on the autobiography as such:

> In autobiography we encounter the highest and most instructive form of the understanding of life. Here a life-course stands as an external phenomenon from which understanding seeks to discover what produced it within a particular environment. The person who understands it is the same as the one who creates it. (…) The power and scope of our own lives and the energy with which we reflect on them provide the basis of historical vision. Self-reflection alone enables us to give a second life to the bloodless shadows of the past. In combination with a boundless need to surrender to, and lose oneself in, the existence of others, it makes the great historian.[156]

The 'existence of others' is in *Die deutsche Katastrophe* Meinecke's own and Germany's history. The self-reflection and insight which follows from this gave Meinecke a new idea. In a letter to Eduard Spranger he explained this idea and suggested replacing the topos *Historia vitae magistra* with *Historiae vita magistra*.[157] Instead of considering history as the mentor of life, Meinecke suggests viewing life as the mentor of history. That is, 'life' is central, thus history should be considered from the perspective of (the individual) life.[158] History, in this sense, is a mirror in which one sees oneself; self-knowledge is the result. Meinecke's insight fits in with Dilthey, but it is Burckhardt's outlook in particular that Meinecke had in mind. A letter to Kaehler proves this point; Meinecke writes: 'Instead of asking primarily for the

---

155 Friedrich Meinecke, 'Wilhelm Dilthey' in: idem, *Zur Theorie und Philosophie der Geschichte* (Stuttgart 1965) 358-363, there 361. 'Er [Dilthey, RK] ist kein Schriftsteller und Philosoph für Jedermann (…) Um ihn zu lieben, muß man schon Neigungen mitbringen, die den seinigen verwandt sind (…)'; also: Meinecke, 'Erlebtes 1862-1901', 72, 80. Hofer, *Geschichtsschreibung*, 12, 242-250, 411.

156 Wilhelm Dilthey, *The Formation of the Historical World in the Human Sciences.* eds. Rudolf A. Makreel and Frithjof Rodi (Princeton and Oxford 2002) 221. Original: Wilhelm Dilthey, 'Der Aufbau der geschichtlichen Welt in den Geisteswissenschaften' in: idem, *Gesammelte Schriften* VII (Göttingen 1958) 199-201.

157 Meinecke, *Ausgewählter Briefwechsel*, 609; Also: Berg, *Der Holocaust*, 85.

158 This is obviously not Nietzsche's understanding of 'life', since his view contains a denial of the past. He claimed that too much history would numb or paralyse us in the present, which is why we should break with the past. This denial of the past is, in Nietzsche's view, the (only) way to still be able to act in the present, to live. This is the theme of his well-known *Vom Nutzen und Nachteil der Historie für das Leben* (1874). Meinecke anticipates here a hermeneutics à la Gadamer. Cf. my final conclusions.

deeds and works of men, we ask: how did deeds and works affect the soul of man himself!'.[159] To Burckhardt, this is the heart of historical writing.

In a lecture of 1948 entitled 'Ranke und Burckhardt', Meinecke set Burckhardt's (above mentioned) conception of history against Ranke's. Meinecke argues in this lecture that Ranke and Burckhardt depart from different questions. For example, Ranke wondered what the significance of the historically acting person was for the suprapersonal whole of the objective mind that man initially tried to understand. Whereas Burckhardt, as mentioned, wondered what the significance of the whole and of world historical events was for the 'culture-creating' individual.[160] In other words, this is about a choice between the two topoi just mentioned: a choice between *Historia vitae magistra* and *Historiae vita* magistra. According to Meinecke, we eventually have to consider both at the same time. If we fail to do so, we will have a one-sided view of the past; if we only focus on Rankean 'objective idealism', in which the emphasis is on the suprapersonal powers, man is reduced to a (subordinate) function in the service of the 'major forces'. But if we only emphasize the subjective, Meinecke argues, the general historical process loses its necessity and stability and becomes subject to the arbitrariness of the subject.[161] Meinecke therefore suggests a reconciliation of Ranke and Burckhardt.

In his lecture on Ranke and Burckhardt, Meinecke analyses the views of both these historians, and also tries to position himself between them – between Ranke and Burckhardt, between Berlin and Basel.[162] With regard to Burckhardt, Meinecke adjusts his views from before the Second World War; he now claims Burckhardt's conception of history is vital for (German) historical writing.[163] He considers Burckhardt's position no longer as isolated, but rather as complementary to the German outlook.[164] Although Meinecke strongly leans toward Burckhardt's conception of history in this lecture of 1948, he eventually concludes that a synthesis between Ranke and Burckhardt would be the best solution for the historiographical prob-

---

159   Meinecke, *Ausgewählter Briefwechsel*, 487; Berg, *Der Holocaust*, 84. 'Statt wie bisher primär nach Taten und Werken der Menschen zu fragen, fragen wir: Wie wirkten Taten und Werke auf die Seele des Menschen selbst!'.

160   Friedrich Meinecke, 'Ranke und Burckhardt' in: idem, *Zur Geschichte der Geschichtsschreibung* (Munich 1968) 93-121, there 107-108.

161   Meinecke, 'Ranke und Burckhardt', 107-108.

162   Also: Gossman, *Basel in the Age of Burckhardt*, 444 en 449.

163   The lecture was held in 1947 and published in 1948. Meinecke, 'Ranke und Burckhardt', 110; Herkless, 'Meinecke and the Ranke-Burckhardt Problem', 297.

164   Meinecke, 'Ranke und Burckhardt', 108-110.

lems which Germany struggled with.[165] This is also expressed in several letters Meinecke wrote during this period. For example, in 1947 he writes to the English historian Gooch:

> (...) Currently, I am concerned with the major problem of comparing Ranke's and Burckhardt's conceptions of history and weighing their significance for today's world-historical situation. Each of the two points of view has had its considerable justification, but today Burckhardt has almost more to tell us than Ranke. To unite Ranke's and Burckhardt's historical thinking somehow in a higher synthesis – that is what is at stake.[166]

Whether this synthesis could even be realized, Meinecke was not sure, but that it indeed would be the best solution, was no question for him. In the same year, Meinecke wrote a letter to Gustav Meyer claiming the following on this issue: 'Whether one can, beyond both, arrive at a more comprehensive and just overall conception? For neither the one nor the other tells us the very final word! Neither do I!'.[167] Here, Meinecke is too modest regarding his own role and position in the German historical sciences. Of course, he does not want to position himself above Burckhardt and certainly not above Ranke, whom he greatly admired, but when reading Meinecke's account on Ranke and Burckhardt, one cannot escape the impression that he describes both historical ideas in a way that is essentially close to his own conception of history.[168]

With an appeal to Dilthey, Meinecke identifies the optimistic and pessimistic world view of Ranke and Burckhardt as an 'objective' and a 'subjective' world view (*Weltanschauung*). According to Meinecke, Ranke's objective conception of history

---

165   Ibidem, 109-110; Meinecke, *Neue Briefe*, 458: 'Ich werde darin [the lecture, RK] nicht etwa untreu gegen Ranke, aber betonte die wachsende Bedeutung Burckhardts für uns und wünsche für die Zukunft eine Vereinigung der Geister Rankes und Burckhardts in unserer Geschichtsschreibung'.

166   Meinecke, *Ausgewählter Briefwechsel*, 277. '(...) augenblicklich beschäftigt mich das große Problem, die Geschichtsauffassungen Rankes und Burckhardts mit einander zu vergleichen und ihre Bedeutung für die heutige welthistorische Situation abzuwägen. Jeder der beiden Standpunkte hat ja seine tiefe Berechtigung gehabt, aber heute hat Burckhardt uns beinahe mehr zu sagen als Ranke. Rankes und Burckhardts geschichtliches Denken irgendwie in einer höheren Synthese zu vereinigen, – darauf käme es eigentlich an'.

167   Ibidem, 276. 'Ob man, über beide hinaus, zu einer noch umfassenderen und gerechteren Gesamtauffassung wird gelangen können? Denn weder der Eine, noch der Andere sagt uns das *aller*letzte Wort! Ich auch nicht!'.

168   Herkless claims that Meinecke was not aware of how close his thought was to that of Burckhardt: Herkless, 'Meinecke and the Ranke-Burckhardt Problem', 320. Herkless, however, does not mention the possibility that Meinecke himself was in fact or could be the synthesis of both. He merely points at some similarities between Meinecke and Burckhardt and between Ranke and Burckhardt.

assumes that the world is 'an explaining of God', that is, Ranke thought God's revelation would be (in part) revealed throughout history by means of historical writing
Ranke himself practiced. Ranke claimed it merely revealed outward appearances
for God in itself is unfathomable. So, throughout history – 'horizontally' – God can
only be divined or intuited (*erahnen*). Burckhardt's subjective conception of history
starts, in Meinecke's view, from 'the subjective consciousness of the moral freedom
of the individual (...)'.[169] The creative spontaneity of the individual is key in this case,
not the affirmation of a divine providence as in the case of Ranke. As mentioned
before, at the heart of Burckhardt's conception of history there is a 'cultural creative'
(not directed by a God) spontaneous human being.[170] According to Burckhardt,
Kultur is created from within the spirit for its own sake: 'What we call culture is
the whole sum of those developments of the mind which happen spontaneously
and do not claim universal or coercive validity'.[171] As mentioned, Burckhardt considered these spontaneous cultural creations as *Verpuppungen* (metamorphoses) of
the eternal Spirit, which can be experienced by the individual through *Anschauung*
– Meinecke would call it a 'vertical' view.[172]

Meinecke claims Ranke's objective, horizontal conception of history, and Burckhardt's subjective, vertical conception of history are two sides of the same historicist coin. When considering both views on the past as such, it is Ranke who
represents the 'principle of development' and Burckhardt the 'principle of individuality'. Ranke, of course, considers the individual also as important, but, according to Meinecke, Ranke ultimately fuses the single individualities together in
a *Gesamtentwicklung*, an overall development.[173] In Burckhardt the development of
individualities is ultimately withdrawn in favour of a cross section (*Querschnitt*) of
a certain condition or characterization of a people or period.[174] With regard to the
task of the historian, Meinecke is on the one hand close to Burckhardt's idea of the

---

169   Meinecke, 'Ranke und Burckhardt', 101. 'dem subjectiven Bewußtsein der sittlichen Freiheit
        des einzelnen (...)'.

170   Ibidem, 106; Gossman, *Basel in the Age of Burckhardt*, 259-260.

171   Cited in: Meinecke, 'Ranke und Burckhardt', 104. 'Kultur nennen wir die ganze Summe
        derjenigen Entwicklungen des Geistes, welche spontan geschehen und keine universale oder
        Zwangsgeltung in Anspruch nehmen'.

172   I have made clear already that Meinecke's definition of *Kultur* was very close to Burckhardt's:
        'Kultur tritt erst da ein, wo der Mensch mit seiner ganzen Innerlichkeit, nicht nur mit dem
        Willen und Verstande den Kampf mit der Natur aufnimmt, wo er wertend im höheren Sinne
        handelt, das heißt, wo er etwas Gutes oder Schönes um seiner selbst willen schafft oder
        sucht oder das Wahre um seiner selbst willen sucht'. Meinecke, 'Kausalitäten und Werte',
        76. *Kultur* is, according to Burckhardt and Meinecke, created from the spononaeous inner self
        of the individual. This reminds us of the unfathomable individual: the spontaneous x, which
        Meinecke borrowed from Droysen. Cf. Chapter 3.

173   Meinecke, 'Ranke und Burckhardt', 106.

174   Ibidem, 106-107.

creative acting historian, and on the other, he holds on to the Rankean ideal of representing the past as faithfully as possible. Meinecke attempts to reconcile Ranke's objective *Weltanschauung* with Burckhardt's subjective *Weltanschauung*. Meinecke put this aptly by means of the Goethean metaphor of the so-called 'creative mirror' (*schaffender Spiegel*), which he equates with the task of the historian:

> The historian may use the term [creative mirror, RK] as a parable for the purpose of his own work. It is not intended to mirror the past in an inanimate manner, but creatively mirror it, to merge the subjective and the objective in such a way that the picture of history thus reproduced represents the past as faithfully and honestly as possible, and yet remains completely animated by the creative individuality of the researcher.[175]

Meinecke takes the view that the (individual creative Burckhardtian) historian should thus mirror, as far as possible, 'the' (Rankean objective) past.[176] In this way, subject (the historian) and object (the past) are reconciled. By means of the metaphor of the 'creative mirror', Meinecke proposed to unite a *historical* with an *ahistorical* view.[177] Ranke personifies the historical view (development), Burckhardt's cross section equals the ahistorical view, since the heart of his conception of history is to unravel the characteristic (individual) cultural values which refer to a 'suprahistorical' level – the good, true and beautiful. This is exactly what Meinecke considered to be the high point of historicism: a reconciliation between the historical and the ahistorical.[178] And this high point he encountered in Goethe's harmonious world view.[179] So Meinecke's conception of history is, in

---

175   'Der Historiker aber darf das Wort [schaffender Spiegel, RK] als Gleichnis für das Ziel seiner eigenen Arbeit anwenden. Sie soll das einst Gewesene nicht mechanisch, sondern schaffend spiegeln, Subjektives und Objektives in sich so verschmelzen, daß das dadurch gewonnene Geschichtsbild zugleich die Vergangenheit, soweit sie zu fassen ist, getreu und ehrlich wiedergibt und dabei doch ganz durchblutet bleibt von der schöpferischen Individualität des Forschers'. Friedrich Meinecke, *Schaffender Spiegel. Studien zur deutschen Geschichtsschreibung und Geschichtsauffassung* (Stuttgart 1948) 7. The notion *schaffender Spiegel* is, according to Meinecke's preface, derived from a deleted scene in Goethe's *Faust*. However, Dobbek claims Herder was early on familiar with this notion which, according to Dobbek, he became aware of through Plotinus' philosophy, via Leibniz' „miroir actif vivant". Dobbek, *J. G. Herders Weltbild*, 15. This of course fits well with Meinecke's panentheism, which is also partly derived from Plotinus. Also: Nisbet, *Goethe and the Scientific Tradition*, 6-22.

176   Hinde, *Jacob Burckhardt*, 201-202.

177   Ankersmit, *De sublieme historische ervaring*, 165-166.

178   The vertical or ahistorical conception of history, by the way, presupposes the horizontal or 'historical' conception of history. Cf. the previous chapter.

179   Cf. Meinecke, *Die Entstehung des Historismus*, 445-584. On countless occasions Meinecke refers to statements in Goethe's writings that point at such a reconciliation of the historical and ahistorical. This next statement of Goethe, as reproduced by Eckermann is exemplary: 'Jeder

that sense, a Goethean-panentheistic reconciliation of Ranke and Burckhardt. It is a reconciliation of politics and culture, of the objective and subjective, of the 'horizontal' and the 'vertical', of the historical and ahistorical, of the general and individual; in sum: it is a reconciliation of historical writing and autobiography, which Meinecke realized at the end of his career.

Conclusion

During the Second World War Meinecke had already begun his research on the causes of the *Schicksal* which had befallen Germany or which Germany had caused itself. In particular, Meinecke tried to find political causes and concluded ultimately that the evil, *kulturwidrige* characteristics of nationalism and socialism had united in National Socialism. Nevertheless, he struggled to bridge the gap between that and the causes for the 'emergence' of Hitler and at the end of the Second World War he was left with the question of 'where to go from here?'. Germany was politically devastated and also Meinecke's beloved *Kultur* was about to collapse. All this was an extremely traumatic experience for Meinecke.[180] The only thing left for him was to urge a regeneration of the humanism of the *Goethezeit*. It was Goethe the Olympian in particular who personified the identity of Germany and historicism. And in following Goethe's *Dichtung und Wahrheit*, Meinecke tried to capture his own life and the history of Germany in one narrative. After the Second World War, Meinecke appointed himself the 'autobiographer' of the history of Germany. In that way, Meinecke managed to defend himself against Germany's fate. Furthermore, by means of his panentheistic conviction he protected himself against a pessimism à la Burckhardt. His vertical, individual, autobiographical, subjective and culture-directed view was indeed important to Meinecke, during and after the Second World War. Meinecke even suggested combining it with Ranke's horizontal, general, objective conception of history, which was aimed at (a power political) development. Meinecke thus deepened his conception of history by means of reconciling these

---

Zustand, ja jeder Augenblick ist von unendlichem Wert, denn er ist der Repräsentant einer ganzen Ewigkeit' (Cited in Meinecke, *Die Entstehung des Historismus*, 565). In this case Goethe is in fact indebted to Leibniz' philosophical system. The 'moment' represents here, on the one hand, the *historical*, for it assumes a development (in time), on the other, it is in its individuality and uniqueness (monad) a reflection of the eternal, timeless form, and in that way ahistorical. Goethe and Meinecke refer to an experience (of the historian) in which the normal duration of a 'moment' is raised in intensity and meaning. Cf: Anglet, *Der "ewige" Augenblick*, 1. Also see the previous chapter.

180   With regard to the theme of trauma and history: Ankersmit, *De sublieme historische ervaring*, 348-407, especially 360-364.

two eminent historians, but also by identifying himself with Germany's history as a whole.

In his last public address in 1951, three years before his death, Meinecke calls to mind Goethe one more time. Consolation for the present and hope for the future, this is what speaks from the Goethe quotation he cites in this lecture: 'If only the eternal remains present to us at any moment, we do not suffer from the passing of time'.[181] The defence of eternal values is at the heart of Meinecke's last address. As the (last) authority of the Rankean historical scientific tradition, Meinecke addresses the future generation of historians. He urges the defence of the highest and most sacred values, for after the destruction of the war this demand was imperative more than ever: 'Freedom, honour, rights and dignity of the personality'[182], these are the most important values in life, in history. They form the basis for a reconciliation of opposites.[183] It is this wisdom, obtained from his own experience and insight, which Meinecke passes on to the new generation. In that sense, he is Germany's conscience.

---

181   'Bleibt uns nur das Ewige jeden Augenblick gegenwärtig so leiden wir nicht an der vergänglichen Zeit'.

182   'Freiheit, Ehre, Recht und Würde der Persönlichkeit'.

183   Friedrich Meinecke, 'Geschichte und Politik' in: idem, *Politische Schriften und Reden* (Darmstadt 1958) 494-497, there 496-497.

Final conclusions

Friedrich Meinecke died on February 6, 1954 at the age of 91. The official obituary of the *Freie Universität Berlin* states that, with the passing of Meinecke, Germany lost 'a historical-political thinker', who 'embodied the voice of its conscience'.[1] With this, the university calls to mind Meinecke's repeated warnings about National Socialism. For that reason, he was appointed *Ehrenrektor* (honorary rector) of the newly established *Freie Universität* in 1948. Meinecke was considered a representative of 'the good Germany', but he obviously represented far more than 'only' this 'voice of conscience'. During every political and intellectual crisis he called for reflection. After the First World War he warned against the veiling of power politics; he publicly resigned himself to the Weimar Republic in becoming a *Vernunftrepublikaner*, he called to mind (after the Second World War) Goethean humanism; he suggested a less idealistic conception of history; he reconciled Ranke with Burckhardt, and above all, he proposed the *individual* conscience as an authority akin to God, which enabled man to approximate or intuit 'God', the true, the good and the beautiful. Put differently, with Meinecke's passing Germany lost a *Mahner* – as the headline of the *Frankfurter Allgemeine* of 8 February 1954 put it.[2]

The obituary of the *Freie Universität* also states that with Meinecke's death the (historical) science lost 'The most important German historian of the first half of our century'. And with the death of Meinecke historical science lost the last great historicist in the tradition of Leopold von Ranke. What makes Meinecke great in this tradition is not only the affirmation of, but also the moulding of this tradition. Ever since his first major work, *Weltbürgertum und Nationalstaat*, he constructed different traditions by means of great historical personalities. His choice of personalities to analyze always showed his own field of interest. This way he created a series of thinkers who confirmed the legitimacy of the unification in 1871, embodied the history of *raison d'état* thinking, and represented the origins of historicism. But he also searched for the causes of National Socialism in such movements as

---

1   In the possession of the author. '(...) einen historisch-politischen Denker', '(...) der Stimme seines Gewissens verkörperte'.

2   In English it is close to 'reminder', 'urger' or an 'hortatory' historian. 'Der Mahner', *Frankfurter Allgemeine Zeitung*, 8 February 1954.

the Enlightenment, nationalism and socialism, and especially in the *kulturwidrige* aspects of these movements. Ultimately, he constructed a synthesis of Burckhardt and Ranke, which he essentially embodied himself. And above all, he identified himself, at the end of his career, in a Goethean way with the (political) history of Germany. All these constructions and grand narratives aimed at reconciling contradictions. In this book I have shown that this idea of reconciliation is the *Leitmotiv* of Meinecke's oeuvre.

Due to the many crises Germany encountered, Meinecke's reconciliatory view came under pressure and was more and more difficult to defend. Initially, he confirmed in *Weltbürgertum und Nationalstaat* (1908) the German political and cultural unification of 1871, as shown in the first chapter; *Macht* and *Kultur*, according to Meinecke, were reconciled. In reality however, a synthesis was out of the question. On a different level, this was also apparent in the contrasts between the political and social groups in Germany. Meinecke, following Friedrich Naumann, wished for these groups to reconcile. This all was in vain, for Germany's society was divided. As a reaction to the failing of this reconciliation, Meinecke, in *Weltbürgertum und Nationalstaat*, falls back on (the idea of) the synthesis between *Macht* and *Kultur* of 1871.

At the outbreak of the First World War, hope grew for an actual reconciliation of power and culture. This again was vain hope. The German Empire collapsed. After the First World War, Meinecke brought to light the dangers of a 'spiritualisation of power' and a 'veiling of evil'. He demonstrated that ethics and power – *Ethos* and *Kratos* – were closely tied together. In this context, he claimed that good and evil were no longer separate but interwoven notions. Ranke and Hegel, according to Meinecke, underestimated the dangerous dimensions of power politics because they concealed it by means of a belief in a divine harmony and the cunning of reason, respectively; in both views evil can have a good result (and vice versa). Meinecke could no longer agree with this – for good reason he considered the history of *raison d'état* thinking a tragic history.

After the First World War historicism fell into a deep crisis. As a result of the war, among other things, man lost his moral footing. The (nineteenth-century) idea of a continuous, harmonious 'stream of becoming' turned out to be untenable; the stream overflowed its banks. Moreover, the idea of a consistent historicization led to a relativization of apparently fixed values. The Second World War had the same, or an even stronger effect on Meinecke and especially on his reconciliatory view. Nevertheless, he maintained that a reconciliation of power and culture could still be possible, albeit that power should first be purged of the shame of the Third Reich before a reunification with *Kultur* was once again possible.

Despite the many political, ideological and intellectual crises, Meinecke kept on devoting himself to reconciling contradictions. I have shown that this is apparent in Meinecke's intellectual development. He advocated a more and more explicit panentheism, which formed the philosophical basis of his reconciliatory view. Initially, in *Weltbürgertum und Nationalstaat* – and mainly in line with the idealism of Humboldt and Ranke, and the legitimizing historical writings of Droysen – Meinecke propagated an implicit panentheism. After the upheaval of the First World War, Meinecke kept on searching for the reconciliatory, and distanced himself from Ranke's all too optimistic attitude towards power politics. In Meinecke's second major work, *Die Idee der Staatsräson* (1924), it was Schiller's aesthetics which seemed a good solution to the conflicts between *Ethos* and *Kratos*. In the process of political decision-making however, ethics are constantly defeated by power. To restore a balance, Meinecke, by means of a '*Staatsräson* as aesthetic category', emphasized ethics. Since the synthesis between power and ethics is ultimately based on an ideal and therefore in reality the conflict between the two remains, Meinecke introduced the statesman's conscience as the last solution. Conscience, as the *Abglanz* of the absolute or divine morality, gives the statesman, in Meinecke's view, the ability to strive for the good, true, and beautiful, and as a result he can make a morally just decision. The statesman's conscience, says Meinecke, functions as the foundation of political decision making.

After this study on political theory, Meinecke could finally fulfill his long-cherished wish: to research the rise of historicism. In a strange way Hitler's take-over came at a convenient moment for Meinecke, for it gave him room to completely dive into his research. After all, forced by the circumstances of National Socialism, he could no longer publish in newspapers and he resigned as editor-in-chief of the *Historische Zeitschrift*, because he disagreed with the new course of the journal. So, during the early 1930s Meinecke, by now emeritus, immersed himself in Goethe and especially his philosophy of nature and its importance for (the rise of) historicism. It was in Goethe's ideas that Meinecke regained harmony. Already during the First World War Meinecke had found comfort in the works of Goethe, but especially during and after the Second World War Goethe was to him a paragon of humanism, a beacon of hope, consolation and love.

Indeed, it was Goethe's philosophy of nature which gave Meinecke the means to reconcile historicism with itself. For, it was only historicism itself that could heal the self-inflicted ethical relativistic wounds. This meant a confirmation of the tradition, but also a historicization of historicism – the heart of *Die Entstehung des Historismus*. Further, Meinecke reinterpreted (*Umdeuten*) the ideas of Goethe in order to fit his own view of historicism, and he next managed to reconcile Enlightenment with historicism. On the basis of Goethe's views and Meinecke's own panentheistic philosophy he also managed to reconcile contradicting notions like the individual and the general, subject and object, and also Ranke and Burckhardt, the historical and the ahistorical. He carried this the furthest in the identification of his own person with the history of Germany. In that sense, he identified himself with 'his own' historicism; he had become the tradition he created. Goethe expressed it aptly:

'(…) that in the present, and even more so in memory, man modulates the outer world according to his peculiarities'.[3]

In sum, at the end of his life Meinecke, in following Goethe, had himself become an Olympian. But like Goethe, Meinecke also remained in contact with reality. And from this reality he strove for higher things in which the dissonances of daily life and history were in harmony with each other. It is this harmonious panentheistic conception of history which is the fundamental principle of Meinecke's oeuvre.

Meinecke's panentheist world view can explain the coherence of his work – a coherence that also leaves room and does justice to the versatile character of his oeuvre. In this sense, this interpretation of Meinecke offers the 'best possible' explanation for what he had in mind and allows for a more convincing explanation for the problem of (dis)continuity in Meinecke's oeuvre than has been the case in the Meinecke-*Forschung* so far.

'What is left of all this work?' asked the Dutch historian Kossmann in an article on Meinecke. Is it outdated? Kossmann thought it was indeed partially the case, but he immediately posed the question what 'outdated' really means in the historical sciences.[4] To stick to the first question: what is the enduring importance of Meinecke's work? In this book I have shown that Meinecke's oeuvre offers a panoramic view of the nineteenth-century historical sciences. Thanks to his historical insight, Meinecke was able to maintain an overall picture despite the changing, and most of all, complex political and intellectual situations in the present. His work offers insight into the key issues of diverse historical-theoretical and political problems, many of whom are still important today.[5] The historical context has obviously changed, but the insight into such matters still offers a starting point for (political) actions, a view on history, insight into the problems of historical writing, and much more. In sum, to think through historicism in all its aspects, Meinecke's works offers wisdom 'forever' – that is the significance of his oeuvre.[6]

---

3   Johann Wolfgang von Goethe, 'Tag- und Jahreshefte' in: idem, *Goethes Werke. Hamburger Ausgabe in 14 Bänden*, ed. Erich Trunz (Munich 1981) volume X, 429-528, there 510. '(…) daß der Mensch in der Gegenwart, ja vielmehr noch in der Erinnerung die Außenwelt nach seinen Eigenheiten bildend modele'.

4   Kossmann, 'Friedrich Meinecke', 223.

5   Such as the problem of (ethical) relativism, subjectivity and objectivity in historical writing, the issue of the unintended consequences of our actions, prudence in politics. The theme of 'unity and difference', which played an important role for Meinecke in nineteenth-century Germany, but is essentially also apparent in the European Union.

6   Burckhardt's famous expression on the meaning of historical knowledge: 'Wir wollen durch Erfahrung nicht sowohl klug (für ein andermal) als weise (für immer) werden'. Burckhardt, *Weltgeschichtliche Betrachtungen*, 10.

And yet, most of the time Meinecke is considered a nineteenth-century historicist à la Ranke, whose faith in a divine or panentheistic harmony is considered extremely unpractical. Moreover, critics agree that Meinecke's view on the state and power are outdated. In that sense, his conception of history is superseded, especially because nowadays the emphasis in the historical sciences is on social and economic changes, political processes, globalization and the transfer of many different (material) expressions of culture. Meinecke (would) never (have) denied the importance of these matters, but he would have emphasized that the key to understanding all these issues consists essentially of 'human thought, will, and feeling'[7], which in these modern areas of research are secondary to the more general processes. Political theory is an exception here, for Meinecke is of course very close to this field; this sub-discipline was for him closely related to the historicist world view.

Meinecke's conception of history is for the most part oriented towards the nineteenth century. Yet, his ideas on how we can have knowledge of the past anticipate some developments of the second half of the twentieth century. In the last chapter I have shown that Meinecke suggested replacing the topos 'history is the teacher of life' with 'life is the teacher of history', thus putting 'human existence' at the heart of his conception of history. According to some commentators, Meinecke should be considered the 'realization' of Dilthey's hermeneutics, in which human experience was central.[8] Nevertheless, Meinecke's view on the 'task of the historian' and his own compliance with this task point in the direction of Gadamer's ontological hermeneutics. After all, Gadamer was interested in the question of how or in what way historical knowledge and experience are essential to human existence.[9] To him every hermeneutic understanding of the past is related to individuality, in this case to the historian. The subject, the historian, is therefore *not* interchangeable, as Dilthey thought. He indeed claimed that every form of understanding (*verstehen*) was the same for every researcher. Gadamer, on the other hand, took the view that an interpretation of the past is closely related to the position of the individual researcher. One can only interpret from one's own position in history, and in that sense all understanding of the past is simultaneously an understanding of one's own position in history; historical insight takes shape in a 'conversation' with

7   Meinecke, 'Willensfreiheit und Geschichtswissenschaft', 29. 'menschliches Denken, Wollen und Empfinden'.

8   Hofer, *Geschichtsschreibung*, 12; more subtle is: Walter Goetz, 'Friedrich Meinecke. Leben und Persönlichkeit' in: idem, *Historiker in meiner Zeit. Gesammelte Aufsätze* (Cologne 1957) 329-350, there 347-348.

9   Ankersmit, *Denken over geschiedenis*, 139.

the tradition – Gadamer speaks of the so-called fusion of horizons (*Horizontver-schmelzung*).[10]

Meinecke's notion of the 'creative mirror', which I discussed in the last chapter, fits perfectly with Gadamer's ontological hermeneutics. The historian must 'creatively mirror' the past, that is: the historian should on the one hand represent the past as faithfully as possible, and on the other, the historian – the individual, non-interchangeable historian – shapes his historical writing in a dialogue with the tradition. As Meinecke so aptly put it in the quote mentioned earlier: 'The historian (...) should not intend to mirror the past in an inanimate manner, but creatively mirror it, to merge the subjective and the objective in such a way that the image of history thus reproduced represents the past as faithfully and honestly as possible, and yet remains completely animated by the creative individuality of the researcher'.[11] Meinecke has elaborated in a grand manner on this 'understanding oneself' with one's tradition in his oeuvre. In essence, his whole career he was in conversation with the historicist tradition to gain insight into (the meaning of) life, into human existence. He once expressed it as follows (in a slightly Gadamerian way): 'This means to enter into the very souls of those who acted, to consider their works and cultural contributions in terms of their own premises and, in the last analysis, through artistic intuition to give new life to life gone by – which cannot be done without a transfusion of one's own life blood'.[12] The similarities with Gadamer's 'fusion of horizons' between object and subject are striking. Meinecke's thought thus represents more the transition from Dilthey to Gadamer than the realization of Dilthey's philosophy.

Perhaps it is also this Gademerian nature that characterizes Meinecke's work, which makes it difficult for some to comprehend. The all too personal or subjective character of his work might convey the *impression* that it was only important to Meinecke himself. According to this view, such subjectivity leads the individual and science in the end to an 'abyss of individuality'.[13] Yet, this notion of the abyss of individuality has two angles. On the one hand, the abyss refers to ethical relativism, which is inherent in (extreme) individualism – this is what Meinecke considered 'life-weakening' (*lebenschwächend*) relativism. On the other, the abyss could be conceived of as 'life-enhancing' (*lebensteigernd*), and thereby emphasize the dynamics and plenitude that the individuality gives to history and life. To Meinecke, as I have

---

10   Gadamer, *Wahrheit und Methode*, 289, 356, 375.

11   Meinecke, *Schaffender Spiegel*, 7.

12   Meinecke, 'Values and Causalities in History', 283. Meinecke, 'Kausalitäten und Werte in der Geschichte', 82. 'Es gilt, sich in die Seelen der Handelnden dabei selbst zu versetzen, von ihren Voraussetzungen aus ihr Werk und ihre Kulturleistung zu betrachten und letzten Endes durch künstlerische Intuition ihr vergangenes Leben neu zu beleben, was ohne Transfusion eigenen Lebensblutes nicht möglich ist'.

13   Dilthey's 'Abyssus von Individualität'.

shown, this last point of view was most valuable, for considering the individual in its historicity is the basis of everything: it is a relativization of perspectives, but also the last ground of being.[14] It is not an easy task to grasp hold of anything from such a point of view; perhaps Meinecke was the last who managed to reconcile himself with these contrasts.

---

14   Not to be confused with Paul Tillich's reference to God.

Bibliography

Albrecht, Andrea, *Kosmopolitismus. Weltbürgerdiskurse in Literatur, Philosophie und Publizistik um 1800* (Berlin and New York 2005).

Ameriks, Karl, 'Introduction: "Interpreting German Idealism"' in: idem ed., *The Cambridge Companion to German Idealism* (Cambridge 2000) 1-17.

Anglet, Andreas, *Der „ewige' Augenblick. Studien zur Struktur und Funktion eines Denkbildes bei Goethe* (Cologne, Weimar and Vienna 1991).

Ankersmit, F. R., 'Narrative Logic. A Semantic Analysis of the Historian's Language' (Groningen 1981). Unpublished dissertation; published version: The Hague 1983.

Ankersmit, F. R., *Denken over geschiedenis. Een overzicht van moderne geschiedfilosofische opvattingen* (Groningen 1986).

Ankersmit, F. R., *History and Tropology. The Rise and Fall of Metaphor* (Berkeley 1994).

Ankersmit, F.R., *Aesthetic Politics. Political Philosophy Beyond Fact and Value* (Stanford 1996).

Ankersmit, F. R., *Macht door representatie. Exploraties III: politieke filosofie* (Kampen 1997).

Ankersmit, F. R., *Historical Representation* (Stanford 2001).

Ankersmit, F. R., *Political Representation* (Stanford 2002).

Ankersmit, F. R., *De sublieme historische ervaring* (Groningen 2007).

Ankersmit, Frank, *Meaning, Truth, and Reference in Historical Representation* (Ithaca 2012).

Ankersmit, Frank, 'Politieke stijl: Schumann en Schiller' in: Dick Pels and Henk te Velde ed., *Politieke stijl. Over presentatie en optreden in de politiek* (Amsterdam 2000) 15-42.

Ankersmit, F. R., 'What is Wrong with World History from a Cosmopolitical Point of View?', Conference paper, Shanghai, November 2008.

Ankersmit, Frank, 'Jacobi. Realist, Romanticist, and Beacon for Our Time', *Common Knowledge* 14 (2008) 221-243.

Ankersmit, F. R., 'Aftermaths and 'Foremaths'' (Groningen 2010) Unpublished lecture.

Antoni, Carlo, *From History to Sociology. The Transition in German Historical Thought* translation Hayden White (Detroit 1959).

Assmann, Aleida, *Arbeit am nationalen Gedächtnis. Eine kurze Geschichte der deutschen Bildungsidee* (Frankfurt, New York and Paris 1993).

Bahners, Patrick, 'Literaturkenntnis schützt vor Denkmalsturz. Völlig losgelöst: der Historismus nach Annette Wittkau', *Frankfurter Allgemeine Zeitung* (17 May 1993).

Baldacchino, Joseph F., 'The Value-Centered Historicism of Edmund Burke', *Modern Age* 27 (1983) 139-145.

Bambach, Charles, *Heidegger, Dilthey and the Crisis of Historicism* (Ithaca and London 1995).

Baneke, David, *Synthetisch denken. Natuurwetenschappers over hun rol in een moderne maatschappij, 1900-1940* (Hilversum 2008).

Bartuschat, Wolfgang, *Baruch de Spinoza* (Munich 1996).

Baumann, Gerhart, *Goethe. Dauer im Wechsel* (Munich 1977).

Beck, Lewis White, *Early German Philosophy. Kant and his Predecessors* (Cambridge Mass. 1969).

Beiser, Frederick C., *The Fate of Reason. German Philosophy from Kant to Fichte* (Cambridge Mass. and London 1987).

Beiser, Frederick C., *Enlightenment, Revolution, and Romanticism: The Genesis of Modern German Political Thought 1790-1800* (Cambridge Mass. 1992).

Beiser, Frederick C., *German Idealism. The Struggle against Subjectivism, 1781-1801* (Cambridge Mass. 2002).

Beiser, Frederick C., *The German Historicist Tradition* (Oxford 2011).

Beiser, Frederick C., 'Hegel's Historicism' in: idem ed., *The Cambridge Companion to Hegel* (Cambridge 1993) 270-300.

Bell, David, *Spinoza in Germany from 1670 to the Age of Goethe* (London 1984).

Benjamin, Walter, 'Schicksal und Charakter' in: idem, *Gesammelte Schriften* II.I, ed. Rolf Tiedemann and Hermann Schweppenhäuser (Frankfurt am Main 1977) 171-179.

Berg, Nicolas, *Der Holocaust und die westdeutschen Historiker. Erforschung und Erinnerung* (Göttingen 2003).

Berg, W. van den, 'De preromantiekconceptie in de Nederlandse literatuurgeschiedenis' in: idem, *Een bedachtzame beeldenstorm. Beschouwingen over letterkunde van de achttiende en negentiende eeuw* (Amsterdam 1999) 13-39.

Berlin, Isaiah, *The Age of Enlightenment. The Eighteenth-Century Philosophers* (Oxford 1979).

Berlin, Isaiah, *Political Ideas in the Romantic Age. Their Rise and Influence on Modern Thought*, ed. Henry Hardy (Princeton and Oxford 2006).

Berlin, Isaiah, 'The Originality of Machiavelli' in: idem, *Against the Current. Essays in the History of Ideas*, (Oxford, Toronto, and Melbourne 1981) 25-79.

Besson, Waldemar, 'Friedrich Meinecke und die Weimarer Republik. Zum Verhältnis von Geschichtsschreibung und Politik', *Vierteljahreshefte für Zeitgeschichte* 7 (1959) 113-129.

Beyerhaus, Gisbert, 'Notwendigkeit und Freiheit in der deutschen Katastrophe. Gedanken zu Friedrich Meineckes jüngstem Buch', *Historische Zeitschrift* 169 (1949) 73-87.

Biernacki, L. and Philip Clayton, eds., *Panentheism across the World's Traditions* (Oxford 2014).

Blanke, Horst Walter and Jörn Rüsen ed., *Von der Aufklärung zum Historismus. Zum Strukturwandel des historischen Denkens* (Paderborn etc. 1984).

Blanke, Horst Walter, *Historiographiegeschichte als Historik* (Stuttgart-Bad Cannstatt 1991).

Blanke, Horst Walter, 'Aufklärungshistorie und Historismus: Bruch und Kontinuität' in: Otto Gerhard Oexle and Jörn Rüsen eds., *Historismus in den Kulturwissenschaften. Geschichtskonzepte, historische Einschätzungen, Grundlagenprobleme* (Cologne, Weimar and Vienna 1996) 69-97.

Blanke, Horst Walter, 'Aufklärungshistorie und "Historismus" im Denken Friedrich Meineckes' in: W. Bialas and G. Raulet ed., *Die Historismusdebatte in der Weimarer Republik* (Frankfurt am Main 1996) 142-160.

Blanke, Horst Walter, 'Vereinnahmungen: „Schiller als Historiker" in der Historiographiegeschichte der letzten 150 Jahre' in: Michael Hofmann, Jörn Rüsen and Mirjam Springer ed., *Schiller und die Geschichte* (Munich 2006) 104-123.

Blaufuß, Dietrich, 'Pietism' in: W.J. Hanegraaff ed., *Dictionary of Gnosis and Western Esotericism* (Leiden 2005) 955-960.

Bock, Gisela and Daniel Schönpflug ed., *Friedrich Meinecke in seiner Zeit. Studien zu Leben und Werk* (Stuttgart 2006).

Bock, Gisela and Daniel Schönpflug, 'Vorwort' in: Gisela Bock and Daniel Schönpflug ed., *Friedrich Meinecke in seiner Zeit. Studien zu Leben und Werk* (Stuttgart 2006) 7-8.

Bock, Gisela, 'Meinecke, Machiavelli und der Nationalsozialismus' in: Gisela Bock and Daniel Schönpflug ed., *Friedrich Meinecke in seiner Zeit. Studien zu Leben und Werk* (Stuttgart 2006) 145-174.

Bock, Gisela, 'Friedrich Meinecke und seine Briefe. Eine Einführung' in: *Friedrich Meinecke, Neue Briefe und Dokumente* (Munich 2012) 1-23.

Bödeker, Hans Erich, Georg Iggers, Jonathan Knudsen and Peter Hans Reill ed., *Aufklärung und Geschichte. Studien zur deutschen Geschichtswissenschaft im 18. Jahrhundert* (Göttingen 1986).

Böge, Ulrike, 'Die Inbesitznahme Goethes durch die Philosophie. Goetherezeption bei deutschsprachigen Philosophen in der ersten Hälfte des 20. Jahrhunderts' (Kiel 2001) unpublished dissertation.

Böhme, Klaus, 'Meinecke als politischer Historiker', *Neue Politische Literatur* (1975) 110-114.

Boos, Stephen, 'Rethinking the Aesthetic: Kant, Schiller, and Hegel' in: Dorota Glowacka and Stephen Boos ed., *Between Ethics and Aesthetics. Crossing the Boundaries* (New York 2002) 15-27.

Börnsen, Hans, *Leibniz' Substanzbegriff und Goethes Gedanke der Metamorphose* (Stuttgart 1985).

Borowski, Audrey, 'Friedrich Meinecke's *Historism* or the Defeat of German Historicism' in Herman Paul and Adriaan van Veldhuizen eds., *Historicism: A Travelling Concept* (London 2020) 165–185.

Bossenbrook, William J., 'Justus Möser's Approach to History' in: James Lea Cate ed., *Medieval and Historiographical Essays in Honor of James Westfall Thompson* (Chicago 1938) 397-422.

Boterman, Frits, *Moderne geschiedenis van Duitsland 1800-heden* (Amsterdam and Antwerp 2005).

Boterman, Frits, *Duitse dichters en denkers. Het belang van cultuur in de moderne Duitse geschiedenis* (Amsterdam and Antwerp 2008).

Boyle, Nicholas, *Goethe. The Poet and the Age. Volume I. The Poetry of Desire (1749-1790)* (Oxford 1991).

Boyle, Nicholas, 'Geschichtsschreibung und Autobiographik bei Goethe (1810-1817)', *Goethe Jahrbuch* 110 (1993) 163-172.

Brands, Maarten C., *Historisme als ideologie. Het 'onpolitieke' en 'anti-normatieve' element in de Duitse geschiedwetenschap* (Assen 1965).

Brands, Maarten, *Het arsenaal van de geschiedenis. Over theorie en geschiedschrijving* (Amsterdam 2013).

Brands, Maarten C., 'Meinecke between "Macht" and "Innerlichkeit"' in: Michael Erbe, ed., *Friedrich Meinecke heute. Bericht über ein Gedenk-Colloquium zu seinem 25. Todestag am 5. und 6. April 1979* (Berlin 1981) 176-185.

Bremi, Willy, *Was ist das Gewissen? Seine Beschreibung, seine metaphysische und religiöse Deutung, seine Geschichte* (Zürich 1934).

Brillenburg Wurth, Kiene, *The Musically Sublime. Infinity, Indeterminacy, Irresolvability* (New York 2009).

Bruch, Rüdiger vom, 'Ein Gelehrtenleben zwischen Bismarck und Adenauer' in: Gisela Bock and Daniel Schönpflug ed., *Friedrich Meinecke in seiner Zeit. Studien zu Leben und Werk* (Stuttgart 2006) 9-19.

Bubner, Rüdiger, 'Die Gesetzlichkeit der Natur und die Willkür der Menschheitsgeschichte. Goethe vor dem Historismus', *Goethe Jahrbuch* 110 (1993) 135-145.

Buck, Theo, 'Dämonisches' in: Bernd Witte ed., *Goethe Handbuch in vier Bänden* (Stuttgart and Weimar) Volume 4/1, 179-181.

Burckhardt, Jacob, *Briefe*, Fritz Kaphan ed. (Leipzig).

Burckhardt, Jacob, *Historische Fragmente*, ed. Emil Dürr (Stuttgart 1957).

Burckhardt, Jacob, *Weltgeschichtliche Betrachtungen*, ed. Rudolf Marx (Stuttgart 1978).

Burckhardt, Jacob, *Judgements on History and Historians* (London and New York 2007 first 1959).

Butterfield, H., 'Review', *The English Historical Review* 86 (1971) 337-342.

Cassirer, Ernst, *Freiheit und Form. Studien zur deutschen Geistesgeschichte* (Berlin 1916).

Cassirer, Ernst, *Goethe und die geschichtliche Welt* (Hamburg 1995, First 1932).

Cassirer, Ernst, 'Die Philosophie der Aufklärung' in: idem, *Gesammelte Werke. Hamburger Ausgabe*, ed. Birgit Recki (Hamburg 2003) volume 15.

Chickering, Roger, *Karl Lamprecht. A German Academic Life* (1856-1915) (Atlantic Highlands 1993).

Clark, Christopher M., *Kaiser Wilhelm II. A Life in Power* (London 2009).

Clark, Mark W., *Beyond Catastrophe. German Intellectuals and Cultural Renewal after World War II 1945-1955* (New York 2006).

Collingwood, R. G., *The Idea of History* (Oxford 1972; first 1945).

Cooper, John W., *Panentheism. The Other God of the Philosophers. From Plato to the Present* (Michigan 2006).

Courtney, C. P., *Montesquieu and Burke* (Oxford 1963).

Croce, Benedetto, *Die Geschichte als Gedanke und als Tat* (Bern 1944).

Croce, Benedetto, *Goethe. Studien zu seinem Werk* (Düsseldorf 1949).

Dann, Otto, Norbert Oellers and Ernst Osterkamp ed., *Schiller als Historiker* (Stuttgart 1995).

Danto, Arthur C., *Nietzsche as Philosopher* (New York and London 1965).

Dassen, Patrick, *De onttovering van de wereld. Max Weber en het probleem van de moderniteit in Duitsland, 1890-1920* (Amsterdam 1999).

Dauner, Paul, 'Das Gewissen' (Stuttgart 2008) unpublished dissertation.

Dehio, Ludwig, *Friedrich Meinecke. der Historiker in der Krise. Festrede, gehalten am Tage des 90. Geburtstags* (Berlin 1953).

Demandt, Alexander, 'Geschichte bei Goethe', *Merkur. Deutsche Zeitschrift für europäisches Denken* (1947) 317-327.

Deursen, A. Th. van, *Geschiedenis en toekomstverwachting. Het onderwijs in de statistiek aan de universiteiten van de achttiende eeuw* (Kampen 1971).

Dilthey, Wilhelm, 'Der Aufbau der geschichtlichen Welt in den Geisteswissenschaften' in: idem, *Gesammelte Schriften* VII (Göttingen 1958).

Dilthey, Wilhelm, 'Leibniz und sein Zeitalter' in: idem, *Gesammelte Schriften* III (Stuttgart and Göttingen 1969) 3-80.

Dilthey, Wilhelm, 'Das Erlebnis und die Dichtung. Lessing, Goethe, Novalis, Hölderlin' in: idem, *Gesammelte Schriften* XXVI (Göttingen 2005).

Dobbek, Wilhelm, *J. G. Herders Weltbild. Versuch einer Deutung* (Cologne 1969).

Dunk, H.W. von der, *De glimlachende sfinx. Kernvragen in de geschiedenis* (Amsterdam 2011).

Dunk, H. W. von der, 'Friedrich Meinecke en het historisme; een blik terug', *Tijdschrift voor Geschiedenis* 79 (1966) 24-37.

Eckermann, Johann Peter, *Gespräche mit Goethe in den letzten Jahren seines Lebens* (Berlin and Darmstadt 1958).

Eckert, G. and G. Walther, 'Die Geschichte der Frühneuzeitforschung in der Historische Zeitschrift', *Historische Zeitschrift* 289 (2009) 149-197.

Eich, Stefan and Adam Tooze, 'The allure of dark times: Max Weber, politics, and the crisis of historicism', *History and Theory* 56 no. 2 (June 2017) 197-215.

Engelhardt, Dietrich von, 'Natur und Geist, Evolution und Geschichte. Goethe in seiner Beziehung zur romantischen Naturforschung und metaphysischen Naturphilosophie' in: Peter Matussek ed., *Goethe und die Verzeitlichung der Natur* (Munich 1998) 58-74.

Erauw, Willem, 'De relatie tussen cultuur en politiek tegen de achtergrond van de Duitse natievorming: een inleiding', *De bijdragen tot de eigentijdse geschiedenis* 11 (2003) 7-19.

Erbe, Michael ed., *Friedrich Meinecke heute. Bericht über ein Gedenk-Colloquium zu seinem 25. Todestag am 5. und 6. April 1979* (Berlin 1981).

Erbe, Michael, 'Das Problem des Historismus bei Ernst Troeltsch, Otto Hintze und Friedrich Meinecke' in: Horst Renz and Friedrich Wilhelm Graf ed., *Umstrittene Moderne. Die Zukunft der Neuzeit im Urteil der Epoche Ernst Troeltsch* (Troeltsch Studien) (Gütersloh 1987) 73-91.

Erbe, Michael, 'Zur Meinecke-Rezeption im Ausland' in: idem ed., *Friedrich Meinecke heute. Bericht über ein Gedenk-Colloquium zu seinem 25. Todestag am 5. und 6. April 1979* (Berlin 1981) 147-165.

Ermatinger, Emil, 'Goethe und die Natur. Rede zur Goethe-Feier der Universität Zürich am 22. Februar 1932', *Wege zur Dichtung. Zürcher Schriften zur Literaturwissenschaft* 13 (1932) 7-33.

Erpenbeck, John, 'Urphänomen' in: Bernd Witte ed., *Goethe Handbuch in vier Bänden* (Stuttgart and Weimar) Volume 4/2, 1080-1082.

Flasch, Kurt, *Die Geistige Mobilmachung. Die deutschen Intellektuellen und der Erste Weltkrieg. Ein Versuch* (Berlin 2000).

Forster, Michael N., 'Herder and Spinoza' in: Eckart Förster and Yitzhak Y. Melamed ed., *Spinoza and German Idealism* (Cambridge 2012) 59-84.

Franz, Erich, *Goethe als religiöser Denker* (Tübingen 1932).

Fresco, Marcel F., 'Platonisme. Naar hoger Honing?' in: Marcel F. Fresco and Rudi van der Paardt ed., *Naar hoger honing? Plato en platonisme in de Nederlandse literatuur* (Groningen 1998) 9-51.

Fricke, Gerhard, 'Schiller und die geschichtliche Welt' in: idem, *Studien und Interpretationen. Ausgewählte Schriften zur deutschen Dichtung* (Frankfurt am Main 1956) 95-118.

Friese, Heidrun ed., *The Moment. Time and Rupture in modern thought* (Liverpool 2001).

Frey, Hugo and Stefan Jordan, 'Inside-Out: The Purposes of Form in Friedrich Meinecke's and Robert Aron's Explanations of National Disaster' in: Stefan

Berger and Chris Lorenz eds., *Nationalizing the Past. Historians as Nation Builders in Modern Europe* (Basingstoke etc 2010) 282-297.

Fulda, Daniel, *Wissenschaft aus Kunst. Die Entstehung der modernen deutschen Geschichtsschreibung 1760-1860* (Berlin and New York 1996).

Gadamer, Hans-Georg, *Wahrheit und Methode. Grundzüge einer philosophischen Hermeneutik* (Tübingen 1960).

Gay, Peter, *Weimar Culture. The Outsider as Insider* (New York and London 1974).

Geiss, Imanuel, 'Kritischer Rückblick auf Friedrich Meinecke' in: idem, *Studien über Geschichte und Geschichtswissenschaft* (Frankfurt am Main 1972) 89-107.

Gjesdal, Kristin, 'Bildung' in: Michael N. Forster and Kristin Gjesdal ed., *The Oxford Handbook of German Philosophy in the Nineteenth Century* (2015) DOI: 10.1093/oxfordhb/9780199696543.013.0035

Gilbert, Felix, 'Friedrich Meinecke' in: idem, *History. Choice and Commitment* (Cambridge Mass. and London 1977) 67-87.

Gilbert, Felix, 'Review Essays', *History and Theory. Studies in the Philosophy of History* 13 (1974) 59-64.

Glaser, Hermann, *Deutsche Kultur. Ein historischer Überblick von 1945 bis zur Gegenwart* (Munich 1997).

Goering, D. Timothy, 'Einleitung. Ideen- und Geistesgeschichte in Deutschland – eine Standortbestimmung' in: idem ed., *Ideengeschichte heute. Traditionen und Perspektiven* (Bielefeld 2017) 7-53.

Goethe, Johann Wolfgang von, 'Aus meinem Leben. Dichtung und Wahrheit' in: idem, *Goethes Werke. Hamburger Ausgabe in 14 Bänden*, ed., Erich Trunz (Munich 1982) Volume IX-X.

Goethe, Johann Wolfgang von, 'Die Leiden des jungen Werther' (1774) in: idem, *Goethes Werke. Hamburger Ausgabe in 14 Bänden*, ed. Erich Trunz (Munich 1982) Volume VI, 7-124.

Goethe, Johann Wolfgang von, 'Erläuterung zu dem aphoristischen Aufsatz „Die Natur"' (1828) in: idem, *Goethes Werke. Hamburger Ausgabe in 14 Bänden*, ed. Erich Trunz (Munich 1982) Volume XIII, 48-49.

Goethe, Johann Wolfgang von, 'Faust' in: idem, *Goethes Werke. Hamburger Ausgabe in 14 Bänden*, ed. Erich Trunz (Munich 1981) Volume 3.

Goethe, Johann Wolfgang von, 'Glückliches Ereignis' (1817) in: idem, *Goethes Werke. Hamburger Ausgabe in 14 Bänden*, ed. Erich Trunz (Munich 1982) Volume X, 538-542.

Goethe, Johann Wolfgang von, 'Materialien zur Geschichte der Farbenlehre' in: idem, *Goethes Werke. Hamburger Ausgabe in 14 Bänden*, ed. Erich Trunz (Munich 1982) Volume XIV, 7-269.

Goethe, Johann Wolfgang von, 'Maximen und Reflexionen' (posth. 1833) in: idem, *Goethes Werke. Hamburger Ausgabe in 14 Bänden*, ed. Erich Trunz (Munich 1982) Volume XII.

Goethe, Johann Wolfgang von, 'Tag- und Jahreshefte' in: idem, *Goethes Werke. Hamburger Ausgabe in 14 Bänden*, ed. Erich Trunz (Munich 1981) Volume X, 429-528.

Goethe, Johann Wolfgang von, *The Autobiography of Goethe. Truth and Poetry: from my own life*, two volumes, Transl. A.J.W. Morrison (London 1881).

Goethe, Johann Wolfgang von, 'Vermächtnis' in: idem, *Goethes Werke. Hamburger Ausgabe in 14 Bänden*, ed. Erich Trunz (Munich 1981) Volume I, 369.

Goethe, Johann Wolfgang von, 'Versuch einer Witterungslehre' in: idem, *Goethes Werke. Hamburger Ausgabe in 14 Bänden*, ed. Erich Trunz (Munich 1982) Volume XIII, 305-313.

Goethe, Johann Wolfgang von, 'Zur Farbenlehre' (1808) in: idem, *Goethes Werke. Hamburger Ausgabe in 14 Bänden*, ed. Erich Trunz (Munich 1982) Volume XIII, 314-523.

Goethe, Johann Wolfgang von, *Maxims and Reflections* transl. Elisabeth Stopp (London etc 1998).

Goethe, Johann Wolfgang von, *Goethe's Theory of Colours* transl. Charles Lock Eastlake (London 1840).

Goethe. UNESCO's Homage on the occasion of the two hundredth anniversary of his birth (Paris 1949).

Goetz, Walter, 'Die Entstehung des Historismus' in: idem, *Historiker in meiner Zeit. Gesammelte Aufsätze* (Cologne 1957) 351-360.

Goetz, Walter, 'Friedrich Meinecke. Leben und Persönlichkeit' in: idem, *Historiker in meiner Zeit. Gesammelte Aufsätze* (Cologne 1957) 329-350.

Goldman, Harvey, *Max Weber and Thomas Mann. Calling and the Shaping of the Self* (Berkeley, Los Angeles and London 1988).

Gossman, Lionel, *Basel in the Age of Burckhardt. A Study in Unseasonable Ideas* (Chicago 2000).

Gregersen, Niels Henrik, 'Three Varieties of Panentheism' in: Philip Clayton and Arthur Peacocke ed., *In Whom We Live and Move and Have Our Being. Panentheistic Reflections on God's Presence in a Scientific World* (Cambridge 2004) 19-35.

Groeneveld, Vasco, 'Friedrich Meinecke en de Goethetijd', *Theoretische Geschiedenis* 22 (1995) 254-278.

Habermas, Jürgen, *Zur Verfassung Europas – Ein Essay* (Berlin 2011).

Hanssen, Léon, *Huizinga en de troost van de geschiedenis. Verbeelding en rede* (Amsterdam 1996).

Hegel, Georg Wilhelm Friedrich, 'Grundlinien der Philosophie des Rechts' in: idem, *Hauptwerke in sechs Bänden* (Hamburg 1999) Volume 5.

Heraclitus, Fragments. *The Collected Wisdom of Heraclitus*, Transl. Brooks Haxton (Harmondsworth 2001).

Herder, Johann Gottfried, 'Vom Erkennen und Empfinden der menschlichen Seele' in: idem, *Sämtliche Werke* 33 Bd (Berlin 1877-1913) Volume 8.

Herder, Johann Gottfried, 'Vom Geist der Ebräischen Poesie: Eine Anleitung für die Liebhaber derselben und der ältesten Geschichte des menschlichen Geistes' (vol. 2) *Sämtliche Werke* 33 Bd (Berlin 1877-1913) Volume 12.

Herkless, J. L., 'Meinecke and the Ranke-Burckhardt Problem', *History and Theory* 9 (1970) 290-321.

Hermand, Jost, *Kultur im Wiederaufbau. Die Bundesrepublik Deutschland 1945-1965* (Munich 1986).

Herzfeld, Hans, 'Friedrich Meinecke. Zu seinem 90. Geburtstage', *Geschichte in Wissenschaft und Unterricht* 3 (1952) 577-591.

Herzfeld, Hans, 'Friedrich Meinecke. Der Historiker, der Politiker und der Mensch nach seinem Briefwechsel' in: idem, *Ausgewählte Aufsätze. Dargebracht als Festgabe zum siebzigsten Geburtstage von seinen Freunden und Schülern* (Berlin 1962) 26-48.

Herzfeld, Hans, 'Friedrich Meinecke: Der Geschichtsdenker' in: Richard Dietrich ed., *Historische Theorie und Geschichtsforschung der Gegenwart* (Berlin 1964) 99-115.

Herzfeld, Hans, 'Friedrich-Meinecke-Renaissance im Ausland' in: *Festschrift für Hermann Heimpel. Zum 70. Geburtstag am 19. September 1971. Erster Band*, ed. Max-Planck-Instituts für Geschichte (Göttingen 1971) 42-62.

Hessing, Jakob, 'Friedrich Meinecke: Naturbegriff und Goethebild. Zur Problematik der konservativen Goetherezeption in Deutschland', *Jahrbuch des Instituts für Deutsche Geschichte* 12 (1983) 317-351.

Heussi, Karl, *Die Krisis des Historismus* (Tübingen 1932).

Hilgers, Klaudia, 'Schicksal' in: Bernd Witte ed., *Goethe Handbuch in vier Bänden* (Stuttgart and Weimar) Volume 4/2, 941-943.

Hinde, John R., *Jacob Burckhardt and the Crisis of Modernity* (Montreal 2000).

Hinrichs, Carl, 'Der Historismus als ein Lebensproblem Friedrich Meineckes' in: idem, *Preußen als historisches Problem. Gesammelte Abhandlungen*, ed. Gerhard Oestreich (Berlin 1959) 360-397.

Hinrichs, Carl, *Ranke und die Geschichtstheologie der Goethezeit* (Göttingen, Frankfurt and Berlin 1954).

Hinrichs, Carl, *Preußentum und Pietismus. Der Pietismus in Brandenburg-Preußen als religiös-soziale Reformbewegung* (Göttingen 1971).

Hofer, Walther, *Friedrich Meinecke als geschichtlicher Denker. Untersuchungen über die Bedeutung der Weltanschauung für die historische Begriffsbildung* (Munich 1949).

Hofer, Walther, *Geschichtsschreibung und Weltanschauung. Betrachtungen zum Werk Friedrich Meineckes* (Munich 1950).

Hofer, Walther, 'Friedrich Meinecke als politischer Denker' in: idem, *Geschichte zwischen Philosophie und Politik. Studien zur Problematik des modernen Geschichtsdenkens* (Basel 1956) 69-97.

Hofer, Walther, 'Einleitung des Herausgebers' in: Friedrich Meinecke, *Die Idee der Staatsräson in der neueren Geschichte* (Munich 1976) ix-xxxii.

Hofer, Walther, "'Keine Fahnenflucht vor der Schlacht". Friedrich Meineckes Warnungen vor dem Nationalsozialismus' in: Catherine Bosshart-Pfluger ed., *Nation und Nationalismus in Europa. Kulturelle Konstruktion von Identitäten. Festschrift für Urs Altermatt* (Frauenfeld, Stuttgart and Vienna 2002) 323-345.

Hölderlin, Friedrich, *Hyperion* Transl. Ben Schomakers (Amsterdam 1987).

Hölderlin, Friedrich, 'Hyperion oder der Eremit in Griechenland' in: idem, Friedrich Hölderlin *Sämtliche Werke und Briefe in drei Bänden*, ed. Jochen Schmidt (Frankfurt am Main 1994) Volume II, 9-175.

Hollander, J. C. den, 'Conservatisme en historisme', *Bijdragen en Mededelingen betreffende de Geschiedenis der Nederlanden* 102 (1987) 380-402.

Horstmann, Rolf-Peter, 'The early philosophy of Fichte and Schelling' in: Karl Ameriks ed., *The Cambridge Companion to German Idealism* (Cambridge 2000) 117-140.

Huber, Peter, 'Systole/Diastole' in: Bernd Witte ed., *Goethe Handbuch in vier Bänden* (Stuttgart and Weimar) Volume 4/2, 1034-1035.

Hübsch, Stefan, *Philosophie und Gewissen. Beiträge zur Rehabilitierung des philosophischen Gewissensbegriffs* (Göttingen 1995).

Hughes, H. Stuart, *Consciousness and Society. The reorientation of European social thought 1890-1930* (Brighton 1979).

Huizinga, Johan, 'Het aesthetische bestanddeel van geschiedkundige voorstellingen' in: idem, *Verzamelde Werken* VII (Haarlem 1950) 3-28.

Humboldt, Wilhelm von, 'Ideen zu einem Versuch, die Gränzen der Wirksamkeit des Staats zu bestimmen' in: idem, *Werke in fünf Bänden*, ed. Andreas Flitner and Klaus Giel (Stuttgart 1960-1981) Volume I, 56-223.

Humboldt, Wilhelm von, 'Über die Aufgabe des Geschichtsschreibers' in: idem, Wilhelm von Humboldt *Werke in fünf Bänden*, ed. Andreas Flitner and Klaus Giel (Stuttgart 1960) Volume I, 585-609.

Humboldt, Wilhelm von, 'On the Historian's Task', *History and Theory* 6 (1967) 57-71.

Humboldt, Wilhelm von, *The Limits of State Action* trans. J.W. Burrow (Cambridge 1969).

Iggers, Georg G., *The German Conception of History. The National Tradition of Historical Thought from Herder to the Present* (Middletown 1968).

Iggers, Georg G., 'The Decline of the Classical National Tradition of German Historiography', *History and Theory* 6 (1967) 382-412.

Iggers, Georg G., 'Review', *History and Theory* 6 (1967) 112-117.

Iggers, Georg G., 'Review: Von der Aufklärung zum Historismus. Zum Strukturwandel des historischen Denkens', *History and Theory* 26 (1987) 114-121.

Iggers, Georg G., 'Historicism: The History and Meaning of the Term', *Journal of the History of Ideas* 56 (1995) 129-152.

Iggers, Georg G., 'Historismus – Geschichte und Bedeutung eines Begriffs. Eine kritische Übersicht der neuesten Literatur' in: Gunter Scholtz ed., *Historismus am Ende des 20. Jahrhunderts. Eine internationale Diskussion* (Berlin 1997) 102-126.

Iggers, Georg G., 'Book reviews', *Central European History* 41 (2008) 151-155.

Irmscher, Hans Dietrich, 'Aspekte der Geschichtsphilosophie Johann Gottfried Herders' in: Marion Heinz ed., *Herder und die Philosophie des deutschen Idealismus* (Amsterdam and Atlanta 1997) 5-47.

Jaeger, Friedrich and Jörn Rüsen, *Geschichte des Historismus. Eine Einführung* (Munich 1992).

Jaeger, Werner, *Paideia. Die Formung des griechischen Menschen* (Berlin and Leipzig 1936) Volume I.

Janz, Rolf-Peter, 'Über die ästhetische Erziehung des Menschen in einer Reihe von Briefen' in: Helmut Koopman ed., *Schiller Handbuch* (Stuttgart 2011) 649-666.

Jensen, Christian Russell, '"Unity in Antinomy". The Hermeneutics of Friedrich Meinecke' (Chicago 1974). Unpublished dissertation.

Jespers, F. P. M., 'Inleiding' in: Gottfried Wilhelm Leibniz, *Monadologie of De beginselen van de wijsbegeerte* transl. F. P. M. Jespers (Kampen 1991) 9-53.

Kämmerer, Wolfgang, *Friedrich Meinecke und das Problem des Historismus* (Frankfurt am Main 2014).

Kampmann, W., *Deutsche und Jude. Studien zur Geschichte des deutschen Judentums* (Heidelberg 1963).

Kant, Immanuel, *Fundering voor de metafysica van de zeden* transl. Thomas Mertens (Amsterdam 1997).

Kelly, Donald R., 'Intellectual History and Cultural History: the Inside and the Outside', *History of the Human Sciences* 15 (2002) 1-19.

Kemper, Dirk, „Ineffabile". *Goethe und die Individualitätsproblematik der Moderne* (Munich 2004).

Kern, Stephen, *The Culture of Time and Space, 1880-1918* (Cambridge Mass. and London 2003, First 1983).

Kessel, Eberhard, 'Einleitung des Herausgebers' in: Friedrich Meinecke, *Zur Theorie und Philosophie der Geschichte* Eberhard Kessel ed., (Stuttgart 1965) vii-xxxiv.

Kessel, Eberhard, 'Friedrich Meinecke in eigener Sicht' in: Michael Erbe ed., *Friedrich Meinecke heute. Bericht über ein Gedenk-Colloquium zu seinem 25. Todestag am 5. und 6. April 1979* (Berlin 1981) 186-195.

Killen, Andreas, *Berlin Electropolis. Shock, Nerves, and German Modernity* (Berkely, Los Angeles and London 2006).

Klapwijk, Jacob, *Tussen historisme en relativisme. Een studie over de dynamiek van het historisme en de wijsgerige ontwikkelingsgang van Ernst Troeltsch* (Assen 1970).

Knudsen, Jonathan B., *Justus Möser and the German Enlightenment* (Cambridge 1986).

Kogon, Eugen, 'Beginn der Geschichtsrevision', *Frankfurter Hefte. Zeitschrift für Kultur und Politik* 8 (1946) 776-779.

Koselleck, Reinhart, *Kritik und Krise. Eine Studie zur Pathogenese der bürgerlichen Welt* (Freiburg and Munich 1976, first 1959).

Koselleck, Reinhart, 'Einleitung – Zur anthropologischen und semantischen Struktur der Bildung' in: idem ed., *Bildungsbürgertum im 19. Jahrhundert. Teil II Bildungsgüter und Bildungswissen* (Stuttgart 1989) 11-46.

Koselleck, Reinhart, 'Goethes unzeitgemäße Geschichte', *Goethe Jahrbuch* 110 (1993) 27-39.

Kossmann, E. H., 'Een kennismaking met Ranke (1795-1886)' in: idem, *Vergankelijkheid en continuïteit. Opstellen over geschiedenis* (Amsterdam 1995) 160-191.

Kossmann, E. H., 'Friedrich Meinecke (1862-1954)' in: idem, *Vergankelijkheid en continuïteit. Opstellen over geschiedenis* (Amsterdam 1995) 209-224.

Kranz, M, 'Schicksal' in: Joachim Ritter ed., *Historisches Wörterbuch der Philosophie* 13 vols. (1971-2007) Volume 8, 1275-1289.

Kraus, Hans-Christof, 'Friedrich Meinecke als Korrespondent. Zu den neuen Briefeditionen', *Historische Zeitschrift* 298 (2014) 89-100.

Kraus, Hans-Christof, 'Ute von Lüpke, Zäsuren – Katastrophen – Neuanfänge. Friedrich Meinecke und die Umbrüche der deutschen Geschichte im 20. Jahrhundert (Review)', *Historische Zeitschrift* 306 (2018) 936-937.

Kreiling, Frederick C., 'Friedrich Meinecke and the Problems of Historicism' (New York 1959, Photomech. Repr. 1967). Unpublished dissertation.

Krol, Reinbert, *Het geweten van Duitsland. Friedrich Meinecke als pleitbezorger van het Duitse historisme* (Groningen 2013).

Krol, Reinbert A., 'Friedrich Meinecke: Panentheism and the Crisis of Historicism', *Journal of the Philosophy of History* 4 (2010) 195-209.

Krol, Reinbert, 'De "scheppende spiegel" van Friedrich Meinecke. Over geschiedschrijving en de taak van de historicus', *Leidschrift* 25 (2010) 73-94.

Krol, R.A., 'Hyper Duitsland. Tijdservaring en nervositeit in het fin de siècle', *Groniek: Historisch Tijdschrift* 219 (voorjaar 2019) 143-156.

Kroll, Frank-Lothar, 'Kultur, Bildung und Wissenschaft im geteilten Deutschland 1949-1989', *Archiv für Kulturgeschichte* 85 (2003) 119-142.

Krul, W. E., *Historicus tegen de tijd. Opstellen over leven en werk van J. Huizinga* (Groningen 1990).

Krul, Wessel, 'Tweemaal fin-de-siècle: van de negentiende naar de twintigste eeuw', *Groniek: Historisch Tijdschrift* 143 (1998) 155-167.

Krul, Wessel, 'Edmund Burke en de oorsprongen van het conservatisme', *Groniek* 164 (2004) 337-348.

Labrie, Arnold, *'Bildung' en politiek, 1770-1830. De "Bildungsphilosophie" van Wilhelm von Humboldt bezien in haar politieke en sociale context* (Amsterdam 1986).

Ladd, Brian, 'Berlin' in: John Merriman and Jay Winter eds., *Europe 1789 to 1914 Encyclopedia of the Age of Industry and Empire* (Detroit etc 2006) 215-220.

Landgrebe, Ludwig, 'Die Geschichte in Goethes Weltbild' in: Gerhard Haselbach and Günter Hartmann ed., *Beiträge zur Einheit von Bildung und Sprache im geistigen Sein. Festschrift zum 80. Geburtstag von Ernst Otto* (Berlin 1957) 371-384.

Langmuir, Gavin I., *History, Religion and Antisemitism* (Los Angeles 1990).

Larmore, Charles, 'Hölderlin and Novalis' in Karl Ameriks ed., *The Cambridge Companion to German Idealism* (Cambridge 2000)141-160.

Laube, Reinhard, *Karl Mannheim und die Krise des Historismus. Historismus als wissenssoziologischer Perspektivismus* (Göttingen 2004).

Leibniz, Gottfried Wilhelm, *The Monadology and other philosophical writings*, transl Robert Latta (London 1948).

Leibniz, Gottfried Wilhelm, *Monadologie of De beginselen van de wijsbegeerte* transl. F. P. M. Jespers (Kampen 1991).

Leibniz, Gottfried Wilhelm, *Monadologie und andere metaphysische Schriften* transl. Ulrich Johannes Schneider (Hamburg 2002).

Leibniz, Gottfried Wilhelm, 'Von der Glückseligkeit' in: idem, *Kleine Schriften zur Metaphysik*, ed. and transl. Hans Heinz Holz (Frankfurt am Main 1965) 390-401.

Loader, Colin T., 'German Historicism and Its Crisis', *The Journal of Modern History* vol.48 No. 3 (September 1976) 85-119.

Lovejoy, Arthur O., *Essays in the History of Ideas* (Baltimore 1948).

Lüpke, Ute von, *Zäsuren – Katastrophen – Neuanfänge. Friedrich Meinecke und die Umbrüche der deutschen Geschichte im 20. Jahrhundert* (Hamburg 2015).

Machiavelli, Niccolò, *The Prince* trans. Luigi Ricci (London 1921).

Mahnke, Dietrich, *Leibnizens Synthese von Universalmathematik und Individualmetaphysik* (Stuttgart and Bad Cannstatt 1964, first 1925).

Maillard, Christine, 'Johann Wolfgang von Goethe' in: W.J. Hanegraaff ed., *Dictionary of Gnosis and Western Esotericism* (Leiden 2005) 432-434.

Mandelkow, Karl Robert, *Goethe in Deutschland. Rezeptionsgeschichte eines Klassikers Band II 1919-1982* (Munich 1989).

Mandelkow, Karl Robert, 'Natur und Geschichte bei Goethe im Spiegel seiner wissenschaftlichen und kulturtheoretischen Rezeption' in: Peter Matussek ed., *Goethe und die Verzeitlichung der Natur* (Munich 1998) 233-258.

Mann, Thomas, *Von deutscher Republik* (Berlin 1923).

Mann, Thomas, *Adel des Geistes. Sechzehn Versuche zum Problem der Humanität* (Stockholm 1945).

Mann, Thomas, 'Lotte in Weimar' in: idem, *Gesammelte Werke* (Berlin 1955) Volume VII.

Mann, Thomas, 'Ansprache im Goethejahr. Gehalten am 25. Juli 1949 in der Paulskirche zu Frankfurt am Main' in: N.N. *Zum Thema: Goethe* (Frankfurt am Main 1982) 63-82.

Mannheim, Karl, *Essays on the sociology of knowledge* ed. Paul Kecskemeti (New York 1952).

Masur, Georg, 'Ethics in a World of Power' (review), *Historische Zeitschrift* 188 (1959) 608-611.

Matussek, Peter, *Goethe zur Einführung* (Hamburg 1998).

Maubach, Franka, '"Wie es dazu kommen konnte". 1933 als Fluchtpunkt deutsch-deutscher Ursachensuche im frühen Kalten Krieg' in: Franka Maubach and Christina Morina ed., *Das 20. Jahrhundert erzählen. Zeiterfahrung und Zeiterforschung im geteilten Deutschland* (Göttingen 2016) 143-189.

Mayer, Hans ed., *Goethe im zwanzigsten Jahrhundert. Spiegelungen und Deutungen* (Frankfurt am Main 1987).

Meinecke, Friedrich, *Das Leben des Generalfeldmarschalls Hermann von Boyen* (Stuttgart I 1896, II 1899).

Meinecke, Friedrich, *Radowitz und die deutsche Revolution* (Berlin 1913).

Meinecke, Friedrich, *Aphorismen und Skizzen zur Geschichte* (Leipzig 1941).

Meinecke, Friedrich, *Schaffender Spiegel. Studien zur deutschen Geschichts-schreibung und Geschichtsauffassung* (Stuttgart 1948).

Meinecke, Friedrich, *Ausgewählter Briefwechsel* (Stuttgart 1962).

Meinecke, Friedrich, *The German Catastrophe. Reflections and Recollections* transl. Sidney Fay (Boston 1963, first 1950).

Meinecke, Friedrich, *Machiavellism. The Doctrine of Raison d'État and its Place in Modern History*, transl. Douglas Scott (New Haven 1962).

Meinecke, Friedrich, *Die Idee der Staatsräson in der neueren Geschichte* (Munich 1963, first 1924).

Meinecke, Friedrich, *Das Zeitalter der deutschen Erhebung* (1795-1815) (Göttingen 1963).

Meinecke, Friedrich, *Die Entstehung des Historismus* (Munich 1965, first 1936).

Meinecke, Friedrich, *Weltbürgertum und Nationalstaat. Studien zur Genesis des deutschen Nationalstaats* (Darmstadt 1969, first 1908).

Meinecke, Friedrich, *Autobiographische Schriften* (Stuttgart 1969) 105.

Meinecke, Friedrich, *Cosmopolitanism and the National State* transl. Robert Kimber (Princeton 1970).

Meinecke, Friedrich, *Historism. The Rise of a New Historical Outlook*, transl. J.E. Anderson (London 1972).

Meinecke, Friedrich, *Geschichte des deutsch-englischen Bündnisproblems 1890-1901* (Munich and Vienna 1972; first 1927).

Meinecke, Friedrich, *Akademischer Lehrer und emigrierte Schüler. Briefe und Aufzeich-nungen 1910-1977*, ed. Gerhard A. Ritter (Munich 2006).

Meinecke, Friedrich, *Neue Briefe und Dokumente* (Munich 2012).

Meinecke, Friedrich, 'Zur Beurteilung Rankes', *Historische Zeitschrift* 111 (1913) 582-600.

Meinecke, Friedrich, 'Pietism as a Factor in the Rise of German Nationalism', *Historische Zeitschrift* 151 (1935) 116-117.

Meinecke, Friedrich, 'Values and Causalities in History' in: Fritz Stern ed., *The Varieties of History. From Voltaire to the present* (New York 1956) 268-288.

Meinecke, Friedrich, 'Geschichte und Politik' in: idem, *Politische Schriften und Reden* (Darmstadt 1958) 494-497.

Meinecke, Friedrich, 'Nationalismus und nationale Idee' in: idem, *Politische Schriften und Reden* (Darmstadt 1958) 83-95.

Meinecke, Friedrich, 'Republik, Bürgertum und Jugend', in: idem, *Politische Schriften und Reden* (Darmstadt 1958) 369-383.

Meinecke, Friedrich, 'Staatsräson', in: idem: *Politische Schriften und Reden* (Darmstadt 1958) 467-470.

Meinecke, Friedrich, 'Von Schleicher zu Hitler. Volksgemeinschaft – nicht Volks-zerreißung' in: idem, *Politische Schriften und Reden* (Darmstadt 1958) 479-482.

Meinecke, Friedrich, 'Zur Selbstbesinnung', in: idem, *Politische Schriften und Reden* (Darmstadt 1958) 484-486.

Meinecke, Friedrich, 'Aphorismen' in: idem, *Zur Theorie und Philosophie der Geschichte* (Stuttgart 1965) 215-263.

Meinecke, Friedrich, 'Deutung eines Rankewortes' in: idem, *Zur Theorie und Philosophie der Geschichte* (Stuttgart 1965) 117-139.

Meinecke, Friedrich, 'Ernst Troeltsch und das Problem des Historismus' in: idem, *Zur Theorie und Philosophie der Geschichte* (Stuttgart 1965) 367-378.

Meinecke, Friedrich, 'Gedanken über Welt- und Universalgeschichte', in: idem, *Zur Theorie und Philosophie der Geschichte* (Stuttgart 1965) 140-149.

Meinecke, Friedrich, 'Geschichte und Gegenwart' in: idem, *Zur Theorie und Philosophie der Geschichte* (Stuttgart 1965) 90-101.

Meinecke, Friedrich, 'Irrwege in unserer Geschichte?' in: idem, *Zur Theorie und Philosophie der Geschichte* (Stuttgart 1965) 205-211.

Meinecke, Friedrich, 'Kausalitäten und Werte in der Geschichte' in: idem, *Zur Theorie und Philosophie der Geschichte* (Stuttgart 1965) 61-89.

Meinecke, Friedrich, 'Klassizismus, Romantizismus und historisches Denken im 18. Jahrhundert' in: idem, *Zur Theorie und Philosophie der Geschichte* (Stuttgart 1965) 264-278.

Meinecke, Friedrich, 'Persönlichkeit und geschichtliche Welt' in: idem, *Zur Theorie und Philosophie der Geschichte* (Stuttgart 1965) 30- 60.

Meinecke, Friedrich, 'Schiller und der Individualitätsgedanke. Eine Studie zur Entstehungsgeschichte des Historismus' in: idem, *Zur Theorie und Philosophie der Geschichte* (Stuttgart 1965) 285-322.

Meinecke, Friedrich, 'Schillers Spaziergang' in: idem, *Zur Theorie und Philosophie der Geschichte* (Stuttgart 1965) 323-340.

Meinecke, Friedrich, 'Über Spenglers Geschichtsbetrachtung' in: idem, *Zur Theorie und Philosophie der Geschichte* (Stuttgart 1965) 181-195.

Meinecke, Friedrich, 'Von der Krisis des Historismus', in: idem, *Zur Theorie und Philosophie der Geschichte* (Stuttgart 1965) 196-204.

Meinecke, Friedrich, 'Wilhelm Dilthey' in: idem, *Zur Theorie und Philosophie der Geschichte* (Stuttgart 1965) 358-363.

Meinecke, Friedrich, 'Willensfreiheit und Geschichtswissenschaft' in: idem, *Zur Theorie und Philosophie der Geschichte* (Stuttgart 1965) 3-29.

Meinecke, Friedrich, 'Zur Entstehungsgeschichte des Historismus und des Schleiermacherschen Individualitätsgedankens' in: idem, *Zur Theorie und Philosophie der Geschichte* (Stuttgart 1965) 341-357.

Meinecke, Friedrich, 'Antrittsrede in der Preußischen Akademie der Wissenschaften' in: idem, *Zur Geschichte der Geschichtsschreibung* (Munich 1968) 1-4.

Meinecke, Friedrich, 'Carl Neumann über Jacob Burckhardt' in: idem, *Zur Geschichte der Geschichtsschreibung* (Munich 1968) 88-92.

Meinecke, Friedrich, 'Jacob Burckhardt, die deutsche Geschichtsschreibung und der nationale Staat', in: idem, *Zur Geschichte der Geschichtsschreibung* (Munich 1968) 83-87.

Meinecke, Friedrich, 'Johann Gustav Droysen, sein Briefwechsel und seine Geschichtsschreibung' in: idem, *Zur Geschichte der Geschichtsschreibung* (Munich 1968) 125-167.

Meinecke, Friedrich, 'Ranke und Burckhardt' in: idem, *Zur Geschichte der Geschichtsschreibung* (Munich 1968) 93-121.

Meinecke, Friedrich, 'Rankes "Große Mächte"' in: idem, *Zur Geschichte der Geschichtsschreibung* (Munich 1968) 66-71.

Meinecke, Friedrich, 'Rankes "Politisches Gespräch"' in: idem, *Zur Geschichte der Geschichtsschreibung* (Munich 1968) 72-82.

Meinecke, Friedrich, 'Zur Beurteilung Rankes' in: idem, *Zur Geschichte der Geschichtsschreibung* (Munich 1968) 50-65.

Meinecke, Friedrich, 'Die deutsche Katastrophe. Betrachtungen und Erinnerungen' in: idem, *Autobiographische Schriften* (Stuttgart 1969) 323-445.

Meinecke, Friedrich, 'Erlebtes 1862-1901' in: idem, *Autobiographische Schriften* (Stuttgart 1969) 1-134.

Meinecke, Friedrich, 'Lebenströster. Betrachtungen über zwei Goethesche Gedichte' in: idem, *Autobiographische Schriften* (Stuttgart 1969) 492-508.

Meinecke, Friedrich, 'Straßburg – Freiburg – Berlin. Erinnerungen' in: idem, *Autobiographische Schriften* (Stuttgart 1969) 135-320.

Meinecke, Friedrich, 'Die deutschen Erhebungen von 1813, 1848, 1870 und 1914' in: idem, *Brandenburg, Preußen, Deutschland. Kleine Schriften zur Geschichte und Politik* (Stuttgart 1979) 509-531.

Meinecke, Friedrich, 'Staatskunst und Leidenschaften' in: idem, *Brandenburg, Preußen, Deutschland. Kleine Schriften zur Geschichte und Politik* (Stuttgart 1979) 578-585.

Meinecke, Friedrich, 'Wilhelm von Humboldt und der deutsche Staat' in: idem, *Brandenburg, Preußen Deutschland. Kleine Schriften zur Geschichte und Politik* (Stuttgart 1979) 279-296.

Meineke, Stefan, *Friedrich Meinecke. Persönlichkeit und politisches Denken bis zum Ende des Ersten Weltkrieges* (Berlin and New York 1995).

Meineke, Stefan, 'Parteien und Parlamentarismus im Urteil von Friedrich Meinecke' in: Gisela Bock and Daniel Schönpflug ed., *Friedrich Meinecke in seiner Zeit. Studien zu Leben und Werk* (Stuttgart 2006) 51-93.

Menke-Glückert, E., *Goethe als Geschichtsphilosoph und die geschichts-philosophische Bewegung seiner Zeit* (Leipzig 1907).

Mensch, Jennifer, 'Intuition and Nature in Kant and Goethe', *European Journal of Philosophy* (2009) 1-23.

Möller, Horst, *Exodus der Kultur. Schriftsteller, Wissenschaftler und Künstler in der Emigration nach 1933* (Munich 1984).

Mooij, J. J. A., *Het morele domein. Over meervoudigheid in de moraal* (Amsterdam 2012).

Muhlack, Ulrich, *Geschichtswissenschaft im Humanismus und in der Aufklärung. Die Vorgeschichte des Historismus* (Munich 1991).

Muhlack, Ulrich, 'Gibt es ein "Zeitalter" des Historismus? Zur Tauglichkeit eines wissenschaftsgeschichtlichen Epochenbegriffs' in: Otto Gerhard Oexle and Jörn Rüsen eds., *Historismus in den Kulturwissenschaften. Geschichtskonzepte, historische Einschätzungen, Grundlagenprobleme* (Cologne, Weimar and Vienna 1996) 201-219.

Müller, Thomas ed., *Philosophie der Zeit. Neue analytische Ansätze* (Frankfurt am Main 2007).

Mullins, R.T., 'The Difficulty with Demarcating Panentheism', *Sophia* vol 55 issue 3 (September 2016) 325-346. DOI 10.1007/s11841-015-0497-6

Nassar, Dalia, 'From a Philosophy of Self to a Philosophy of Nature: Goethe and the Development of Schelling's Naturphilosophie', *Archiv für Geschichte der Philosophie* vol 92 issue 3 (2011) 302-321.

Nicholls, Angus, *Goethe's Concept of the Daemonic: After the Ancients* (New York 2006).

Nietzsche, Friedrich, *Over nut en nadeel van de geschiedenis voor het leven. Tweede traktaat tegen de keer* Transl. Vertalerscollectief Historische Uitgeverij (Groningen 1983).

Nipperdey, Thomas, *Deutsche Geschichte 1800-1866. Bürgerwelt und starker Staat* (Munich 1984).

Nipperdey, Thomas, 'Historismus und Historismuskritik heute' in: idem, *Gesellschaft, Kultur, Theorie. Gesammelte Aufsätze zur neueren Geschichte* (Göttingen 1976) 59-73.

Nipperdey, Thomas, 'Probleme der Modernisierung in Deutschland' in: idem, *Kann Geschichte objektiv sein? Historische Essays* (Munich 2013) 84-104.

Nisbet, Hugh Barr, *Goethe and the Scientific Tradition* (London 1972).

Nisbet, Hugh Barr, 'Goethes und Herders Geschichtsdenken', *Goethe Jahrbuch* 110 (1993) 115-133.

Nordau, Max, *Entartung*, 2 vols (Berlin 1892-1893).

Novalis, Werke, *Tagebücher und Briefe* 3 vols. (Munich 1978).

Obst, Helmut, *August Hermann Francke und sein Werk* (Halle 2013).

Oexle, Otto Gerhard, '"Historismus". Überlegungen zur Geschichte des Phänomens und des Begriffs' (1986) in: idem, *Geschichtswissenschaft im Zeichen des Historismus. Studien zu Problemgeschichte der Moderne* (Göttingen 1996) 41-72.

Oexle, Otto Gerhard, 'Meineckes Historismus. Über Kontext und Folgen einer Definition' in: idem, *Geschichtswissenschaft im Zeichen des Historismus. Studien zu Problemgeschichte der Moderne* (Göttingen 1996) 95-136.

Oexle, Otto Gerhard, 'Die Geschichtswissenschaft im Zeichen des Historismus. Bemerkungen zum Standort der Geschichtsforschung' in: idem, *Geschichtswissenschaft im Zeichen des Historismus: Studien zu Problemgeschichten der Moderne* (Göttingen 1996) 17-40.

Oexle, Otto Gerhard, 'Krise des Historismus – Krise der Wirklichkeit. Eine Problemgeschichte der Moderne' in: idem ed., *Krise des Historismus – Krise der Wirklichkeit. Wissenschaft, Kunst und Literatur 1880-1932* (Göttingen 2007) 11-116.

Offringa, C., 'Friedrich Meinecke herdacht', *Theoretische Geschiedenis* 14 (1987) 220-223.

Otterspeer, Willem, *Huizinga voor de afgrond. Het incident-Von Leers aan de Leidse universiteit in 1933* (Utrecht 1984).

Pachter, Henry, 'Masters of Cultural History II: Friedrich Meinecke and the Tragedy of German Liberalism', *Salmagundi* No. 43 (Winter 1979) 12-42.

Paddock, Troy R. E., 'Rethinking Friedrich Meinecke's historicism', *Rethinking History* 10 (2006) 95-108.

Paul, Herman, *Het moeras van de geschiedenis. Nederlandse debatten over historisme* (Amsterdam 2012).

Paul, Herman, 'A Collapse of Trust. Reconceptualizing the Crisis of Historicism', *Journal of the Philosophy of History* 2 (2008) 63-82.

Paul, Herman, 'Historisme op een procrustusbed', *Tijdschrift voor Geschiedenis* 126.1 (2013) 134-136.

Peters, Rik, *History as Thought and Action. The Philosophies of Croce, Gentile, de Ruggiero, and Collingwood* (Exeter and La Vergne 2013).

Peters, Rik, 'Italian Legacies', *History and Theory* 49 (2010) 115-129.

Pinkard, Terry, *German Philosophy 1760-1860. The Legacy of Idealism* (Cambridge 2002).

Pinson, Koppel S., *Pietism as a Factor in the Rise of German Nationalism* (New York 1934).

Pois, Robert A., *Friedrich Meinecke and German Politics in the Twentieth Century* (Berkeley and Los Angeles 1972).

Popkin, Jeremy D., *History, Historians and Autobiography* (Chicago and London 2005).

Popkin, Jeremy D., *From Herodotus to H-Net. The Story of Historiography* (New York and Oxford 2016).

Pree, Carla du, 'Intellectueel in tijden van crisis. Friedrich Meinecke en het publieke debat in Duitsland van 1914 tot 1954', *Streven: cultureel maatschappelijk maandblad* 79 (2012) 226-236.

Prüfer, Thomas, *Die Bildung der Geschichte. Friedrich Schiller und die Anfänge der modernen Geschichtswissenschaft* (Cologne, Weimar and Vienna 2002).

Pyper, Jens Fabian, 'Meinecke, Croce, and the Individual: The Moral Foundations of the Study of History, 1918-1946' (Florence 2008) unpublished dissertation.

Radkau, Joachim, *Das Zeitalter der Nervosität: Deutschland zwischen Bismarck und Hitler* (Munich and Vienna 1998).

Radkau, Joachim, 'Die wilhelminische Ära als nervöses Zeitalter, oder: Die Nerven als Netz zwischen Tempo- und Körpergeschichte', *Geschichte und Gesellschaft* 20 No. 2 (April/June 1994) 211-241.

Ranke, Leopold von, 'Anhang über Machiavelli' in: idem: *Sämtliche Werke* 54 vols. (Leipzig 1874) Volume 34, 151-174.

Ranke, Leopold von, 'Über die Verwandtschaft und den Unterschied der Historie und der Politik' in: idem, *Sämtliche Werke* 54 vols. (Leipzig 1877) Volume 24, 280-293.

Ranke, Leopold von, *Über die Epochen der neueren Geschichte* (Darmstadt 1959).

Ranke, Leopold von, *The Theory and Practice of History*, Georg G. Iggers ed. New translations by Wilma A. Iggers (London and New York 2011).

Raulet, Gérard, 'Strategien des Historismus' in: Wolfgang Bialas and Gérard Raulet ed., *Die Historismusdebatte in der Weimarer Republik* (Frankfurt am Main 1996) 7-38.

Reed, T.J., *Goethe* (Oxford and New York 1984).

Reed, T.J., *Light in Germany. Scenes from an Unknown Enlightenment* (Chicago and London 2015).

Reich-Ranicki, Marcel, *Goethe noch einmal. Reden und Anmerkungen* (Munich and Stuttgart 2002).

Reill, Peter Hanns, 'Aufklärung und Historismus: Bruch oder Kontinuität?' in: Otto Gerhard Oexle and Jörn Rüsen eds., *Historismus in den Kulturwissenschaften. Geschichtskonzepte, historische Einschätzungen, Grundlagenprobleme* (Cologne, Weimar and Vienna 1996) 45-68.

Reill, Peter Hanns, 'Schiller, Herder, and History' in: Michael Hofman, Jörn Rüsen and Mirjam Springer ed., *Schiller und die Geschichte* (Munich 2006) 68-78.

Reill, Peter Hanns, *The German Enlightenment and the Rise of Historicism* (Berkeley 1975).

Richards, Robert J., *The Romantic Conception of Life. Science and Philosophy in the Age of Goethe* (Chicago and London 2002).

Ritter, Gerhard, 'Die Idee der Staatsräson', *Neue Jahrbücher für Wissenschaft und Jugendbildung* 1 (1925) 101-114.

Roberts, David D., *Benedetto Croce and the Uses of Historicism* (Berkeley, Los Angeles and London 1987).

Rossi, Pietro, '"Historismus" und "Storicismo": zwei Denktraditionen' in: Arnold Esch and Jens Petersen, ed., *Geschichte und Geschichtswissenschaft in der Kultur Italiens und Deutschlands. Wissenschaftliches Kolloquium zum hundertjährigen Bestehen des Deutschen Historischen Instituts in Rom (24. – 25. Mai 1988)* (Tübingen 1989) 39-69.

Rothfels, Hans, *Friedrich Meinecke – Ein Rückblick auf sein wissenschaftliches Lebenswerk. Trauerrede, gehalten in Berlin am 27. Februar 1954* (Berlin 1954).

Ruddies, Hartmut, '"Geschichte durch Geschichte überwinden". Historismuskonzept und Gegenwartsdeutung bei Ernst Troeltsch' in: Wolfgang Bialas and Gérard Raulet ed., *Die Historismusdebatte in der Weimarer Republik* (Frankfurt am Main 1996) 198-217.

Rumpf, Max, 'Friedrich Meinecke, Die Idee der Staatsräson in der neueren Geschichte', *Archiv des öffentlichen Rechts* 48 (1925) 340-349.

Rüsen, Jörn, 'Friedrich Meineckes "Entstehung des Historismus". Eine kritische Betrachtung' in: Michael Erbe ed., *Friedrich Meinecke heute. Bericht über ein Gedenk-Colloquium zu seinem 25. Todestag am 5. und 6. April 1979* (Berlin 1981) 76-100.

Rüsen, Jörn, 'Historismus als Erkenntnisprinzip und Wissensform – einige Gesichtspunkte', in: idem, *Konfigurationen des Historismus. Studien zur deutschen Wissenschaftskultur* (Frankfurt am Main 1993) 17-28.

Rüsen, Jörn, 'Historismus als Wissenschaftsparadigma. Leistung und Grenzen eines strukturgeschichtlichen Ansatzes der Historiographiegeschichte' in: Otto Gerhard Oexle and Jörn Rüsen eds., *Historismus in den Kulturwissenschaften. Geschichtskonzepte, historische Einschätzungen, Grundlagenprobleme* (Cologne, Weimar and Vienna 1996) 119-137.

Saler, Michael, 'Introduction', in: idem ed., *The Fin-De-Siècle World* (London and New York 2015) 1-8.

Sandkühler, Hans Jörg ed., *Handbuch Deutscher Idealismus* (Stuttgart 2005).

Schicketanz, Peter, *Der Pietismus von 1675 bis 1800* (Leipzig 2001).

Schieder, Theodor, *Frederick the Great*, ed. and transl. Sabina Berkeley and H.M. Scott (London and New York 2000).

Schiller, Friedrich, 'Letters Upon the Aesthetic Education of Man' in: Charles W. Eliot ed., *Literary and Philosophical Essays. French, German, And Italian* Volume 32 (New York 1910) 221-313.

Schiller, Friedrich and Johann Wolfgang von Goethe, 'Xenien: 'Das Deutsche Reich'' in: Friedrich Schiller, *Sämtliche Werke in fünf Bänden* (Munich 2004) Volume 1, 267.

Schiller, Friedrich, 'Die Götter Griechenlandes': Friedrich Schiller, *Sämtliche Werke in fünf Bänden* (Munich 2004) Volume 1, 169-173.

Schiller, Friedrich, 'Über die ästhetische Erziehung des Menschen in einer Reihe von Briefen' in: Friedrich Schiller, *Sämtliche Werke in fünf Bänden* (Munich 2004) Volume 5, 570-669.

Schmidt, Alfred, 'Natur' in: Bernd Witte ed., *Goethe Handbuch in vier Bänden* (Stuttgart and Weimar) Volume 4/2, 755-776.

Schmidt, Georg, 'Friedrich Meineckes Kulturnation. Zum historischen Kontext nationaler Ideen in Weimar-Jena um 1800', *Historische Zeitschrift* 284 (2007) 597-621.

Schmidt, Gustav, *Deutscher Historismus und der Übergang zur parlamentarischen Demokratie. Untersuchungen zu den politischen Gedanken von Meinecke / Troeltsch / Weber* (Lübeck 1964).

Schmitt, Carl, 'Zu Friedrich Meineckes "Die Idee der Staatsräson"' in: idem, *Positionen und Begriffe im Kampf mit Weimar – Genf – Versailles 1923-1939* (Hamburg 1940) 45-52.

Schmitz, Hermann, 'Das Ganz-Andere. Goethe und das Ungeheure' in: Peter Matussek ed., *Goethe und die Verzeitlichung der Natur* (Munich 1998) 414-435.

Schnädelbach, Herbert, *Geschichtsphilosophie nach Hegel. Die Probleme des Historismus* (Freiburg and Munich 1974).

Schneider, Hans, '"Mit Kirchengeschichte, was hab' ich zu schaffen?" Goethes Begegnung mit Gottfried Arnolds Kirchen- und Ketzerhistorie' in: Hans-Georg Kemper and Hans Schneider ed., *Goethe und der Pietismus* (Tübingen 2001) 79-110.

Schönpflug, Daniel, 'Revolution und "Erhebung". Friedrich Meinecke über 1789 und die deutsche Geschichte' in: Gisela Bock and Daniel Schönpflug ed., *Friedrich Meinecke in seiner Zeit. Studien zu Leben und Werk* (Stuttgart 2006) 21-49.

Schulin, Ernst, 'Friedrich Meinecke', in: Hans-Ulrich Wehler ed., *Deutsche Historiker* 9 vols. (Göttingen 1971) Volume 1, 39-57.

Schulin, Ernst, 'Robert A. Pois, Friedrich Meinecke and German Politics in the Twentieth Century', *Historische Zeitschrift* 217 (1973) 454-456.

Schulin, Ernst, 'Das Problem der Individualität. Eine kritische Betrachtung des Historismus-Werkes von Friedrich Meinecke' in: idem, *Traditionskritik und Rekonstruktionsversuch. Studien zur Entwicklung von Geschichtswissenschaft und historischem Denken* (Göttingen 1979) 97-116.

Schulin, Ernst, 'Meineckes Leben und Werk. Versuch einer Gesamtcharakteristik' in: idem, *Traditionskritik und Rekonstruktionsversuch. Studien zur Entwicklung von Geschichtswissenschaft und historischem Denken* (Göttingen 1979) 117-132.

Schulin, Ernst, 'Friedrich Meinecke und seine Stellung in der deutschen Geschichtswissenschaft' in: Michael Erbe ed., *Friedrich Meinecke heute. Bericht über ein Gedenk-Colloquium zu seinem 25. Todestag am 5. und 6. April 1979* (Berlin 1981) 25-49.

Schulin, Ernst, 'Aufklärung und Geschichte', *Storia della storiografia* 12 (1987) 108-113.

Schulin, Ernst, 'Neue Diskussionen über Historismus', *Storia della storiografia* 33 (1998) 109-118.

Schulin, Ernst, 'Zeitgemäße Historie um 1870. Zu Nietzsche, Burckhardt und zum „Historismus"', *Historische Zeitschrift* 281 (2005) 33-58.

Schulin, Ernst, 'Treitschke und Max Weber, Meinecke und Gerhard Ritter. Politik und Geschichte in Freiburg zwischen 1863 und 1967' in: Achim Aurnhammer and Hans-Jochen Schwier ed., *Poeten und Professoren. Eine Literaturgeschichte Freiburgs in Porträts* (Freiburg 2009) 193-220.

Schultz, Werner, 'Die Bedeutung des Tragischen für das Verstehen der Geschichte bei Hegel und Goethe', *Archiv für Kulturgeschichte* 38 (1956) 92-115.

Schulz, Gerhard, 'Chaos und Ordnung in Goethes Verständnis von Kunst und Geschichte', *Goethe Jahrbuch* 110 (1993) 173-183.

Schulze, Winfried, *Deutsche Geschichtswissenschaft nach 1945* (Munich 1989).

Shanahan, William O., 'Germany: Realities of Power and Reconstruction', *The Review of Politics* 10 (1948) 502-509.

Shanahan, William O., 'Friedrich Naumann: A Mirror of Wilhelmian Germany', *The Review of Politics* 13 (1951) 267-301.

Sigurdson, Richard F., 'Jacob Burckhardt: The Cultural Historian as Political Thinker', *The Review of Politics* 52 (1990) 417-440.

Skinner, Quentin, *Machiavelli. A Very Short Introduction* (Oxford, New York 2000; first 1981).

Smith, Leonard S., *The Expert's Historian: Otto Hintze and the Nature of Modern Historical Thought* (Eugene 2017).

Snow, Dale Evarts, 'F. H. Jacobi and the Development of German Idealism', *Journal of the History of Philosophy* 25 (1987) 397-415.

Snyder, Louis L., *German Nationalism. The Tragedy of a People. Extremism contra Liberalism in Modern German History* (Harrisburg and Pennsylvania 1952).

Sontheimer, Kurt, 'Thomas Mann als politischer Schriftsteller' in: Helmut Koopman ed., *Thomas Mann* (Darmstadt 1975) 165-226.

Sösemann, Bernd ed., Friedrich Meinecke, *Die deutsche Katastrophe. Betrachtungen und Erinnerungen. Edition und internationale Rezeption* (Berlin 2018).

Spranger, Eduard, *Der gegenwärtige Stand der Geisteswissenschaften und die Schule* (Leipzig and Berlin 1922).

Spranger, Eduard, 'Das Historismusproblem an der Universität Berlin seit 1900' in: idem, *Gesammelte Schriften V. Kulturphilosophie und Kulturkritik* (Tübingen 1969) 430-446.

Srbik, Heinrich Ritter von, *Geist und Geschichte vom deutschen Humanismus bis zur Gegenwart* 2 vols. (Munich and Salzburg 1951).

Stark, Werner, 'Introduction' in: Friedrich Meinecke, *Machiavellism. The Doctrine of Raison d'état and its Place in Modern History* (New Haven 1957) xi-xlvi.

Stark, Werner, 'Friedrich Meinecke' in: idem, *Social Theory and Christian Thought. A Study of Some Points of Contact. Collected Essays Around a Common Theme* (London 1958) 201-245.

Steiner, Georg, *Heidegger* (Stanford Terrace 1978).

Sterling, Richard W., *Ethics in a World of Power. The Political Ideas of Friedrich Meinecke* (Princeton 1958).

Sterling, Richard, 'Political necessity and moral principle in the thought of Friedrich Meinecke', *The Canadian Journal of Economics and Political Science* 26 (1960) 205-214.

Stiewe, Barbara, *Der "Dritte Humanismus". Aspekte deutscher Griechenrezeption vom George-Kreis bis zum Nationalsozialismus* (Berlin and New York 2011).

Stirk, S.D., 'Review: The German Catastrophe through German Eyes', *International Journal* 2 (1947) 343-353.

Stoffers, Manuel, *Het nerveuze tijdperk en zijn historici: de opkomst van de mentaliteitsgeschiedenis in Duitsland 1889-1915* (Maastricht; dissertation 2007).

Stolleis, Michael, 'Friedrich Meineckes Die Idee der Staatsräson und die neuere Forschung' in: Michael Erbe ed., *Friedrich Meinecke heute. Bericht über ein Gedenk-Colloquium zu seinem 25. Todestag am 5. und 6. April 1979* (Berlin 1981) 50-75.

Strauss, Gerald, 'Meinecke, Historismus, and the Cult of the Irrational', *German Quarterly* 26 (1953) 107-114.

Strauss, Leo, *Natural Right and History* (Chicago 1953).

Sturma, Dieter, 'Politics and the New Mythology: The Turn to Late Romanticism' in: Karl Ameriks ed., *The Cambridge Companion to German Idealism* (Cambridge 2000) 219-238.

Talmon, Jacob L., *The Origins of Totalitarian Democracy* (London 1952).

Talmon, Jacob L., *The Myth of the Nation and the vision of Revolution. The Origins of Ideological Polarisation in the Twentieth Century* (London 1980).

Tessitore, Fulvio, *Kritischer Historismus. Gesammelte Aufsätze* (Cologne, Weimar and Vienna 2005).

Tessitore, Fulvio, 'Ernst Troeltschs "Kompromiß" und Friedrich Meineckes "neuer Dualismus"' in: idem, *Kritischer Historismus. Gesammelte Aufsätze* (Cologne 2005) 45-50.

Tessitore, Fulvio, 'Friedrich Meinecke und die Auflösung der onto-naturalistischen Hypothese' in: idem, *Kritischer Historismus. Gesammelte Aufsätze* (Cologne, Weimar and Vienna 2005) 37-44.

Tessitore, Fulvio, 'Meinecke in Italien' in: Gisela Bock and Daniel Schönpflug ed., *Friedrich Meinecke in seiner Zeit. Studien zu Leben und Werk* (Stuttgart 2006) 227-256.

Thielicke, Helmut, *Glauben und Denken in der Neuzeit. Die großen Systeme der Theologie und Religionsphilosophie* (Tübingen, 1988 first 1983).

Tollebeek, Jo, 'Het Duitse debat. Geschiedenis rond 1900' in: Herman Beliën and Gert Jan van Setten ed., *Geschiedschrijving in de twintigste eeuw. Discussie zonder eind* (Amsterdam 1991) 15-40.

Troeltsch, Ernst, *Der Historismus und seine Probleme. Erstes Buch: Das logische Problem der Geschichtsphilosophie* (Tübingen 1922).

Troeltsch, Ernst, *Der Historismus und seine Überwindung. Fünf Vorträge von Ernst Troeltsch. Eingeleitet von Friedrich von Hügel-Kensington* (Berlin 1924).

Troeltsch, Ernst, *Glaubenslehre. Nach Heidelberger Vorlesungen aus den Jahren 1911 und 1912* (Munich and Leipzig 1925).

Troeltsch, Ernst, 'Die Zukunftsmöglichkeiten des Christentums im Verhältnis zur modernen Philosophie' in: idem, *Zur religiösen Lage, Religionsphilosophie und Ethik* (Tübingen 1913) 837-862.

Troeltsch, Ernst, 'Wesen der Religion und der Religionswissenschaft' in: idem, *Zur religiösen Lage, Religionsphilosophie und Ethik* (Tübingen 1913) 452-499.

Trunz, Erich, 'Das Vergängliche als Gleichnis in Goethes Dichtung' in: idem, *Ein Tag aus Goethes Leben* (Munich 2006, first 1999) 167-187.

Ullrich, Volker, *Die Nervöse Grossmacht 1871-1918: Aufstieg und Untergang des deutschen Kaiserreichs* (Frankfurt am Main 1997).

Vermeulen, E. E. G., *Waarden en geschiedwetenschap. Een vergelijking van de standpunten ingenomen door H. W. von der Dunk, A. G.Weiler, M. C. Brands. Met notities over die van J. M. Romein, G. Harmsen, J. H. J. van der Pot* (Assen 1978).

Vierhaus, Rudolf, 'Goethe und der Historismus', *Goethe Jahrbuch* 110 (1993) 105-114.

Vierkandt, Alfred, 'Friedrich Meinecke, Die Idee der Staatsräson in der neueren Geschichte', *Kant-Studien* 33 (1928) 299-300.

Wagner, Fritz, *Geschichtswissenschaft* (Freiburg im Breisgau 1951).

Wallmann, Johannes, *Der Pietismus* (Göttingen 2005).

Walther, Peter Th., 'Die Zerstörung eines Projektes. Hedwig Hintze, Otto Hintze und Friedrich Meinecke nach 1933' in: Gisela Bock and Daniel Schönpflug ed., *Friedrich Meinecke in seiner Zeit. Studien zu Leben und Werk* (Stuttgart 2006) 119-143.

Wecker, Regina, *Geschichte und Geschichtsverständnis bei Edmund Burke* (Bern 1981).

Wehrs, Nikolai, 'Demokratie durch Diktatur? Friedrich Meinecke als Vernunftre-publikaner in der Weimarer Republik' in: Gisela Bock and Daniel Schönpflug ed., *Friedrich Meinecke in seiner Zeit. Studien zu Leben und Werk* (Stuttgart 2006) 95-118.

Wertz, Marianna and William F. Wertz Jr. ed., *Friedrich Schiller. Poet of Freedom* Volume IV (Washington 2003).

Wertz, Marianna (13 July 2016). http://www.schillerinstitute.org/transl/trans_schil_2poems.html

Weston, John C., 'Edmund Burke's View of History', *Review of Politics* 23 (1961) 203-229.

White, Hayden, *Metahistory. The Historical Imagination in Nineteenth-Century Europe* (Baltimore and London 1973).

White, Hayden, 'Historism: The Rise of a New Historical Outlook by Friedrich Meinecke (review)', *Pacific Historical Review* vol43, No. 4 (November 1974) 597-598.

Willey, Thomas E., *Back to Kant. The Revival of Kantianism in German Social and Historical Thought*, 1860-1914 (Detroit 1978).

Wippermann, Wolfgang, 'Friedrich Meineckes Die deutsche Katastrophe. Ein Versuch zur deutschen Vergangenheitsbewältigung' in: Michael Erbe ed., *Friedrich Meinecke heute. Bericht über ein Gedenk-Colloquium zu seinem 25. Todestag am 5. und 6. April 1979* (Berlin 1981) 101-121.

Wippermann, Wolfgang, '"Deutsche Katastrophe" Meinecke, Ritter und der erste Historikerstreit', in: Gisela Bock and Daniel Schönpflug ed., *Friedrich Meinecke in seiner Zeit. Studien zu Leben und Werk* (Stuttgart 2006) 177-191.

Wittkau, Annette, *Historismus. Zur Geschichte des Begriffs und des Problems* (Göttingen 1994).

Wittkowski, Wolfgang ed., *Friedrich Schiller. Kunst, Humanität und Politik in der späten Aufklärung. Ein Symposium* (Tübingen 1982).

Wolfson, Philip J., 'Friedrich Meinecke (1862-1954)', *Journal of the History of Ideas* 17 (1956) 511-525.

Wyman Jr., Walter E., *The Concept of Glaubenslehre. Ernst Troeltsch and the Theological Heritage of Schleiermacher* (Chico, California 1983).

Zabka, Thomas, 'Ordnung, Willkür und die "wahre Vermittlerin". Goethes ästhetische Integration von Natur- und Gesellschaftsidee' in: Peter Matussek ed., *Goethe und die Verzeitlichung der Natur* (Munich 1998) 157-177.

Zammito, John H., *The Genesis of Kant's Kritik der Urteilskraft* (Chicago 1992).

Zammito, John H., *Kant, Herder, and The Birth of Anthropology* (Chicago and London 2002).

Zammito, John H., 'Herder, Kant, Spinoza und die Ursprünge des deutschen Idealismus' in: Marion Heinz ed., *Herder und die Philosophie des deutschen Idealismus* (Amsterdam and Atlanta 1997) 107-144.

Zammito, John, '(Re)Discovering Johann Gottfried Herder. A Personal Memoir', *Groniek* 171 (2006) 191-214.

Ziegler, Klaus, 'Zu Goethes Deutung der Geschichte', *Deutsche Vierteljahrsschrift für Literaturwissenschaft und Geistesgeschichte* 30 (1956) 232-267.

Index

GPSR Authorized Representative: Easy Access System Europe, Mustamäe tee 50, 10621 Tallinn, Estonia, gpsr.requests@easproject.com